PLANT TAXONOMY
PAST, PRESENT, AND FUTURE

PLANT TAXONOMY
PAST, PRESENT, AND FUTURE
Dr Prithipalsingh Festschrift

Editor
Rajni Gupta

The Energy and Resources Institute

© The Energy and Resources Institute, 2012
First reprint 2014

ISBN 978-81-7993-359-6

All rights reserved. No part of this publication may be reproduced, stored in a retrieval system, or transmitted in any form or by any means, electronic, mechanical, photocopying, recording or otherwise, without the prior permission of the publisher.

All export rights for this book vest exclusively with The Energy and Resources Institute (TERI). Unauthorized export is a violation of terms of sale and is subject to legal action.

Suggested citation
Gupta R. 2011. *Plant Taxonomy: past, present, and future*. New Delhi: The Energy and Resources Institute (TERI)

Published by
The Energy and Resources Institute (TERI)
TERI Press
Darbari Seth Block
IHC Complex, Lodhi Road India
New Delhi – 110 003
India

Tel. 2468 2100 or 4150 4900
Fax 2468 2144 or 2468 2145
+91 • Delhi (0)11
E-mail teripress@teri.res.in
Website www.teriin.org

Printed in India

Foreword

Taxonomy is considered "dead" by some so-called "reformists of botany syllabi". If this thinking predominates, students of botany would not be able to identify even common plants such as *Tagetes*. This would be a sad day for botany in general.

Taxonomy, directly or indirectly, is a part of all branches of botany; hence it cannot die. The majority of botanists, unfortunately, consider people knowing only the names of plants as taxonomists. In fact, the names of plants represent phyletic, phylogenetic, and phytogeographic features, and also (directly or indirectly) have many other important implications. Like other sciences, taxonomy is based on principles and has many characteristic theories. Therefore, if taxonomy is to be understood in its entirety, each botany department should have a teacher like Dr Prithipalsingh (to whom this book is dedicated) on its faculty. I am sure the readers of this volume shall be able to appreciate the scope of taxonomy in its entirety.

Professor (Retd) K M M Dakshini
Department of Botany
University of Delhi

Message

It is a great pleasure to write about Dr Prithipalsingh, an affable, intelligent, and diligent student and equally hardworking and painstaking teacher. He has interacted with several generations of young minds in the pursuit of knowledge and truth. In fact, Dr Prithipalsingh is a brilliant taxonomist. His doctoral thesis was based on the study of the genus *Blumea* (Family: Asteraceae) under the supervision of the well-known taxonomist, Professor K M M Dakshini, my colleague in the Department of Botany, University of Delhi.

Now, I would like to develop a theme within the domain of Dr Prithipalsingh's interest, which is plant taxonomy. In doing so, I will largely draw upon my own research works. During my PhD programme at the Department of Botany, University of Delhi, I was given material of *Adonis flammea* fixed in FAA in 1961. Dr B M Johri provided this to me for investigations in the family Ranunculaceae (he had obtained the material thanks to Dr van der Heel from Leiden). When I investigated the embryology of this plant I found many unique features, some of which I would like to describe below.

There are numerous carpels in the flower and each carpel has many ovules, one of which is fertile while the rest are sterile. Uniquely, the development of the embryo sac is similar in the sterile ovules as well as the fertile ovule. At the stage of megaspore, the mother cell cuts off a cell which starts dividing multiple times to produce a large number cells resulting in a "parietal tissue". One of these cells is placed on top of the megaspore mother cell, simulating a megaspore. But, this soon starts dividing and contributes to the parietal tissue. The megaspore mother cell divides and results in two megaspores. The nuclei in each of the two cells further divide, resulting in the formation of two binucleate megaspores. Subsequently, the upper megaspore degenerates and the lower one gives rise to a binucleate embryo sac.

At the insistence and direction of my teacher and guru, Professor Panchanan Maheshwari, I went to the interiors of Shimla in search of a rare species of the genus, *Adonis aestivalis*. Needless to say, it was a Herculean task going deep into the woods all alone, with no one to share my thoughts and difficulties in finding the rare species. After about five days of rigorous labour and frustration, I finally found a single plant. After Professor Maheshwari confirmed the identity of the plant, satisfaction dawned on me, and I relaxed because my pursuit had ended. After more searching, I found a patch of plants in flowering stage. It was an ideal spot. In addition, I also received fixed material of *Adonis chrysocyanthus* collected by Dr C P Malik and Dr G L Dhar from Kashmir.

Fixing research material for embryological studies of this species (*A. aestivalis*) helped me confirm whether this species also showed bisporic development. So I carried out detailed embryological research, which showed that this species as well as *A. chrysocyanthus* had bisporic embryo sac development. I concluded that, probably, all the species of *Adonis* have bisporic embryo sac development. It was therefore suggested that a new taxonomic unit be created within the family Ranunculaceae.

This reminded me of the statement of Sir J D Hooker who, in the *Introductory Essay to Flora indica* in 1855, remarked: "A knowledge of the relative importance of characters can only be acquired by long study...". In this context, the importance of embryological characters in plant taxonomy could be understood because of the extensive studies carried out by Professor Maheshwari and his students. Interestingly, Professor Maheshwari interpreted embryology in a very broad sense to include all events, taking into account the sequence from ovule and anther formation to fertilization as well as embryo and endosperm development. The observations on the development of the embryo sac in *Adonis* formed the basis of the suggestion about the taxonomic position of the genus. It gives me great satisfaction to emphasize that Armen Takhtajan (*Diversity and Classification of Flowering Plants*, 1997) has recognized Adonaideae as a distinct taxonomic unit in the family Ranunculaceae.

The importance of careful observation of minute details and the correct interpretation of the information has always been of great significance in plant taxonomy. Embryological characters have provided many interesting examples of their utility in plant taxonomy.

It is a matter of great joy that we are able to dedicate this volume to Dr Prithipalsingh. I wish him the very best in his life and hope that his work can inspire a new generation of taxonomists to enter this deeply enriching and widely interesting field of research.

Professor (Retd) N N Bhandari
Department of Botany
University of Delhi

Preface

Plant taxonomy is a fundamental science. It is recognized as the "focal point of biology" based on the fact that it is related to morphology, anatomy, embryology, cytology, chemistry, and evolution, as well as the classification of plants. In order to understand the developments of this science, the present volume attempts to provide a full scenario of plant taxonomy. Thus, topics related to nomenclature, species and evolutionary aspects, methods of identification, anatomical features, palynology, and molecular systematics have been discussed.

Dr Prithipalsingh has devoted his life to academic pursuits. Ever since he joined the Department of Botany, University of Delhi, he has been engaged in the study of different aspects of plant taxonomy. He has been teaching the subject for almost 40 years. This Festschrift commemorates his long span as a teacher who inspired generations of students by his depth of knowledge and the easy manner in which he explained the fundamental aspects of plant taxonomy. I have had the good fortune of interacting with Dr Prithipalsingh for more than 10 years and learning many concepts of angiosperm taxonomy from him.

In compiling this volume, it has been my endeavour to focus on different aspects of plant taxonomy. Nomenclature is a significant aspect of plant taxonomy. Different aspects of this science of naming plants have been developed in some of the contributions. It is also an established fact that "species is a fundamental unit in biology". The species concept and the process of speciation have been elaborately discussed. The importance of proper identification of the species has been recognized and modern tools and techniques used for this purpose have been elaborated. The wisdom of traditional knowledge and "folklore taxonomy" as well as the importance of protecting this ancient science needs elaborate legal knowledge and has been described in detail. The significance of plant taxonomy for managing "genetic resources" has been analysed in depth. The use of evidence from anatomy, cytology, chemistry, and palynology for taxonomic purposes leads to a better understanding of taxonomic relationships. These have been evaluated in great detail in this volume.

Since understanding phylogeny and phylogenetic relationships has always been a challenge to students of plant taxonomy, these have been explained in simple terms. The importance of herbaria and data storage systems, as well as the floristics of the future, have been elaborately described. Thus, this volume shall serve as a useful source of information for graduate and postgraduate students in plant science courses.

I am highly obliged to all the contributors for their cooperation. In spite of such short notice, they prepared their manuscripts and generously accepted the suggestions made by the reviewers. Without their support and cooperation, the task of editing the volume would have been impossible. The citation of references in most of the contributions follows a basic pattern, but in the case of one contribution (*Indigenous Knowledge of Plants and Biopiracy in India*) a different pattern has been adopted. This was necessary in view of the legal aspects covered in this contribution.

It is an honour to have the foreword written by Professor K M M Dakshini, a teacher who has been the source of inspiration for generations of students. Due to indifferent health, he was unable to provide a complete article for inclusion in this Festschrift. Special thanks are also extended to Professor N N Bhandari for a very useful message on the importance of observations in plant taxonomy. He has provided some interesting facts about the embryology of the family Ranunculaceae and the taxonomic significance of these observations.

I thank all the contributors and my colleagues in the Department of Botany, Kirori Mal College, for their valuable cooperation and encouragement, which helped me complete the task. Last, but not the least, I remain grateful to my family members for their valuable support.

Contents

Foreword	*v*
Message	*vii*
Preface	*ix*
List of Contributors	*xiii*
The Life and Training of a Plant Taxonomist: *Dr Prithipalsingh*	*xv*

1. **Ethnobotanical Noah's Ark** — 1
   Sudhir Chandra

2. **Plant Nomenclature: an Overview** — 29
   Bharati Bhattacharyya

3. **Plants of Delhi: Scientific Names and their Meaning** — 61
   Neelam Pari Malkani

4. **Species and Speciation** — 91
   M A Khalid

5. **Modern Tools for Identification of Plants** — 115
   Rajni Gupta and Ruchitra Gupta

6. **Plant Taxonomy in Plant Genetic Resource Management** — 129
   Anjula Pandey, D C Bhandari, and K Pradheep

7. **Indigenous Knowledge of Plants and Biopiracy in India** — 141
   Piyush K Sharma

8. **Herbaria and Data Information Systems in Plant Taxonomy** — 167
   Satish K Aggarwal

9. **Phylogenetic Systematics** — 175
   Manoj M Lekhak, Anil Kumar, and Shrirang R Yadav

10. **Plant Anatomy in Relation to Taxonomy**   211
   Rajni Gupta and Kusum Shukla

11. **Chemotaxonomy**   231
   Anand Sonkar and Sharda Mahilkar Sonkar

12. **Cytotaxonomy and its Evolutionary Significance in the Evaluation of Orchidaceae and Cyperaceae**   273
   Prabha Sharma and P L Uniyal

13. **Palynology: Timeline**   289
   Meenakshi Prajneshu

14. **Role of Molecular Markers in Evaluation of Plant Diversity**   309
   Mandeep Kaur, Gurveen Kaur, Rajneet Kour Soodan, Jatinder Kaur Katnoria, and Avinash Nagpal

15. **E-flora: the Future of Floristic Documentation**   329
   Gurcharan Singh

About the Editor   *351*

List of Contributors

Anand Sonkar, Assistant Professor, Department of Botany, Hansraj College, University of Delhi, Delhi – 110 007; <anandsonkar@gmail.com>

Anil Kumar, Assistant Professor, Department of Botany, Swami Shraddhanand College, University of Delhi, New Delhi – 110 036; <crazygrassdel@gmail.com> (Kirori Mal College alumnus)

Anjula Pandey, Sr Scientist, Division of Plant Exploration, National Bureau of Plant Genetic Resources, Indian Agricultural Research Institute, New Delhi – 110 012; <pandeyanjula@nbpgr.ernet.in>

Avinash Nagpal, Professor, Department of Botanical and Environmental Sciences, Guru Nanak Dev University, Amritsar – 143 005, Punjab; <avnagpal@rediffmail.com>

Bharati Bhattacharyya, Associate Professor (Retd), Department of Botany, Gargi College, University of Delhi, New Delhi – 110 049; <bbharati46@ gmail.com> (40/5 Chittaranjan Park, New Delhi – 110 019)

D C Bhandari, Head, Division of Plant Exploration, National Bureau of Plant Genetic Resources, Indian Agricultural Research Institute, New Delhi – 110 012; <bhandaridc@nbpgr.ernet.in>

Gurcharan Singh, Associate Professor (Retd), Department of Botany, SGTB Khalsa College, University of Delhi, Delhi – 110 007; <singhg@sify.com> (932 Anand Kunj, Vikaspuri, New Delhi – 110 018)

Gurveen Kaur, Department of Botanical and Environmental Sciences, Guru Nanak Dev University, Amritsar – 143 005, Punjab

Jatinder Kaur Katnoria, Lecturer, Department of Botanical and Environmental Sciences, Guru Nanak Dev University, Amritsar – 143 005, Punjab

K Pradheep, Sr Scientist, Division of Plant Exploration, National Bureau of Plant Genetic Resources, Indian Agricultural Research Institute, New Delhi – 110 012

Kusum Shukla, Department of Botany, Kirori Mal College, University of Delhi, Delhi – 110 007; <kusum1753@gmail.com>

M A Khalid, Field Director, India Regional Climate Centre, Earthwatch Institute, Yellapur Road, Sirsi, Karnataka; <makhalid2007@gmail.com>

Mandeep Kaur, Department of Botanical and Environmental Sciences, Guru Nanak Dev University, Amritsar – 143 005, Punjab

Manoj Lekhak, Assistant Professor, Department of Botany, Shivaji University, Kohlapur – 416 004, Maharashtra; <mlekhak@gmail.com> (Kirori Mal College alumnus)

Meenakshi Prajneshu, Associate Professor, Department of Botany, Deshbandhu College, University of Delhi, New Delhi – 110 019; <prajneshum@gmail.com>

Neelam Malkani, Associate Professor, Department of Botany, Daulat Ram College, University of Delhi, Delhi – 110 007; <n.malkani@yahoo.co.in>

N N Bhandari, Head (Retd), Department of Botany, University of Delhi, Delhi – 110 007; <nn_bhandari@yahoo.co.uk>_(BB 4 B, Janakpuri, New Delhi – 110 058)

Piyush Kumar Sharma, Department of Botany, Kirori Mal College, University of Delhi, Delhi – 110 007; <piyushsharma2005@rediffmail.com>

P L Uniyal, Associate Professor, Department of Botany, University of Delhi, Delhi – 110 007, <uniyalpl@rediffmail.com>

Prabha Sharma, Department of Botany, University of Delhi, New Delhi – 110 007; <prabha3@gmail.com>

Rajneet Kour Soodan, Department of Botanical and Environmental Sciences, Guru Nanak Dev University, Amritsar – 143 005, Punjab

Rajni Gupta, Department of Botany, Kirori Mal College, University of Delhi, Delhi – 110 007; <rajnigupta68@rediffmail com>

Roshini E Nayar, Principal Scientist, Division of Plant Exploration, National Bureau of Plant Genetic Resources, Indian Agricultural Research Institute, New Delhi – 110 012, <roshinienayar@yahoo.com>

Ruchitra Gupta, Department of Botany, Kirori Mal College, University of Delhi, Delhi – 110 007; <ruchitragupta85@gmail.com>

Sharda Mahilkar Sonkar, Department of Chemistry, Miranda House, University of Delhi, Delhi – 110 007

S K Aggarwal, Associate Professor, Department of Botany, Deshbandhu College, University of Delhi, New Delhi – 110 019; <skaggarwal_102@yahoo.co.in>

S R Yadav, Department of Botany, Shivaji University, Kohlapur – 416 004, Maharashtra; <yadavdu@rediffmail.com>

Sudhir Chandra, Professor (Retd), Department of Botany, Kumaun University, Nainital – 263 001, Uttaranchal; <chandras46@yahoo.com> (G8 Biotech Complex, The Mall, near Shalimar Hotel)

The Life and Training of a Plant Taxonomist: Dr Prithipalsingh

Roshini E Nayar

My association with Dr Prithipalsingh spans over four decades. I was first acquainted with him in 1971 when I was a BSc (Honours) third-year student and Dr Prithipalsingh participated in the conduction of practical classes in plant taxonomy. He was the first student to submit his PhD thesis under the guidance of Professor K M M Dakshini of the Department of Botany, University of Delhi, while I was among the last students who submitted their theses under the same guide. Our association has continued over the years as I have regularly sought Dr Prithipalsingh's participation in the teaching faculty for training programmes on plant taxonomy. Dr Prithipalsingh is so well known in the plant genetic resources (PGR) scenario that he is now an honorary member of the PGR library at the National Bureau for Plant Genetic Resources (NBPGR).

Dr Prithipalsingh joined the Department of Botany in the University of Delhi in 1967 for post-graduation after graduating from St Joseph's College, Bengaluru. When we first met him, his claim to fame was being a Punjabi fluent in South Indian languages (Kannada and Tamil). He was born in Murad Nagar, Uttar Pradesh, but his family first moved to Secunderabad, Andhra Pradesh, and then to Bengaluru, Karnataka, where his father served in the Ministry of Defence, Government of India. He grew up in Bengaluru, completing his schooling from the Bishop Cotton School in 1963, when the Indian School Certificate Examination was first conducted under the aegis of the University of Cambridge (Local Examinations Syndicate).

The foundation for his interest in plant taxonomy was laid as a graduate student when Fr (Dr) Cecil Saldanha taught the first batch of students to be awarded the BSc degree by the Bangalore University (earlier, all colleges in Bengaluru were affiliated to the University of Mysore). He learnt the basics of plant taxonomy and experienced hands-on training during several field excursions in a city like Bengaluru, from "a well-travelled field botanist" and "a committed conservationist"; he has taken a lead in passing on the love of plants and their diversity even to the common man. Thereafter, when he joined the University of Delhi to pursue master's programme, Dr Prithipalsingh consciously chose plant taxonomy as his specialization and submitted his dissertation on Compositae of Delhi in partial fulfilment of the MSc degree and later his PhD thesis titled *Morphology and Chemotaxonomy in the Systematics of Blumea (Asteraceae: Inuleae) in India*, both under the guidance of Professor Dakshini. This understanding of the fundamentals of plant taxonomy and love for the subject remained with him during his teaching and research career.

He had valuable experience as a student, and later as a lecturer, of plant taxonomy in the Department of Botany. When he was pursuing his PhD, he was a "teaching assistant for conducting practical classes". This was followed by a stint of teaching in the Department of Botany (1973), followed by teaching graduates in Deshbandhu College (1973–78), and finally in Kirori Mal College (1978 onwards). This was an important time for botany in general and taxonomy in particular. Dr Prithipalsingh knew and interacted closely with students working on taxonomy, ecology, systematics, and other subjects for more than three decades. This was also a time when the inclinations of a student were taken into consideration and all disciplines of botany were considered equally important. Dr Prithipalsingh's strong foundations in taxonomy and his interest in the subject became evident when he won a competition organized by the Delhi University Botanical Society (1967/68). A quote published in one of the older issues of *Bulletin of the Torrey Botanical Club* (later renamed *Journal of the Torrey Botanical Society*), "Come one, come all, win a big cash prize. Say in 500 words or less why you like taxonomy", was the basis for his presentation. Professor B M Johri (then Head, Department of Botany, University of Delhi) recognized this and said that Dr Prithipalsingh would devote his life to plant taxonomy. This prediction has come true.

The Department of Botany generally hosted events in which a diverse range of scientists from different parts of the world participated. Dr J Hutchinson, Dr K R Sporne, Professor V H Heywood, Mr B L Burtt, Dr Heslop-Harrison, Dr Armen Takhtajan, Dr J K Maheshwari, Fr K M Mathew, Dr K Subramaniam, Dr M A Rau, Dr Peter Bell, and Dr Harbhajan Singh were some of the visitors to the department. Among the taxonomists, mention must be made of Professor V H Heywood, Mr B L Burtt, Professor Peter Endress, Dr Dan Nicolson, Dr Armen Takhtajan, Dr J K Maheshwari, Fr K M Mathew, Dr K Subramaniam, Dr R S Rao,

Dr M A Rau, Dr T A Rao, Dr A S Rao, Dr C L Malhotra, Dr A N Henry, and Dr B A Razi (from the University of Mysore). Dr Prithipalsingh recalls many interesting aspects of plant taxonomy that he learned from these interactions. They helped him in the learning process and are a significant part of the background of Dr Prithipalsingh's entry into the field of plant taxonomy as a student and a teacher. His PhD thesis was evaluated by stalwarts of taxonomy like Professor V H Heywood (University of Reading, UK), Mr B L Burtt (Royal Botanic Garden, Edinburgh, UK), and Dr J K Maheshwari (Director, Botanical Survey of India). One of the comments by the examiners was that the thesis presented "a refreshing look at Indian plant taxonomy".

Identification of plants is an activity taxonomists, by training, carry out all through their career. It not only gives them the satisfaction of being able to relate to the diversity of plants in an area, but also provides them with the opportunity to add to their repertoire of knowledge. Some of Dr Prithipalsingh's important experiences to date have been recalled here. During his early years as a student, and later as well, there was a close collaboration between botany and chemistry departments for identifying plants used for phytochemical analysis. It was also the period when there was considerable focus on phytochemistry and chemotaxonomy at the global level.

When he was a young student, Dr Prithipalsingh was summoned by Professor N S Rangaswamy to identify a sample of plant material received from the Department of Chemistry. This is a case in point of the difficulties involved in identification. After careful analysis of the fragmentary plant material (consisting of inflorescence only), he concluded that it was "probably *Centella asiatica*". Professor N S Rangaswamy counselled him that he should always be sure and should express opinion with full confidence, never using the word "probably". This was an important lesson in plant identification for the young Dr Prithipalsingh. He later wrote on this aspect in his PhD thesis by quoting from the book *Principles of Angiosperm Taxonomy* by Davis and Heywood. This is a guiding principle relevant to both taxonomists and non-taxonomists: "Descriptions are abstract for practical convenience. They provide some of the characters that are easily recognizable and comparable, but not all those features that may be used in making a decision about the status. In this activity, taxonomists not only gather information (that is, characters), but also present them in simple terms that can be easily understood. This is not often realized, especially by non-taxonomists who tend to assume that a particular classification stands or falls by the evidence quoted. However, it creates an ambiguity regarding the amount of data that should be presented in any particular context, and the purpose of the publication is a major consideration".

The process of identification in plant taxonomy requires proper characterization of the specimen, and "expressing an opinion is similar to pronouncing a judgement". Dr Dakshini's method of teaching plant identification was similar; he would present a single leaf or a flower stripped of leaves and give equal emphasis on knowing both the botanical and common names of the plant. This was also the theme of lectures and practical demonstrations when Dr Prithipalsingh was invited to serve as a resource person for "plant taxonomy and biodiversity training programmes" organized by the NBPGR in New Delhi. These programmes were for the benefit of a large number of officers working in different NBPGR stations and other similar institutes. Besides interacting with the trainees during the lectures, he insisted that it was "more important to provide a hands-on training in herbarium". Many herbarium specimens of the same plant species (collected from different parts of the country) were used for these training programmes. Each trainee was encouraged to carefully analyse the characteristics of specimens so as to correctly identify them. This is a major aspect of training in taxonomic identification. There are several interesting incidents that reflect this. Dr Prithipalsingh recalls identifying plant material based on information provided telephonically by Dr Savithri Singh (now Principal, Acharya Narendra Dev College), who called from Jodhpur for the identification of a common weed. On the basis of a few select characters, the plant was identified as *Tribulus terrestris*. Dr Tej Pal Singh (then Senior Fellow, TERI, New Delhi; now with IUCN in Bangkok) once requested him for information on *Salvadora persica*, and Dr Prithipalsingh gathered information about the characters of this plant by consulting his library. Professor Dakshini mentioned this while recalling his trips from the University of Reading, UK, to parts of the Alps and his first exposure to the temperate and alpine flora. He was a versatile botanist, familiar with plants from different parts of India, and he trained Dr Prithipalsingh in the proper identification of plants based on their characters. However, it was a surprise to Dr Prithipalsingh when, on his visits to Nottingham, New Jersey, and Niagara Falls, he was able to identify some plants that he had never seen earlier by analysing their characters.

Meticulous attention to detail and a systematic approach to the subject are important in the training of a taxonomist. Dr Prithipalsingh feels that this was in part learnt "at home", and he was proud that he shared this trait with his father. During World War II, Dr Prithipalsingh's father was decorated by the commander-in-chief of the British Army because he meticulously prepared an "inventory of material" (both major and minor items offloaded by a British warship at Karachi; the detailed papers had not been sent with the ship for security reasons), which completely matched the original list.

Collecting and collating data is also a crucial aspect of plant taxonomy, which was evident in Dr Prithipalsingh's thesis. He not only collected plants from different locations across India, but also examined a large number of herbarium specimens. Each specimen was treated as a representative of a naturally occurring population for analysing the characters.

On an average, Dr Prithipalsingh has examined about 1000 capitula of each of the 22 species of *Blumea* (in India) to collect data on floral characters. Thus, he actually dissected about 3 188 500 florets and used the information to group the species into different categories to understand the pollen presentation mechanisms in this genus. This information was used by Mr B L Burtt in a detailed article published in the monograph *The Biology and Chemistry of Compositae*. This aspect of training of a plant taxonomist needs to be inculcated by all students aspiring to establish themselves in this discipline. This trait also helped Dr Prithipalsingh undertake the task of preparing *The Flora of Faridabad District* (an inventory), under the District Flora Scheme of the Botanical Survey of India. This scheme was funded by the Ministry of Environment and Forests, Government of India, and Dr Prithipalsingh was one of the very few college teachers and the only botanist from the University of Delhi assigned for such a project.

Teaching and learning go hand in hand, more so in taxonomy, which is needed in any activity related to plants. Dr Prithipalsingh considered it a challenge when he served as guest faculty at The School of Planning and Architecture, New Delhi, and taught the basics of botany and plant ecology to students pursuing Masters in Landscape Architecture. He is presently a guest faculty at the Indira Gandhi National Open University (IGNOU) and teaches plant taxonomy to students pursuing Masters in Life Sciences.

The science of plant taxonomy requires a basic understanding of Latin, and Dr Prithipalsingh studied this language informally when he analysed the original literature on the taxonomy of *Blumea*. This helps in simplifying the science of practical taxonomy to students, not only for identifying plants in the field but also for providing the correct spellings of botanical names. This was the theme of a much appreciated presentation on plant nomenclature at the NBPGR, where Dr Prithipalsingh explained the broad aspects and finer details of the process of naming plants occurring in the wild as well as cultivated species.

Dr Prithipalsingh always attempts to understand the etymology of plant names and to learn the correct spellings as published by the author. This aspect of learning plant taxonomy was initiated during his graduation at St Joseph's College. In fact, Professor Sudhir Chandra of Kumaon University, Nainital, a close associate of Dr Prithipalsingh, has spent almost 35 years in compiling a multi-volume work *Dictionary of Commemorative Plant Generic Names*. Dr Chandra attributes this endeavour to a presentation made by Dr Prithipalsingh during a competition of the Delhi University Botanical Society.

It is crucial for a taxonomist to understand classification. Dr Prithipalsingh has lucidly explained the fundamental aspects of the process of classification in his book *An Introduction to Biodiversity*. Though his specialization is in angiosperm taxonomy, he has not neglected the classification of other groups of organisms routinely studied by botany students. As regards angiosperms, he keenly follows recent developments, especially the APG III version on relationships among orders and families.

Taxonomists should always present their work in a concise and precise manner. This requires strict discipline, which comes easily to Dr Prithipalsingh, for he is "a hard taskmaster" according to many students. He insists on careful analysis of plants during practical classes, both in the field and the laboratory. This was affirmed in 1971 when one of my classmates could not get her practical record signed by him. He brought a plant of the same species and asked her to compare the specimen with the drawing in the practical book.

Dr Prithipalsingh emphasizes the importance of taxonomic discipline even in his daily life. This is echoed in the words of Bill McKelvey (Professor, Organizational Science and Management, Graduate School of Management, University of California, Los Angeles), "No scientific progress can be achieved without applying the basic principles of systematics".

Dr Prithipalsingh has amassed a vast collection of books on botany, especially taxonomy; a large number of students have benefitted by having unlimited access to these writings. At the NBPGR, Dr Prithipalsingh's donation of his prized volumes of the journal *Taxon* and a large number of books for use by the scientific community has been much appreciated.

Besides botany, Dr Prithipalsingh has shown interest in different aspects of the Sikh religion and history. To mark the 300 years of the Khalsa Panth, he participated in a seminar organized by the Kalindi College (January 2000) and presented a paper titled *Transformation in Sikhism: the contributions of women*. He has also published several articles in different journals and magazines (*The Sikh Review*, Kolkata; *Atam Science/Atam Paratam*, Dagshai; *Sachkhand Patra*, Nanded; and *Sampark*, Karnataka Sikh Welfare Society, Bengaluru) on varied topics. Some of the articles are *Shadian de Sartaj: Guru Arjan Dev*; *Sri Guru Granth Sahib: a unique scripture*; *The Tradition of Guruship in Sikh Religion*; *Guru Manyo Granth: the Granth be thy Guru*; *Baisakhi: a universal festival*; *Saint, Sainthood and Sikhism*; *Gurbani and Science*; and *Aartee: the evening worship*.

Dr Prithipalsingh epitomizes the training of a taxonomist through teaching and training under experts, besides meticulous observation and rigorous study of plants in the field and the laboratory. A taxonomist is in the happy position of always having "the tools of his trade" at hand. I look forward to continuing my interactions with Dr Prithipalsingh since he will always be a practising taxonomist.

1

Ethnobotanical Noah's Ark

Sudhir Chandra

INTRODUCTION

When I was a PhD student at the Department of Botany, University of Delhi, in 1968, a student symposium entitled "What Interests Me Most in Biology" was held under the auspices of the Delhi University Botanical Society. Dr Prithipalsingh (then an MSc student) presented his summation on the role of taxonomy and systematics (prepared under the supervision of Dr Bharti Bhattacharya, nee Chakravarty, a PhD student of Dr J K Maheshwari, author of *Flora of Delhi*). Among his many arguments in support of taxonomy, Dr Prithipalsingh said that many renowned botanists had been honoured in plant names, including Professor B M Johri [*epon*: *Piptocephalis brijmohanii* (Fungi: Zygomycetes: Mucorales)] and Professor Panchanan Maheshwari [*Maheshwariella* D. D. Pant et D. D. Nautiyal 1963 (fossil seed)]. This started my voyage on the discovery of plant names in honour of botanists. The odessey is far from complete after 43 years. Five of the projected 20 volumes of the *Dictionary of Commemorative Plant Generic Names* have been published by Bishen Singh Mahendra Pal Singh, Dehradun. Volumes six to eight are in press in various stages of publication.

In an article entitled *Ethnobotanical ark*, I said that like the "Ark of Genesis", Linnaeus, through his binomial system, built a "Nomenclature Ark" for the systematization of plant names (Chandra 1995). This Ark became the "Ethnobotanical and Ethnobiological Ark" through the incorporation of primeval environment in Latin plant and animal names rather dealt under arbitrary 19 categories: plant names based on human organs or parts, animals, languages and vernacular names, household goods and needs, gods, musical instruments, alphabets, slavery, educational centres, agricultural terms, relationship, human values and sentiments, medicinal uses, tribals and tribes, celestial bodies, colours, weapons and warfare, geographical terms, architectural terms, and mathematical terms. In this chapter, I have elaborated on household goods and needs and how different terms find their way into Latin plant names.

I dedicate this chapter to my friend Dr Prithipalsingh for inspiring me to start this work and his continuous support. He has made outstanding contributions to botany in the fields of chemotaxonomy (*Blumea*), floristics (flora of Faridabad district), and biodiversity (for the Tata Energy Research Institute, now The Energy and Resources Institute [TERI]). He has also edited volumes, including a Festschrift in honour of his teacher Professor K M M Dakshini and wrote incisive and authorative reviews for *Phytomorphology*, a textbook for students, and "Units of Study Material" for the Indira Gandhi National Open University. He has made an enriching contribution to botany in India by training students in taxonomy, who are dispersed throughout India and abroad, like *Blumea/Bidens* and other composite fruits get dispersed with the help of pappus. His gift of *Taxon* volumes to the National Bureau of Plant Genetic Resources (NBPGR) will be an asset forever to the institution and will be of immense use to generations of students.

PLANTS NAMED AFTER HOUSEHOLD GOODS AND NEEDS

The following is a presentation of how different terms find their way into plant names.

Arch	*Camarotis* Lindl. Orchidaceae: Greek *kamarotos* = arch; indicating the orchid lip
Aroma	*Osmorhiza* Rafin, 1818: Greek *osme* = aroma, *rhiza* = root; referring to the aroma of the root, whether "fresh or dried"
	Smyrnium L., 1753. Smyrnion, ancient Greek name for *Smyrnium perfoliatum* or *Smyrnium olusatrum*: From *smyrna* = myrrh (gum resin), referring to the plant's pleasant aroma; reminiscent of myrh
Arrow	*Beloperone*. Acanthaceae. *belos* = arrow, *perone* = buckle
	Sagittaria L. Alismataceae: Latin *sagitta* = arrow, *sagittarius* = archer; referring to the shape of leaves
Axe	*Pelecyphora*: Greek *pelekys* = axe or hatchet, *phorein* = to carry; as the tubercles resemble the blade of a hatchet

	Securidaca L. Polygalaceae: Latin *secures* = axe or hatchet; referring to the shape of the wing at the end of the pod
	Securinega Comm. ex Juss. Euphorbiaceae
Bag	*Saccopetalum* Benn. Annonaceae: Greek *saccos* = bag, *petalon* = petal
Band	*Cryptotaenia* DC: Greek *cryptos* = hidden, *taenia* = band, stripe; referring to the canals hidden in the depths of the fruit
	Desmodium Desv: Greek *desmos* = band; because of the shape of the inflorescence
Basin	*concavus, -a, -um*; basin-shaped
Basket	*Calathea* G F W Mey. Marantaceae: Greek *kalathos* = basket or cup; referring to the form of the stigma
	Canistrum: Latin *canistra* = flat basket; in allusion to the appearance of inflorescence of the Brazilian plants
	Calathrinus, -a, -um; basket-like
Bed	*Clinogyne* Salisb. ex Benth. Marantaceae: Greek *klinion* = little bed, *gyne* = ovary
	Cyathocline Cass. Compositae: Greek *kyathos* = ladle, *kline* = bed; indicating the shape of leaves
Bed of straw	*Stramentopappus* H Robinson and V Funk. *stramentum* = bed of straw combined with pappus; referring to the pappus formed by the numerous short deciduous pappus bristles in the mature heads
Bell	*Campanula*: Latin *campana* = bell; in allusion to the form of the flowers
	Campanumoea Blume. Campanulaceae: Latin *campana* = bell, *homoios* = similar; indicating similarity with the genus *Campanula*
	Codonacanthus Nees. Acanthaceae: Greek *codon* = bell, *acanthos* = acanthaceous; referring to the shape of the flowers
	Codonopsis Wall. Campanulaceae: Greek *codon* = bell; referring to the shape of the corolla
	Cryptocodon Fed. Campanulaceae: Greek *cryptos* = hidden, *codon* = bell; akin to hidden bell

	Leptocodon Lem. Campanulaceae: Greek *leptos* = slender, *codon* = bell; referring to the slender bell-shaped flowers
	Megacodon (Hemsl.) H Smith. Gentianaceae: Greek *megas* = great, *codon* = bell; referring to the large bell-shaped flowers
	Platycodon A. DC. Campanulaceae: Greek *platos* = broad, *codon* = bell
	campanulatus, a, um
	campanularius, a, -um
	campanuloides
Bench	*Synedrella* Gaertn. Compositae: Greek *synedrella* = little bench; alluding to the naked receptacle
Birth	*Aristolochia* L. Aristolochiaceae: Greek *aristos* = best, *lochia* = birth; referring to the flower resembling a human foetus
Bitter	*Peucedanum* L: Greek *peukedanos* = bitter, sharp; referring to the sharp taste and smell of the plant
Bitumen (coal)	*bituminosus, -a, -um*
Boat	*Cymbidium* Sw. Orchidaceae: Greek *kymbe* = boat; referring to the hollow, boat-shaped recess in the labellum
	Cymbopogon Spreng. Gramineae: Greek *kymbe* = boat, *pogon* = beard; referring to the boat-shaped glumes
	Lembotropis Griseb: Greek *lembos* = boat or its prow, *tropis* = keel; implying a short keel
	Phaseolus L. Papilionaceae: Greek *phaseolus* = little boat, canoe; referring to the shape of the bean
Bolt	*Gomphandra* Qwall. ex Lindl. Icacinaceae: Greek *gomphos* = bolt or nail, *andra* = androecium; referring to the nature of stamens
Boot	*Crepis* L. Compositae: Greek *krepis* = boot
Bow	*Toxocarpus* Wt and Arn. Asclepiadaceae: Greek *toxon* = bow, *karpos* = fruit; allusion to the follicles in the form of a bow
Bowl	*Buxus* L: Greek *pyxos* = the name of *Buxussempervirens*, the Latin *buxus* = boxwood; classical name used by Virgil

Box	*Phialacanthus* Benth. Acanthaceae: Greek *phiale* = bowl or flat vessel
	Scaphium Schott and Endl. Sterculiaceae: Latin *scaphium* = hollow vessel, Greek *scaphe* = a bowl
	Capsella Medic. Cruciferae: Latin *capsella* = little box; referring to the fruit
	Cibotium: Greek *kibotos* = small box; from the appearance of sori of these ferns
	Cistanche Hoffm. et Link. Orobanchaceae: Greek *kiston* = box or capsule, *anchis* = son of Capys and father of Aeneas of Greek mythology
	Pachylarnax Dandy. Magnoliaceae: Greek *pachys* = thick, *larnax* = box or chest
	Thecagonum Babu. Rubiaceae: Greek *theke* = box or case for something, *gonia* = a joint, a knee
Boundary	*Terminthia* Bernh. Anacardiaceae: Greek *termin* = boundary, *thia, theio* = to run; referring to gum exudates from the branches
Bracelet (collar)	*armillaris, -e* = encircled; as with a bracelet or collar
	armillatus, -a, -um; like a bracelet or collar
Bridle	*Bifrenia* L. *bis* = twice, *frenum* = bridle; allusion to the two caudicles by which the pollen masses are connected with their gland in these orchids
Broom	*Calluna* Salisb: Greek *kallunein* = to sweep; brooms are made of the twigs of this plant
	Corona: Greek *koroma* = broom
	Sarothamnus Winm: Greek *saros* = broom, *thamnos* = shrub; implying a broom-like shrub; *cytisoides*, resembling a broom
	Scoparia L. Scrophulariaceae: Greek *scope* = a broom
Brush	*Hemigraphis* Nees. Acanthaceae: Greek *hemi* = half, *graphis* = brush or pencil; referring to the dense hair covering the filaments of outer stamens
Buckle	*Beloperone*. Acanthaceae: Greek *belos* = arrow, *perone* = buckle
Bundle	*Amischotolype* Rao et Kamm. Commelinaceae: Greek *a* = not/without, *miskos* = stalk, *phakelos* = bundle; in allusion to the nature of the inflorescence

	Desmanthus Willd. Mimosaceae: Greek *desmos* = bundle or chain, *anthos* = flower; referring to the arrangement of flowers
	Desmodium Desv. Papilionaceae. Alluding to jointed pods
	Phacelea Juss. Hydrophyllaceae: Greek *phakelos* = bundle; in allusion to the bunched flowers
	Soranthes Ldb, 1829: Greek *soros* = bundle, *anthos* = flower
Camphor	*Camphorosma*: Greek *kamphora* = camphor, *osme* = odour, scent; alluding to the shrub that smells like camphor
	Camphoratus, -a, -um; pertaining to or resembling camphor
Candle	*Cephalocereus*: Greek *kephale* = head, *cereus* = wax candle
	Cereus: Latin *cereus* = wax candle
	Chamaecereus: Greek *chamai* = ground, *cereus* = wax candle; referring to the prostrate habit
	Echinocereus: Greek *echinos* = porcupine, *cereus* = wax candle
Cane	*Bactris* Jacq. Palmae: Greek *Bactron* = cane; referring to the small stem used as walking stick
Canoe	*Phaseolus* L: Greek *phaseolus* = boat, canoe; for the shape of the bean
Cap	*Mitracarpum* Zucc. Rubiaceae: Greek *mitra* = head-dress or cap, *karpos* = fruit
	Mitragyna Korth. Rubiaceae; referring to cap-shaped ovary
	Mitrasacme Labill. Spigeliaceae
	Mitrastemon Makino. Rafflesiaceae
	Mitreola L. Spigeliaceae
	Mitrephora (Bl.) Hook. f. and Thoms. Annonaceae
	Pileostegia Hook. f. and Thoms. Hydrangeaceae: Greek *pilos* = felt cap, *stege* = roof
Carpenter	*Tectona* L.f. Verbenaceae: Greek *tekton* = builder or carpenter

Chaff	*Coelachyrum* Hochst. and Nees. Gramineae: Greek *koilos* = hollow, *achuron* = chaff; referring to the nature of the glume
	Coelachyropsis Bor. Gramineae
	Diplachne Beauv. Gramineae: Greek *diploos* = double, *achne* = chaff; referring to the nature of the spikelets
Chain	*Alyxia* Banks ex R. Br: Greek *halusis* = chain; in allusion to the fruits
Chessboard	*Fritillaria* L. Liliaceae: Latin *fritillus* = chessboard; referring to the chequered markings on the perianth of some species
Childbirth	*Aristolochia* L. Aristolochiaceae: Greek *aristos* = straight or upright (like an aristocrat/best), Latin *lochia* = discharge of fluid from the vagina after childbirth; referring to the supposed value in aiding childbirth
Children	*Aerides* Lour. Orchidaceae: Greek *aeri* = air or wind, *des* = children; referring to children of air
Clip	*Colutea* L. Pl; name used by Theophrastos from Greek *coluteon* = to shorten, clip.
Cloak	*Chisocheton* Blume. Meliaceae: Greek *cheton* = covering or cloak; referring to the nature of the flowers
	Hymenochlaena Bremek. Acanthaceae: Greek *hymen* = membrane, *chlaena* = covering or cloak; referring to the membranous covering
	Hymenolaena DC. Umbelliferae
	Notholaena R. Br: Greek *nothos* = spurious, *chlaena* = cloak
	Tricholaena Schrad ex Roem. and Schult. Gramineae: Greek *thrix* = hair, *chlaina* = cloak; referring to the plants with a covering of hairs
Clothing (outer)	*Hymenolaena* DC: Greek *hymen* = membrane, *laena* (for *chlaena*) = outer clothing
Club	*Cleidion* Bl. Euphorbiaceae: Greek *kleidion* = club or clavicle; referring to thickened female pedicels
	Cordyline Comm. ex Juss. Agavaceae: Greek *kordyle* = club; allusion to large fleshy roots of some species

	Cordyloblaste Hensch ex Moritzi. Symplocaceae: Greek *kordyle* = club, *blastos* = embryo or bud
	Cryptocoryne Fisch. ex Wydl. Araceae: Greek *kryptos* = hidden, *koryne* = club; referring to the shape of the flower
	Gomphia Schrad. Ochnaceae: Greek *gomphos* = a club or nail; referring to the shape of the fruit
	Rhopalocarpus Boj. Sphaerosepalaceae: Greek *rhopalon* = club, *karpos* = fruit
	Rhopalocnemis Jungh. Balanophoraceae: Greek *rhopalon* = club, *knemis* = knee
Coat	*Chlamydites* Drumm. Compositae: Greek *chlamys* = cloak, *mites* = suffix denoting "like" or "belonging to"
	Eriolaena DC. Sterculiaceae: Greek *erion* = wool, *chlaina* = cloak; referring to woolly calyx
	Trinia (Hall.) Scop: Latin *tunica* = coat, cover, skin
	Tunica Scop. Caryophyllaceae: Latin *tunica* = coat, referring to the bracts closely filling the calyx
Coin	*Soldanella* L. Name derived from the word *solidus* (Italian *soldo*); referring to the shape of leaf similar to a coin known by this name, or from the Italian word *soldana*; referring to the corolla, which is cut into narrow lobes
Comb	*Cenolophium* Koch: Greek *kenos* = empty, *lophos* = comb; ribs of the fruit inflated hollow
	Ctenolepis Hook. f. Cucurbitaceae: Greek *kteis*, *ktenos* = comb, *lepis* = scale; referring to comb-like arrangement of scales
	Dactyloctenium Willd. Gramineae: Greek *dactylos* = finger, *kteris* = comb; referring to the digitate comb-like appearance
	Scandix L: Greek *xandix* = comb; referring to the appearance of the ripe umbel of *Scandix pecten-veneris* L.
Comb, monkey's	*Pithecoctenium*. Begoniaceae: Greek *pithos* = large, *ctenoid/ktenos* = comb; meaning monkey's comb
	ctenoides: Greek *ktenos* = comb, *eidos* = form; resembling a comb

Companionship	*Hetaeria* Bl. Orchidaceae. Greek *hetaireia* = companionship; indicating the intimate association of plants of this genus with genera of the tribe Neottieae
Cone	*Conocephalus* Bl. Urticaceae: Greek *konos* = cone, *kephale* = head; referring to the form of flowers
Container	*Halletheca* T N Taylor, 1971. Pennsylvanian Pteridosperm Pollen Organ. The generic name Halletheca is proposed in honour of T G Halle in recognition of his work with pteridosperm pollen organs: Greek *theca* = container
	Melissiotheca Brigitte Meyer-Berthaud, 1986. Fossil Pteridosperm Pollen Organ: Greek *melission* = honeycomb, *theca* = container; referring to the appearance of the synangium when cut in transverse section
Cord	*Hedera* L. Araliaceae. Celtic *hedera* = cord
Cork	*Quercus phellos* L. Fagaceae: Greek *phellos* = cork
Cornice	*Geissanthera* Schlechter: Greek *geison* = cornice, *anther* = flower; referring to the peculiar bristle-like secondary bracts opposite each leaf and flowers, which may be likened to ornate cornices
Cosmetic	*Anchusa* L. Boraginaceae: Greek *ankousa* = cosmetic paint for skin (*A. tinctoria* has been used for staining skin since ancient times. In ancient Greece, women used the root paste of this plant as rouge).
Cotton	*Bombax* L. Bombacaceae: Greek *bombyx* = cotton; referring to the cotton in the pods
Couch	*Stromanthe* Marantaceae: Greek *stroma* = spread (like a couch), *anthos* = flower; allusion to the form of inflorescence
Covering	*Adenocaulon* Hook: Greek *aden* = gland, *chlaena* = covering; referring to the dense glandular nature
	Adenocalymma. Bignoniaceae. *Aden* = gland, *calymma* = covering
	Derris Lour. Papilionaceae: Greek *derris* = skin or leathery cover; referring to the pods
	Diplycosia Blume. Ericaceae: Greek *diploos* = double, *kos* = covering

Elymus L. Gramineae: Greek *elyo* = to cover; the leaves of *E. maritimus* are woven into a coarse fabric

Elyonurus Humb. and Bonpl. ex Willd. Gramineae

Elytranthe Bl. Loranthaceae: Greek *elytron* = cover or sheath, *anthos* = flower; referring to long, sheath-like flowers

Epithema Blume. Gesneriaceae: Greek *epithema* = cover; referring to the nature of the flowers

Eucalyptus L' Herit. Myrtaceae: Greek *eu* = well, *kalypto* = to cover; referring to the united calyx lobes and petals forming a lid

Wrightia tomentosa. Apocynaceae: Latin *tomentosa* = densely covered with hairs

Cross

Crucianella: Latin *crux* = cross; the leaves being cross-wise

Stauranthera Benth. Gesneriaceae: Greek *stauros* = cross, *anther* = anther; referring to coherent anthers forming a cross

Staurogyne Wall. Acanthaceae

Stauropsis Reichb. f. Orchidaceae

Cruciatus, -a, -um; in the form of a cross

Crucifer, crucifera, -um

Crux-andrae, meaning St Andrew's Cross

Crux-maltae, meaning Maltese cross

Cup

Calathodes Hook. f. and Thoms. Ranunculaceae: Greek *kalathos* = cup, oides = resembling; referring to the flowers resembling a cup

Cotula. Compositae: Greek = small cup; the base of leaves for cups

Cotyledon L. Crassulaceae: Greek *kotyledon*, from *kotyle* = cavity or small cup; referring to cup-like leaves

Cyathea Cyatheaceae: Greek *kyatheion* = little cup; in allusion to the spore cases

Cyathodes: Greek *kyathodes* = cup-like; in allusion to the cup-shaped, toothed disc

Kollodepas Hassk. Euphorbiaceae: Greek *kollos* = hollow, *depas* = cup.

Poterium L. Rosaceae: Latin *poterium* = cup, from Greek *poterion* = drinking cup

Pterocymbium R. Br. Sterculiaceae: Greek *pteros* = wing, *kymbe* = cupor boat; referring to its fruits

Scyphiphora Gaertn f. Rubiaceae: Greek *skyphos* = cup, *phora* = bearing; referring to oblong fruits with six to eight ridges

Cymbal *Cymbalaria*: Latin *cymbalum* = cymbal; referring to the leaf shape

Dagger *Machaerocereus*: Greek *machaira* = dagger; the longest spine is knife-like

Dart *Ecbolium* Kurz. Acanthaceae: Greek *ec* = out of, *bole* = a throw, dart; referring to ejection of seeds from the capsule

Sporobolus R. Br. Gramineae: Greek *sporos* = seed, *bolis* = dart, *boleo* = to throw; referring to seeds that are let loose

Dirty *Ryparosa* Bl. Flacourtiaceae: Greek *rhyparos* = dirty; refering to the dirty hair

Disc *Corallodiscus* Batalin. Gesneriaceae: Greek *korallion* = coral, *discus* = disk; in allusion to the colour of disc

Cryptodiscus Schrenk. Greek *cryptos* = hidden, *discos* = disc; referring to stylopodium

Diplodiscus Turcz. Tiliaceae: Greek *diploos* = double, *diskos* = disc; referring to the disc present in the flowers

Disocactus Pfeiffer Cactaceae: Greek *diskos* = disc; resembling the round and nearly flat body of the genus

Discocainia J Reid et Funk. Fungi: Heotiales: Greek *diskos* = disc; the surname of Professor Roy F Cain (teacher and colleague of the authors), in whose honour this fungus is named

Dish *Biscutella*: Latin *bis* = twice, *scutella* = small flat dish; in allusion to the form of the fruit

Hydrocotyle L. Umbellifrae: Greek *hydor* = water, *kotyle* = dish, umbilicus; referring to the habitat of *H. vulgaris* L. and to the shape of its leaves

	Lecanthus Wedd. Urticaceae: Greek *lekone* = dish or pot
Door	*Biforea* Hoffm: Latin *biforis* = two-doored (*bis* = twice, *feris* = door); referring to two perforations in the pericarp near the commissure
	Dicliptera Juss. Acanthaceae: Greek *diklis* = double/folding door, two-valved, *pieron* = wing; referring to the wing-like compartments in the valves
Dress	*Apodytes* E Mey. ex Arn. Icacinaceae: Greek *apodutos* = undressed; in allusion to the minute calyx for large flowers
	Tulipa Liliaceae Persian. Persian *Tulipan* = turban; resemblance of the flower
Drinking cup	*Poterium* L: Greek *poterion* = drinking cup; name of a plant by Dioscorides and Pliny
Dung	*Coprotus* Korf and Kimbr. Fungi: Ascomycetes: Greek *kopros* = dung, *otos* = ear, from the generic name *otidea*
	Sterculia L. Sterculiaceae: Latin *stercus* = dung; in allusion to the foetid smell of the flowers.
Dust	*Conyza* Less. Compositae: Greek *konis* = dust; referring to the powdery covering on the plant
Dye	*Baphia*: Greek *baphe* = dye; the tree provides the bar of cam wood of commerce, which yields a red dye
	Baphicacanthus Bremek. Acanthaceae: Greek *bapto* = to dye, *akanthos* = spiny; the spiny nature of the plant yielding a blue dye (*B cusia* [Nees] Bremek)
	Baptisia: Greek *bapto* = to dye; it has sometimes been used as a substitute for true indigo
	Betony. French, Latin *betonica* (Vettonica is a Gaulish tribe that used these plants as medicine and for dyeing); a common British Labiatae plant growing in woods, it was of great repute in ancient and medieval medicine and was also used to dye wool yellow
	Carthamus L. Compositae. Arabic *quarton* = to point; referring to the dye obtained from the flowers for colouring food
	Chrozophora: Greek *chrozo* = to tinge, dye, stain, *phorus* = carrying

	Dracaena draco. Greek *drakaina* = dragon's blood; a red colour used in the varnish industry
	Ficus infectoria. Moraceae: Latin *infectus* (past particple of *inficere)* = to dye, infectoria = dye yielding
	Indigofera L, 1753. Papilionaceae: Latin *indigo* or *indicum* = name of a blue dye originally obtained from India, *ferre* = to bear, yield
	Rubia L. Rubiaceae: Latin *ruber* = red; in allusion to the reddish dye obtained from roots
Dye (red)	*Anchusa*, a plant yielding a red dye called anchisin, similar to the dye obtained from *Rubia*
Earth	*Geodorum* G. Jacks. Orchidaceae: Greek *ge* = earth, *doron* = gift; meaning this plant is a gift of earth
Envelope	*Elytrophorus* P Beauv. Gramineae: Greek *elytron* = envelope, *phoros* = bearing; referring to the inflorescence
Fan	*Coccothrinax* Sargent. Palmae: Greek *kokkos* = berry, *thrinax* = fan; from the fancied morphology of leaves and fruits resembling a berry
	Thrinax. Palmae: Greek *thrinax* = fan; the shape of the leaves
Fence	*Aphragmus* Andrzej. Cruciferae: Greek *a* = without, *phragmos* = fence or partition; referring to the fruit without partititons
	Fraxinus L. Oleaceae: Greek *fraxis* = to divide; referring to the wood being easily cleaved or its being used to erect fences, thus dividing land; it is also possible that the actual derivation is from Latin *frangere*, meaning to bear; in allusion to the fragility of ash branches
Fibre	*Fibraurea* Lour. Menispermaceae. Latin *fibra* = fibre, filament; referring to the climbing nature
Fire	*Pyracantha* Roem: Greek *pyr* = fire, *acanthos* = thorn
Flame	*Paraphlomis* Prain. Labiatae: Greek *para* = beside, *phlogmos* = flame
	Phlogacanthus Nees. Acanthaceae: Greek *phlogos* = a flame, hence reddish, and *akanthos*
	Phlox L. Polemoniaceae

Flask	*Lagenandra* Dalz. Araceae: Greek *lagenos* = flask, *aner* = man
	Lagenaria Ser. Cucurbitaceae; referring to the bottle-shaped fruit of some species
	Lagenifera Cass. Compositae
Flat	*Planea* Karis, 1990. Asteraceae: Latin *planus* = flat; referring to the distinctly flat leaf margins
	Ampullaceus, -a, -um; flask-like
Flax	*Camelina* Crantz: Greek *khamai* = low, *linon* = flax; a plant suppressing flax, not allowing it to grow
Flower	*Chalcanthus* Boiss: Greek *chalcos* = copper, *anthos* = flower
Forest	*Drymaria* Willd. ex Roem and Schult. Caryophyllaceae: Greek *drymos* = forest; referring to the habitat of the plant
Fountain	*Ceropegia* L. Asclepiadaceae: Greek *keros* = wax, *pege* = fountain; referring to the milky exudation from the plant of *C. noorjahaniae* Ansari
Fragrance	*Anethum.* Umbelliferae: Greek *ana* = through, *ethein* = burn; referring to the burning taste of the fruits. Others derive from Greek *anethon*, Aristophane's name for the plant closely similar to *anison*, from the Greek *aeni*, meaning fragrance
Fraternity	*Hemiadelphis* Nees. Acanthaceae: Greek *hemi* = half, *adelphia* = fraternity
Funnel	*Chonemorpha* G. Don. Apocynaceae: Greek *chone* = funnel, *morpha* = form; referring to the funnel-shaped flowers
Garland	*Actinostemma* Griff. Cucurbitaceae: Greek *aktin* = ray, *stemma* = wreath or garland; referring to the radiating branches
	Agrostemma: Greek *agros* = field, *stemma* = crown, garland
	Cyathostemma Griff. Annonaceae: Greek *kyathos* = ladle, *stemma* = garland; indicating ladle-shaped fruits
	Gomphostemma Wall. Labiatae: Greek *gomphos* = nail, *stemma* = crown or garland

	Gynostemma Blume. Cucurbitaceae: Greek *gyne* = female, *stemma* = wreath or garland; indicating that the ovules are arranged in the form of a garland inside the ovary
Garment	*Borago* Boraginaceae: Latin *burra* = hairy garment; in allusion to hairy leaves
Gift	*Armodorum* Breda. Orchidaceae: Greek *harmos* = crack in the wall, *doron* = gift
	Dorema D. Don. Umbelliferae: Greek *dorema* = gift; in allusion to the rich flow of resins
	Pandorea Spach. Bignoniaceae: Greek *pan* = all, *dora* = gift
Girdle	*Brachystelma* R. Br. Asclepiadaceae: Greek *brachys* = short, *stelma* = girdle; referring to the short coronal processes of the flowers
	Zostera L. Zosteraceae: Greek *zoster* = girdle or band
Glove	*Digitalis* L. Scrophulariaceae: Latin *digitus* = finger; referring to the flowers that look like the fingers of a glove
Glue	*Collomia:* Greek *kolla* = glue; the seeds are mucilaginous when wet
	Gluta L. Anacardiaceae: Latinized form of *gluten* = glue; a resin derived from bark
Gum	*Copaifera:* Brazilian *copaiba* = balsam, Latin *fere* = to break; the trees yield copal, a commercial gum used in making varnish
	copallinus, -a, -um; gummy
Goblet	*Caltha* L. Ranunculaceae: Greek *kalathos* = goblet; the form of the corolla is goblet-like
	Coelodepas Hassk. Euphorbiaceae: Greek *koilos* = hollow, *depas* = goblet; referring to the cup-shaped calyx in female flowers
	Cypella: Greek *kypellon* = goblet
Gold	*Bracteola* Jason R. Gramineae: Latin *brattea* = thin metal plate; a bract is a small leaf growing at the base of a flower: Latin *bracteole* = dimunitive form of a bract; a thin gold leaf; referring to the golden, shining compressed spikelets

Chrysalidocarpus Wendl. Palmae: Greek *chrysos* = gold, *karpos* = fruit; referring to the golden fruit

Chrysanthelium Rich. Compositae: Greek *chrysos* = gold, *anthellum* = flower; referring to small flowered chrysanthemum

Chrysanthemum L. Compositae: Greek *chrysos* = gold, *anthos* = flower

Chrysobalanus L. Chrysobalanaceae: Greek *chrysos* = gold, *balanos* = acorn; referring to the yellow fruit

Chrysobraya Hara. Cruciferae: Greek *chrysos* = gold, *bray* = to pound or crush into fine bits/grind into a powder; probably referring to the spread of the golden pollen

Chrysoglossum Fl. Orchidaceae: Greek *chrysos* = gold, *glossa* = tongue; referring to the golden tounge-shaped labellum of the flower

Chrysogonum L. Compositae: Greek *chrysos* = gold, *gonos* = seed; referring to the shiny seeds

Chrysophyllum L. Sapotaceae: Greek *chrysos* = gold, *phyllon* = leaf; referring to the shiny leaves

Chrysopogon Trin. Gramineae: Greek *chrysos* = gold, *pogon* = beard; referring to the silky hairs on the inflorescence

Chrysosplenium L. Saxifragaceae: Greek *chrysos* = gold, Latin *splenius*/Greek *splenion* = bandage; referring to the compressed shiny spike

Aurum = gold (the metal), represented by the symbol Au. It was once believed that gold had botanical origin, till the Swiss metallurgist Agricola refuted it. The reason for this belief was that the metal chiefly occurs embedded in igneous rocks in a crystalline, wiry, branch-like form, which gives it the appearance of a plant in the rocks

Ground

Chamaepericlymenum Graebn: Greek *chamai* = on the ground; *periclymenon*: Greek *peri* = around, *clymene* from Greek Mythology referring to the mother of Atlas and Prometheus, referring to the prostrate stem circling on the ground; Disocorides named it for a prostrate or twining plant

Ethnobotanical Noah's Ark 17

	Chamaesciadium C. A. M: Greek *chamai* = on the ground, *scias* = umbrella; providing a shadow on the ground
Grove	*Alseodaphne* Nees. Lauraceae: Greek *alsos* = grove, *daphne* = bay laurel; referring to the affinity and habitat
Guard	*Diphylax* Hook. f. Orchidaceae: Greek *dis* = double, *phylux* = guard; referring to the two slender spurs of the column one on each side of the anther
Guest	*Kleinhovia hospita*. Sterculiaceae: Latin *hospes* = guest. This name is believed to be used in reference to Kleinhoff's well-known hospitality
Hammer	*Sphyranthera* Hook. f. Euphorbiaceae: Greek *sphyra* = hammer, *anther* = anther
Hair dye	*Xanthium* L: Greek *xanthos* = yellow; referring to the colouring matter obtained from the plant
Handle	*Floscopa* Lour. Commelinaceae: Latin *flos* = flower, *kope* = handle; referring to inflorescence
Hat	*Petasites* Mill. Compositae: Greek *petasos* = hat with a broad rim; referring to the large, broad leaves
	Pilea Lindl. Urticaceae: Greek *pileos* = hat; referrring to the enlarged sepal covering the achene in some species
Hatchet	*Pelecyphora*: Greek *pelekys* = axe or hatchet, *phorein* = to carry; as the tubercles resemble the blade of a hatchet
Headband	*Oxymitra* (Bl.) Hook. f. and Thoms. Annonaceae: Greek *oxys* = acid, *mitra* = headband
Heap	*Agathis* Salisbury. Coniferae: Greek *agathis* = claw, heap; referring to the crowded staminate aments
	Camptosorus Link. Polypodiaceae: Greek *comptein* = to bend, *sorus* = heap, pile (sorus)
	Pleurosoriopsis Fom: Greek *pleura* = film, *soros* = heap, *opsis* = appearance; resembling *Pleurosorus*
	Struchium P. Br. Compositae: Greek *struius* = heap
Hedge	*Haplophragma* Dop. Bignoniaceae: Greek *haplos* = single, *phragma* = hedge; indicating the partition of the capsules

	Phragmites Adans. Gramineae: Greek *phragmos* = hedge; referring to the formation of hedge-like growth along marshy fields
Heat	*Aethionema* R. Br: Greek *aithos*= burning heat, *nema* = thread
Helmet	*Brachycorythiis* Lindl. Orchidaceae: Greek *brachys* = short, *korys* = helmet; referring to the cucullate labellum remotely resembling a helmet
	Corylus L. Corylaceae: Latin *korys* = helmet; referring to the calyx covering the fruit
	Craniostome Reichb. Labiatae: Greek *kraneion* = helmet, *temmo* = to cut; referring to the shape of floral parts
	Galeola Lour. Orchidaceae: Latin *galea* = helmet, *galeola* = helmet-shaped vessel; referring to the form of the lip
Hole	*Fallopia* Adans. Polygonaceae: Greek *fullo* = to deceive, *ope* = opening or hole
	Malope L. Malvaceae: Greek *malos* = delicate, *ope* = hole, slit
	Opilia Roxb. Opiliaceae: Greek *ope* = opening or hole, *eilo* = to close up
	Trema Lour. Ulmaceae: Greek *trema* = a hole indicating pitted seeds
Hollow	*Cicuta* L. Umbelliferae: Greek *klein* = to be hollow; referring to the hollow, fistular stem
	Stenocoelium Ldb: Greek *stenos* = narrow, *coelos* = hollow; referring to the narrow valleculae
Honey	*Melianthus* L. Melianthaceae: Greek *meli* = honey, *anthos* = flower
	Melicope J. R. and G. Forst. Rutaceae
	Melinis Beauv. Gramineae
	Meliosma Bl. Meliosmaceae
	Melodorum Hook. f. and Thoms. Annonaceae
	Melilotus Adans. Leguminosae: Greek *meli* = honey, *lotos* = name applied to many plants, including certain clovers
Hood	*cucullaris, -is, -e*

cucullatus, -a, -um; hood-like, having sides or apex curved inwards to resemble a hood

Hook   *Uncifera* Lindl. Orchidaceae: Latin *uncus* = hook or barb, *fero* = to bear

Horn   *Aceras* R. Br. Orchidaceae: Greek *a* = without, *keras* = horn; from the absence of a spur or calcar

Acroceras Stapf. Gramineae: Greek *akros* = terminal, *keras* = horn; indicating the nature of terminal glume

Aegiceras Gaertn. Myrsinaceae: Greek *keras* = horn; referrring to the nature of fruit

Carpoceras Boiss: Greek *carpos* = fruit, *keras* = horned

Ceratocarpus L. Chenopodiaceae: Greek *ceras* = horn, *carpos* = fruit; referring to the shape of the fruit

Ceratochloa Beauv. Gramineae: Greek *keras* = horn, *chloe* = grass; referring to seeds with horns

Cerastium L. Caryophyllaceae: *keras* = horn; in allusion to the shape of the capsule in some species

Ceratocarpus L.; referring to the shape of the fruit

Ceratocephalus Moench: Greek *cerato* = horny, *cephale* = head

Ceratonia L. Caesalpiniaceae: Greek *keratonia* = name of the tree, *keras* = horn; referring to the shape of the pods

Ceratophyllum L. Ceratophyllaceae: Greek *keras* = horn, *phyllon* = leaf; referring to leaves resembling a stag's antlers

Ceratostigma Bunge. Plumbaginaceae: Greek *keras* = horn, *stigma* = stigma; referring to the shape of stigma

Ceratostylis Blume. Orchidaceae: Greek *keras* = horn, *stylis* = style; referring to the shape of the fleshy column

Ceratotheca Endl. Pedaliaceae: Greek *keras* = horn, *theke* = capsule; referring to the shape of the capsule

Christolea Camb: Greek *okto* = eight, *keras* = horn

Cornus L. Cornaceae. Roman name for the Latin *cornus*, meaning horn; referring to the wood, which is durable and hard as horn

Dittoceras Hook. f. Asclepiadaceae: Greek *dittos* = double, *keras* = horn; referring to the corona lobes

Thelycrania (Dumort.) Fourr: Greek *thelys* = feminine, *kraneia* = horn; a literal translation of "Cornus foemina" of the Romans, who referred to *T. sanguinea* as feminine and "Cornus mascula" as masculine

Diploclisia Miers. Menispermaceae: Greek *diploos* = double, *klisia* = enclosure or hut; referring to the ovary chamber

Jade   *Ushania* K H Keng, 1957. Gramineae: The name *Yushania* (*yu* = jade and *shan* = mountain) is a Latinized form of two components of the Chinese name (in national dialect) of the locality of *Y. nitikamensis* type species of the genus, a renowned geographical feature of Formosa (the island of Taiwan). Yshan is known in Japanese as *Niitakayama* and in English as Mount Morrison.

Knife   *Machaerocereus*: Greek *machaira* = dagger; the longest spine is knife-like

Knife blade   *cultratus, -a, um*; Latin, knife-shaped

Cultriformis, -s, -e; shaped like a knife blade

Knot   *Atylosia* Wt and Arn. Papilionaceae: Greek *a* = without, *tulos* = knot; referring to the absence of callus on the standard petal

Hammada Iljin. Chenopodiaceae: Greek *hamma* = knot

Ladder   *Scaligeria* DC. Umbelliferae: Greek *scala* = ladder, staircase, *gero* = to carry or bear

Ladle   *Cyathocalyx* Champ. ex Hook. f. and Thoms. Annonaceae: Greek *kyathos* = ladle, calyx; referring to the ladle-shaped calyx

	Cyathocline Cass. Compositae: Greek *kyathos* = ladle, *kline* = bed indicating shape of leaves
	Cyathopus Stapf. Gramineae: Greek *kyathos* = ladle, *pous* = foot; referring to the stoloniferous culm in the shape of a ladle
	Cyathostemma Griff. Annonaceae: Greek *kyathos* = ladle, *stemma* = garland; indicates ladle-shaped fruits
	Cyathula Blume. Amaranthaceae: Greek *kyathos* = ladle
	Spathodea Beauv. Bignoniaceae: Greek *spathe* = laddle; allusion to the shape of the calyx
Lamp	*Lychnis* L. Caryophyllaceae: Greek *lychnos* = lamp
Lamp's wick	*Lychnis* L. Name occurring in the writings of Theophrastus, derived from Greek *lychnos*, meaning lamp; as the leaves of *Lychnis coronaria* (L.) Desv. used for wicks
Lancet	*Callistemon lanceolatus.* Myrtaceae: Greek *kallos* = beauty (referring to the beautiful flowers), Latin = *lanceolatus* (referring to the lancet-shaped)
	Dalbergia lanceolaria. Papilionaceae. Lancet-like; referring to the shape of the pod
Lead	*Plumbago* L. Plumbaginaceae: Latin *plumbum* = lead; some species are said to be remedy against lead poisoning
Leather	*Coriaria* L. Coriariaceae: Latin *corium* = leather; indicating that some species like *C. nepalensis* are used in tanning
Lottery	*Clerodendron* L. Verbenaceae: Greek *kleros* = a lot of/many (as in a lottery); name given by Burman, suggested by Hermann's Latin name for *C. infortunatum*
Lyre	*Citharexylum* Mill. Verbenaceae: Greek *kithara* = lyre, *xylon* = wood; referring to the wood used to make musical instruments
Mallet	*Tupidanthus* Hook. f. and T Thoms. Araliaceae: Greek *tupis* = mallet, *anthos* = flower; referring to the shape of the flower bud
	Tupistra Ker-Gawl. Liliaceae

Manure	*Sterculia* L. Sterculiaceae: Latin *stereus* = muck or manure; allusion to the evil smell of the flowers of *S. foetida*
Marriage	*Gamosepalum* Schlechter. Orchidaceae: Greek *sepalum* = sepal, *gamos* = marriage; in allusion to the union of sepals with the petals and the lip into a more or less tubular structure
Mausoleum	*Mausolea* Bunge ex Pojakov: Greek *mausolus* = mausol, the tsar of Karia (about 360 BC), in memory of whom a splendid tomb called "mausoleum" was erected
Measure	*Cotula* L. Compositae: Greek *kotula* = graduated liquid measure used in medicine; referring to the fancied resemblance of the bases of the leaves
	Modiola Moench. Malvaceae; a small measure or nave of a waterwheel; in allusion to the whorled orientation of carpels
Milk	*Astragalus* L. Leguminosae: Greek *astron* = star, *gala* = milk
	Galactia P. Br. Papilionaceae: Greek *galaktos* = milky; referring to the milky juice
	Galium L. Rubiaceae: Greek *gala* = milk; referring to the practice of using *G. verum* for curdling of milk in the process of cheese-making
	Lactuca Linn. Compositae: From the Latin name of the lettuce *Lactuca sativa*, derived from *lac*, meaning milk; in allusion to its milky juice
	Papaver L. Papaveraceae: Latin *papa, pap* = thick milk
	Opopanax Koch. Umbelliferae: *Opos* = milky, *pan* = all, *akos* = remedy
	Polygala L. Polygalaceae: Greek *polys* = much, *gala* = milk; referring to its reputation as good cattle fodder
Monkey's comb	*Pithococtenium* Mart. ex DC. Bignoniaceae: Greek *pithos* = large, *ktenoeides* = comb-shaped
Mound	*Aggericutis*, 1984: Greek *aggeris* = heap, mound; referring to the domed papillae of the epidermal cells

Musical instrument	*Citharexylum* Mill. Verbenaceae: Greek *kithara* = lyre, *xylon* = wood; from the use of wood to make musical instruments
	Jatropha panduraefolia. Euphorbiaceae: Latin *pandûra* = fiddle shaped; referring to the shape of the leaves
	Sambucus L. Sambucaceae: *Sambuke* = stringed musical instrument, parts of which were sometimes made of the elder tree
Nail	*Gomphandra* Wall. ex Lindl. Icacinaceae: Greek *gomphos* = bolt or nail, *andra* = androecium; referring to the nature of stamens
Necklace	*Ormopterum* Schischk: Greek *ormos* = necklace, *pteron* = wing; referring to the unique sculpture of the fruit surface
	Ormosciadium Boiss: Greek *ormos* = necklace, crow, *scias* = umbrella
	Ormosia Jacks. Papilionaceae. *Ormos* = necklace; referring to the scarlet seeds of *O. coccinea* strung as beads in the form of a necklace
Needle	*Acantholimon* Boiss: Greek *acanthi* = needle, *leimon* = meadow, meadow grass
	Chamaeraphis R. Br. Gramineae: Greek *chamai* = dwarf, *raphis* = needle; indicating the nature of the spikelets
	Parkinsonia aculeata. Caesalpiniaceae. Specific epithet *aculeate*, meaning "a small needle"; referring to the small leaflets
	Rhapidophyton Iljin: Greek *raphis* = needle (alluding to the leaf shape), *phyton* = plant
	Triraphis R. Br. Gramineae: Greek *tri* = three, *raphis* = needle; referring to the presence of bristles in the spikelets
	Acicularis = shaped like a needle
	Aciculifer = needle forming
Nest	*Nidularium* Lem. Bromeliaceae: Latin *nidus* = nest
Net	*Dictyosperma* Wendl. and Drude. Palmae: Greek *diction* = net, *sperma* = seed; referring to the nature of the seeds

	Hymenodictyon Wall. Rubiaceae: Greek *hymena* = membrane, *diktyon* = net; referring to the nature of the seeds
Odour	*Olax* L. Olacaceae: Latin *olax* = odorous
Ornament	*Chlidanthus*: Greek *chlide* = luxury, costly ornament
	Dictyocoprotus, 1991. Fungi: Pyrenonemataceae: Greek *diction* = net; the generic name *Coprotus* refers to the reticulate ornamentation on the walls of the ascospores but resembles *Coprotus*, which has been combined to form the new name *Dictyocoprotus*
	Perculticuus Kovach and Dilcher, 1984 (Fossil cuticle): Latin *percultus* = highly adorned; referring to the cuticular ornamentation characteristic of this genus
	anthos, meaning flower
	comptus, -a, -um; meaning adorned, ornamented
Paint	*Carthamus*. From an Arabic word meaning to paint; in allusion to the brilliant colour yielded by the flowers
Paper	*Chartoloma* Bunge: Greek *khartos*, Latin *charta* = paper, and Greek *loma* = joint, silicle
Partition	*Heterophragma* DC. Bignoniaceae: Greek *heteros* = different, *phragma* = a partition; in allusion to the peculiar four-angled divisions of the ovary
Perfume	*Myrtus* L. Myrtaceae: Greek *myron* = perfume
Poison (fish)	*Walsura piscidia*. Piscidia: Latin *piscis* = fish, *caedo* = to kill; in allusion to the use of bark as a fish poison
Polish	*Delima* L: Latin *delimare* = to polish or smoothen; in allusion to the use of rough leaves. The native name has the same meaning. The native name for *D. sarmentosa* is korasawel.
Potash	*Kalidium* Moq; apparently in allusion to the potash content of the plants
Sac	*Thylacospermum* Fenzl: *thylacos* = sac, *sperma* = sac
Saline	*Halocharis* Moq: Greek *hals* = salt, *charis* = beauty, grace

	Halanthium C. Koch: Greek *hals* = salt, *xylon* = tree
	Halimodendron Fisch: *hals* = saline, *dendron* = tree
	Halimocnemis C. AM: Greek *halimos* = saline, *cnemis* = shank
	Halocnemum M. B: Greek *hals* = salt, *cnemis* = sheath
	Halogeton C. AM: Greek *hals* = salt, *geiton* = neighbour
	Halopeplis: Greek *hals* = salt, *peplis* = name of a plant of family Lythraceae
	Halostachys C. AM: Greek *hals* = salt, *stachys* = spike
	Halotis Bunge: Greek *hals* = salt, *anthos* = flower
	Haloxylon Bunge: Greek *halos* = salt, *xylon* = tree
Salty taste	*Salsola* L. A plant name first used by Cesalpino; in allusion to the salty taste, to *Halogeton sativus*, which was earlier included in the genus *Salsola* (Ulbrich)
Sandal	*Crepidatus, -a, um*; shaped like a sandal or slipper
Scent	*Camphorosma* L: Greek *camphora* = camphor, *cosme* = scent
	Chaerophyllum L: Greek name referring to the agreeable scent of the foliage
Sheath	*Coleoptryphe*: Greek *koleos* = sheath, *tryphe* = hole
	Coleus: Greek *koleos* = sheath; in allusion to the way in which the stamens are enclosed
Shuttle	*Cercis* L: Greek *kerkis* = shuttle; a tree name mentioned by Theophrastus and Aristotle; possibly an allusion to pod shape
Silver	*Argyrolobium* Eckl. et Zeyh: Greek *argyros* = silver, *lobos* = pod
Shoe	*Calceolaria*: Latin *calceolus* = slipper; in allusion to the form of the flower
Silk	*Bombax* L: Greek *bombyx* = silk; in allusion to the fluffy, silky hairs filling the seed capsule
Smoke	*Fumaria* L: Latin *fumus* = smoke, fume
	Fumariola Korsh, 1898: Fumariaceae. This is the name of one more genus formed from the Latin term *fumus*.

Soap	*Saponaria* L: Greek *sapon* = soap; the roots of some species being used as substitute for soap
Sour	*Oxyria* Hill: Greek *oxys* = sour
Sovereign	*Hegemone* Bunge: Greek *hegemone* = sovereign
Spoon	*Cochlearia* L: Latin *cochlear* = spoon; from the shape of the basal leaves of the plant
	Cochlearifolius, -a, -um; with leaves like *Cochlearia*
	Cochlearis, -is, -e; spoon-shaped
Stick	*Bacularia*: Latin *baculum* = walking stick
Stone	*Petroselinum* Hoffm: Greek *petros* = stone or rock, *selinon* = wreath; referring to the use of the plant in making wreaths
Strap	*Loranthus* L: *loron* = strap, *anthos* = flower
Sweep, to	*Calluna* Salisb: Greek *kallunein* = to sweep; brooms are made of the twigs of this plant
Sweet	*Glycine* L: Greek *glyk* = sweet; because of the sweetness of the roots
	Glycyrrhiza L: *glycys* = sweet, *rhiza* = root; Pliny's name for the plant
Sweet juice	*Glinus* Loefl: Greek *glinos* = sweet juice
Sweet smell	*Hedysarum* L: Greek *hedys-arum* = sweet smell; the name given to the genus because of the sweet smell of some of its species
Taste, burning	*Caryota urens* L: From the very hot burning taste of the fruit
Thread	*Aethionema*: Greek *aethes* = unusual, *aitho* = to scorch, *nema* = thread
	Caleonema: Greek *koleos* = sheath, *nema* = thread; referring to the filaments of the stamens, which are folded in the petals
Totie	*Ligustrum* L: A name mentioned by Virgil in the Bucolics. The name might have been derived from the Latin word *ligare*, meaning to tie/bind; referring to the use of thin and pliable branchlets in wickerwork
Tunic	*Brachychiton*: Greek *brachys* = short, *chiton* = tunic

Umbrella	*Chamaesciadium* C. AM: Greek *chamai* = on the ground, *scias* = umbrella
	Grammosciadium DC: *gramma* = stripe or line, *scias* = umbrella; presumably referring to the linear subulate lobes of the involucre
	Helosciadium Koch: Greek *helos* = swamp, *scias* = umbrella
	Ormosciadium Boiss: Greek *ormos* = necklace or crown, *scias* = umbrella
Varnish	*Phytolacca* L: Greek *phyton* = plant, *kicca* = varnish
Vessel	*Aerangis*: Greek *aer* = air, *aggos* (*angos*) or *aggeion* (*angeion*)
	Angophora: Greek *aggeion* = vessel, *phoreo* = to carry; in allusion to the form of the fruit
	Hydrangea L: Greek *hydro* = water, *angeion* = vessel; the fruit is shaped like a bowl
Walking staff	*Bactris*: Greek *bactron* = a walking staff; suggesting the way the young stems of palm are used
Water	*Apium* L. Celtic *apon* = water; referring to the plant's preference for damp habitats; or from Latin *apis*, meaning bee, as bees readily gather on it to collect nectar
	Aquilegia L. Russian *vodosbor* is an exact translation of the Latin *aqua* meaning water and *lego* meaning collect.
	Hydrangea L.
	Hydrocotyle L: Greek *hydro* = water, *kotyle* = dish, umbilicus; referring to the habitat of *H. vulgaris* L. and the shape of the leaves
	Salix L: Celtic *sal* = near, *lis* = water; in allusion to the predominant habitat
	Sium L: Celtic *siu* = water
Wedge	*Sphenocarpus* Korov: Greek *sphen* = wedge, *carpon* = fruit
Wheel	*Actinidia*: Greek *aktis* = ray; in allusion to the styles that radiate like the spokes of a wheel
	Rotala L: Latin *rota* = wheel; from the whorled leaves of *R. verticillaris* L.

Web	*Catatrama* Anna E Franco-Molano, 1991 (fungi): Greek *cata* = down, Latin *trama* = web; the hyphae of the trama are oriented downwards.
Wing	*Ormopterum* Schischk: *ormos* = necklace, *pteron* = wing; referring to the unique sculpture of the fruit surface
	Polylophoium Boiss: Greek *poly* = many, *lophos* = elevation, crest, wing
	Tilia L: Greek *pillion* = wing; in allusion to the peduncle furnished with a wing-like bract
Wool	*Bulbocodium*: Greek *bolbos* = bulb, *kodion* = wool
	Eriosynaphe DC: Greek *erion* = wool, *synaphe* = commissure, point of contact
	Lachnoloma Bunge: Greek *lakhnos* = wool, *loma* = segment, silicle
Wash, to	*Laurus* L: Ancient Latin name of the tree, meaning obscure; perhaps from *lavate*, meaning to wash, or from *laus*, meaning pride
Wreath	*Agrostemma* L: Greek *agros* = field, *stemma* = wreath; the plant being useful for inclusion in wreaths of field plants
	Coronaria L: Latin *corona* = wreath, crown; in allusion to the shape of the corolla
	Petroselinum Hoffm: Greek *petros* = stone or rock, *selinon* = wreath; referring to the use of the plant in making wreaths
	Selinum L: Greek *selinon* = an ancient name for an umbellifer, whose leaves are used to make wreaths; Greek *selas* = lustre (lustrous leaves), alternatively from *helisso* = twisting, winding; referring to the above-mentioned use of the leaves

BIBLIOGRAPHY

Chandra S. 1995. **Ethnobotanical ark**. *Ethnobotany* **7**: 105–116

Pant D D and Nautiyal D D. 1963. **On *Maheshwariella bicornuta* gen. et. sp. nov. A compressed seed from lower Gondwana of Karharbari coalfield, India**. *Journal of the Indian Botanical Society* **42A**: 150–158

2

Plant Nomenclature: an Overview

Bharati Bhattacharyya

INTRODUCTION

Nomenclature is the assignment of names to any object, entity or taxon, usually to assist in an easy means of reference. It is a symbol of communication and a reference base for storage, retrieval or simply documentation of information. Human beings have been nomenclaturists since time immemorial. Even in our lives, we search for names for babies, sometimes even before they are born. Of course, there are no rules and recommendations for naming a human being. A family name attached to a suitable first name suggested for a newborn is usually sufficient, although there may be exceptions.

However, naming plants, animals, and other biological species is not an easy task. Plants have always been useful to man; even prehistoric man could probably distinguish between useful, useless, and harmful plants, and in doing so, they must have given them names. There is, however, no record of these names because they would be as unreliable as the records of early civilizations. During the Middle Ages, scientific achievements in Europe were at a very low profile. At the turn of the 12th and 13th centuries, a scientific revival, an intellectual revolution called the Renaissance, began. This was a period of a little over 1000 years, between the fall of the Roman Empire and the Renaissance. During this period, a ninth-century monk, Wahlafrid Strabo, composed the poem, *De Cultura Hortorum,* in which he mentioned the names of the herbs growing in the garden of the monastery and their medicinal uses. This goes to show that plants were considered important and significant enough to make a note of them, even during such early years.

The first written account about plants (and therefore their names) was manuscripts in Greek and Latin. Modern botanical names are scientific names by which taxonomic groups of plants can be recognized. These names are either Latin words or words Latinized from some other language, most often Greek. Botanists all over the world can talk about any plant growing in any part of the world if they know the Latin name for it. In modern nomenclature, each plant can have only one correct scientific name, which should be in Latin. This was not the situation during the medieval period. When scientific nomenclature had not been introduced, long sentences were used to help recognize the plants. For example, today's *Grevillea robusta* might have been named *"Grevillea robusta, grandiflora, australiana"*. In the herbal of Clusius published in 1583, a species of willow (present day *Salix*) was named *Salix pumila angustifolia altera* (Jones and Luchsinger 1987). These names are known as polynomials.

It was not feasible for botanists to use this complex name–description system. The arrival of an increasing number of plants from new areas of exploration caused concern as it became a cumbersome process and expansion of a name became impossible.

But why do we need scientific names at all? The reasons are obvious. Students of taxonomy need to learn the scientific names of plants, without which they cannot progress in their identification, classification, and description, determine correct names or come up with new names. Furthermore, it would not be possible for them to determine which plants are rare, endangered or threatened.

Plants have aesthetic value; they are economically useful, and we are dependent on plants for our very existence and survival, be it food, clothing, shelter, drugs for healing or even water and oxygen. It is imperative to have a globally common symbol of communication or scientific nomenclature that can be understood and used by all, especially taxonomists. As a matter of fact, the scientific names must become the everyday vocabulary of such workers.

Many students are not keen to study taxonomy and sometimes even botany as a whole. The most common reason given is difficulty in "learning and memorizing scientific names for plants", especially since they do not have any knowledge of Greek or Latin. However, if the meanings of a number of scientific terms and their roots are learnt, it is quite easy to pronounce and remember the scientific names of plants and the characteristics of most of them. Table 1 lists some common Latin words used as specific epithets, along with their respective meanings. The Latin used in botanical nomenclature is not the classical Latin of the scholars. Instead, it is a more or less popular Latin spoken by common people during the Middle Ages and thus found its way into plant nomenclature. But why only Latin? The reasons are: Latin is specific and exact in its meaning. Its preciseness and conciseness make it pertinent to the needs of descriptive phrases of natural sciences. Since Latin is

Table 1 Some common Latin specific epithets and their meaning

Latin specific epithet	Meaning
Alba	white
Amabile	lovely
aquatic	in water
Arborea	tree-like
Aureus	golden
Borealis	northern
Bracteata	leafy bracts on flower stalks
Bicolour	two-coloured
Bulbiferum	bulbil-bearing
Cerifera	wax-bearing
Coccineus	scarlet
Communis	gregarious
caespitose	closely tufted
Dichotoma	twice forked
Decumbens	prostrate
Dendroideum	tree-like
Dulcis	sweet
Edulis	edible
echioides	echium-like
erecta	upright
fluitans	floating
fumaraefolia	leaves like those of *Fumaria*
Flava	yellow
Foetida	ill-scented
Granulata	rough
Giganteum	gigantic
Grandiflora	large-flowered
Glabra	smooth
Herbaceum	herbaceous
Humilis	dwarf
Hortensis	of gardens
Indica	of India
Isophylla	equal-leaved
Incana	woolly
Japonica	of Japan
Linearis	narrow, linear
Longifolia	long-leaved
Marginata	margined
Magnus	large
Minuta	very small
mexicana	of Mexico
Nitida	smooth and lustrous
Nigra	black
Ochroleuca	yellowish white
Officinalis	of the shop (plants always kept "in stock" by herbalists)

Contd ...

Table 1 Contd...	
Latin specific epithet	Meaning
Obovata	obovate (leaves)
Prismaticum	cut like a prism
Palustris	of marshes or swamps
Purpureus	purple
Paniculata	flowers in a panicle
Quadrifolia	four leaves or four lobes to the leaves
Repens	creeping
rara	rare
reticulata	netted or striped
Roseus	rose-coloured
Speciosa	showy
Spectabilis	notable
Scapous	having a scape
Sulcatus	furrowed
Tridentatus	with three spines
Tenellus	slender
Terrestris	growing on dry ground
Usitatissimum	most commonly used
Uncinatus	hooked
Vulgaris	common
Virens	green
Velutinous	velvety
Venusta	lovely
Winteri	the aromatic bark (winter's bark)
Zeylanicus	of Ceylon

written in the Roman alphabet, any confusion created by the use of a language with different scripts such as Chinese, Japanese, Russian or Sanskrit, are avoided. Moreover, Latin is a "dead" language now and, hence, cannot arouse political controversy. Objections, however, have been voiced from many quarters against the use of Latin for plant names. Kelsey and Dayton (1942) tried to introduce English nomenclature using common English names or anglicized Latin names for plants. Using any spoken language, especially the national language of any country, is quite difficult for naming plants. It is even more difficult as no world flora has been published with English or any other "vernacular" names.

One may also wonder why common names of plants are not used. There is no dearth of common names in any language. Benson (1962) provided the following reasons as to why vernacular or common names cannot replace Latin or Latinized names of plants.

Names in any common vernacular are ordinarily applicable in only a single language. For example, the common name for a member of family Asteraceae is Chrysanthemum in English, *Guldavari* in Hindi, and *Chandramallika* in Bangla. Which of these is to be selected so that everyone can recognize the plant? Devil's tree (in English) is *Chhatim* in Hindi and Bangla, *Saptaparni* in Sanskrit, *Chatinan* in Odiya, *Satvin* in Marathi, *Mukampalai* in Tamil, and *Palaigh* in Telegu. The Latin name for the plant is *Alstonia scholaris*, which can be used globally without giving any particular vernacular importance.

In most parts of the world, relatively few species have common or vernacular names in any language. Many ornamental plants such as *Aster, Dahlia*, and *Phlox* do not have alternative vernacular names.

Common names are applied indiscriminately to genera, species or varieties. For example, lotus is an English common name for *Nelumbo nucifera* of the family Nelumbonaceae; lotus is a specific epithet in *Nymphaea lotus* (water lily) of Nymphaeaceae; and lotus is a generic name in *Lotus corniculatus* (birds' foot trefoil) of Leguminosae.

Often, two or more unrelated plants are known by the same name. Frequently, even in one language, a single species may have two to several common names applied either in the same or different localities. The common name "lily" is used for many genera of the Liliaceae (*Lilium, Erythronium, Hemerocallis*), Iridaceae (*Belamcanda, Nemastylis*), Amaryllidaceae (*Crinum, Zephyranthes*), and Zingiberaceae (*Hedychium*). *Spathodea campanulata* is known by four different names in English: "squirt tree", "scarlet bell", "fountain tree", and "African tulip" (Bhattacharyya 2009).

For plant nomenclaturists, the two most important pioneering works are *Genera Plantarum* published in 1737 and *Species Plantarum* published in 1753 by Linnaeus. Earlier in 1735, Linnaeus published the well-known "Sexual System", an artificial classification system in *Systema Naturae*. This work classified all known animals, minerals, and plants of that period. Both *Genera Plantarum* (1737) and *Species Plantarum* (1753) were based on this system and were very useful and popular during those days. The *Genera Plantarum* listed and described plant genera recognized by Linnaeus; many names suggested by Bauhin and Tournefort were also given genus rank. Comparatively new names were either descriptive or commemorative names (after the names of distinguished botanists), such as *Bauhinia, Dioscorea, Fuchsia, Lobelia, Magnolia, Tournefortia*, and *Linnaea*. Together, *Genera Plantarum* and *Species Plantarum* include published information for 1105 genera and 7700 species names.

Generic descriptions are given in *Genera Plantarum* but not in *Species Plantarum*, which enumerates the number, name, and a brief description of various species under each genus. References to earlier literature and synonyms (if present) are mentioned in *Species Plantarum*, along with habitats and countries of origin. Thus, the description of each species turned out to be in the form of a "phrase–name" of up to 12 words, commencing with the genus name. Some of these "phrase–names" were new but many were taken from earlier publications of Linnaeus and other botanists. For a new species without any earlier reference, a longer descriptive paragraph was prepared. This polynomial descriptive phrase was intended to serve as a definition of the species. For genera with only one species, no description for the species was written, as the description for the genus (in *Genera Plantarum*) was considered sufficient.

In addition to the "phrase–name", each species was also provided with a "trivial name", later known as the "specific epithet", usually written in the margin. It appears that these names "originated as an indexer's paper-saving device" (Stace 1989). Essentially, it was shorthand for the real name—the polynomial "phrase–name" (Jones and Luchsinger 1987).

The convenience of using the trivial names given in the margin in combination with the generic names became obvious and, very soon, the standard names of each species. This resulted in the "binomial nomenclature". It was not even mandatory to use a keyword from the phrase–name for these specific epithets, such as the third example in Table 2.

Linnaeus, however, was not the first person to use two names comprising a binomial for each species. More than 100 years earlier, Gaspard Bauhin made use of binomials in his publication *Pinax* (1623),

Table 2 Examples from species *Plantarum, Linnaeus* (1737)

Specific epithet	Polynomial description
ovata	PONTEDERIA foliis ovatis, floribus capitatus.
	Narukila. Rheed. mal, 11.p.
	67.34. Raj. Suppl. 573
	Habitat in Malabariae aquafis.
	A part of phrase–name "ovatis" has been used as the specific epithet
europaeum	253, Habitat in Europe, australis
	The specific epithet here is not based on part of phrase–name but the habitat
glauca	SERRATULA foliis ovato-oblongis accuminatus serratis, floribus corymbosis, calycibus subrotundis
	Habitat Marilandia, Virginia, Carolina
	Here the specific epithet is not a part of the phrase–name

Note In all three examples, the generic name is followed by the polynomial descriptive "phrase–name". On the margin is the "trivial name" or "specific epithet". The generic names and specific epithets together form the binomials: *Pontederia ovata, Heliotropium europaeum,* and *Serratula glauca* though not consistently. It is not accurate to state that Linnaeus devised this method, but Linnaeus's nomenclatural pattern survives as most nomenclaturists still follow his method for naming new taxa. *Species Plantarum* (1753) marks the first consistent use of binomial nomenclature. It has subsequently been adopted by botanists all over the world as the starting point for botanical nomenclature. Names published before *Species Plantarum*, that is, before 1 May 1753, are of no significance in nomenclatural priority. However, there are some exceptions where certain earlier names have been conserved. It is a fact that Linnaeus and his contemporaries brought a large degree of order and stability in the nomenclatural confusion that existed till almost the middle of the 18th century.

PRESENT SITUATION

For a long time, there were no generally accepted rules for naming plants except for Linnaeus's "Principles of Nomenclature". By common accord, it was accepted that no two genera could have the same generic name and that no two species within a genus could bear the same species name. Also, when a genus was divided into two or more genera (as a result of some taxonomic research), the original generic name was to be retained for one of these new genera; when a variety was recognized, the binomial ought to be a combination of the parent species and the varietal name. Priority of publication was given importance, but most authors avoid assigning different names to identical species or identical names to different species. During the early 19th century, the discovery of a number of new plants increased and there was much confusion regarding their nomenclature. Moreover, for a long time, there was no published data from which botanists from one part of the world might learn about the names already employed by other botanists elsewhere. The first important index of this nature was *Nomenclator Botanicus* written by Steudel in 1821, for flowering plants only. This was the forerunner of the famous *Index Kewensis Plantarum Phanerogamarum* originally compiled by B D Jackson during 1893–95 under the supervision of Sir J D Hooker.

As time passed, it became apparent that unified nomenclatural rules were necessary. It was Alphonse de Candolle who took the lead and organized the first meeting of about 150 botanists in Paris in 1867. This was the First International Botanical Congress, where the standardization and legislation of nomenclatural practices took root. The rules laid down in this Congress were known as the Paris Code of 1867, also known as de Candolle Rules.

The Rochester Code, 1892, introduced the Type concept (that is, every plant name should be permanently associated with a particular specimen) and more strict rules of priority even if the name was a tautonym (similar "generic name" and "specific epithet", for example, *Sassafras sassafras*).

The Vienna Code, 1905, according to which a tautonym was not accepted and Latin diagnosis was made compulsory for any new species. Moreover, the conservation of generic names was introduced and a list of conserved names was approved (*Nomina generica conservanda*). However, the American Code, 1907, accepted neither the list of conserved names nor the necessity of Latin diagnosis for new species.

It was only at the Cambridge Congress, 1930, that these differences of opinions were sorted out and an International Code of Nomenclature came into being. This Code came up after the Fifth International Botanical Congress and accepted the concept of the Type method, rejected tautonyms, made Latin diagnosis compulsory, and approved the conservation of generic names.

After this agreement, the Code has been amended every six years in following Botanical Congresses. The 16th International Botanical Congress was held at St Louis in 1999. The Code was published in 2000 (Greuter, McNeill, Barrie, *et al.* 2000). The 17th Congress was held at Vienna, Austria, in 2005. The Code was published in 2006 (McNeill, Barrie, Burdet, *et al.* 2006). The 18th International Botanical Congress was held in July 2011 in Melbourne, Australia. The nomenclatural section was held during 18–22 July 2011, before the commencement of the Congress. Changes to the International Code of Botanical Nomenclature were voted upon during this time (Appendix 1).

Modern botanists all over the world assign names using a formal system. The criteria for naming organisms considered plants are based on the rules and recommendations of the International Code of Botanical Nomenclature (ICBN) (McNeill, Barrie, Burdet, *et al.* 2006). Similarly, for animal groups, there is the International Code of Zoological Nomenclature (ICZN) and for prokaryotes, the International Code for Nomenclature of Bacteria (ICNB). One difficulty is that photosynthetic bacteria are named under both ICBN and ICNB. Similarly, some of the so-called protists (itself a paraphyletic assemblage) are named under both ICBN and ICZN (Judd, Campbell, Kellogg, *et al.* 2008). In this process, some organisms have two names originating from two different nomenclatural codes.

In a simple and precise manner, the ICBN deals with (1) terms that denote the ranks of taxonomic groups or units and (2) scientific names

applied to the individual taxonomic groups of plants (Greuter 1988). The ICBN has two basic activities.
1. Naming new taxa hitherto unnamed and probably not yet described.
2. Determining the correct name for previously named taxa, which might have been divided, united, transferred or changed in rank after careful taxonomic revision.

PRINCIPLES OF NOMENCLATURE

The principles of the ICBN from the Vienna Code, 2005 (McNeill, Barrie, Burdet, *et al.* 2006) are listed as follows.

- Botanical nomenclature is independent of zoological and bacteriological nomenclature. The Code applies equally to names of taxonomic groups treated as plants, whether or not these groups were originally so treated.
- Application of names of taxonomic groups is determined by means of nomenclatural types.
- Nomenclature of a taxonomic group is based on priority of publication.
- Each taxonomic group with a particular circumscription, position, and rank can bear only one correct name, the earliest that is in accordance with the Rules, except in specified cases.
- Scientific names of taxonomic groups are treated as Latin regardless of their derivation.
- The rules of nomenclature are retroactive unless expressly limited.

The detailed provisions are divided into rules and recommendations. The objectives of the rules are to put the nomenclature of the past into order. These are organized as "articles" and are binding. The objective of the recommendations is to try and bring about greater uniformity and clarity, particularly in future nomenclature. Recommended names should await the formation of rules for their application. These are non-binding but preferred. The rules and recommendations of the ICBN are applicable to all living organisms treated as plants (including fungi but excluding bacteria and fossils). As already mentioned, the nomenclature of bacteria is governed by the ICNB. Cyanobacteria is a problematic group that was proposed to be named under the ICNB but was already named under the ICBN. As the ultimate goal of taxonomists is to achieve a universal "BioCode" that would encompass all living organisms, it is desirable that both botanical and bacteriological taxonomists use unified rules for the nomenclature of taxa of cyanobacteria, cyanophyta or blue–green algae.

Furthermore, there should be a single taxonomic classification scheme (Oren 2004).

Principle I: Botanical Nomenclature is Independent of Zoological Nomenclature

In basic principles, the two Codes are similar. The independence of the two Codes has often resulted in a similar name assigned to both animals and plants. For example, *Cecropia* is a moth according to zoological nomenclature, although it sometimes refers to a fast-growing pioneer tree from tropical America belonging to the family Cecropiaceae of order Urticales (Mabberley 1997). Another example is *Pieris,* cabbage butterflies according to zoological nomenclature and an ornamental shrub of family Ericaceae according to botanical nomenclature.

In addition to the green plant clade, the botanical Code includes eukaryotic clades such as stramenopiles, some alveolates (the dinoflagellates), rhodophytes (red algae), fungi, euglenoids, and slime moulds. Some euglenoids and dinoflagellates are considered animals and are, hence, governed by the nomenclatural rules of the zoological Code. Therefore, such organisms may have two names—one under the botanical Code and another under the zoological Code (Judd, Campbell, Kellogg, *et al.* 2008).

Principle II: Application of Names to Taxonomic Groups is Determined by Means of Nomenclatural Types

According to this principle, the name of each species is permanently attached or associated with a particular specimen—the nomenclatural type. The following types are recognized.

Holotype: The specimen or other element used or designated by the author in the original publication as the "main nomenclatural type". Any type selected after the original publication is not regarded as a holotype. It is essential that a holotype designated for a newly described species be deposited in a national herbarium (Box 1).

Isotype: A duplicate specimen of a holotype. These are plants forming part of the same gathering as the holotype or growing with it and gathered at the same time, by the same person. Isotypes are important as they are reliable duplicates of the same taxon and may be placed in many other herbaria so that they are easily available to taxonomists of various regions.

Syntype: One, two or more specimens studied and cited by the author, when the holotype is not designated by him/her (Box 1). A duplicate of a syntype is an isosyntype.

Paratype: A specimen cited in the original publication but not designated as holotype, isotype or syntype (Box 1).

Lectotype or *neotype*: When the author fails to designate a holotype or the holotype is missing, a lectotype or a neotype is selected to serve as the nomenclatural type. A lectotype is a specimen selected from material cited by the author: from the original material—isotype, syntype (Box 1) or isosyntype. A neotype is a specimen selected from the material not cited by the author with the original description. A neotype is selected only when all the original specimens collected and cited by the author are missing.

Epitype: A specimen or illustration selected to serve as an interpretative type if the holotype, isotype, lectotype or neotype is ambiguous with respect to the identification of the particular taxon (Simpson 2010). The type for a genus is the name of any one species (usually the first named species); for a family, it is the name of a genus and for an order, it is the name of a family.

BOX 1   **Information on the "TYPE" specimens of *Impatiens thomsonii* Hook. f.**

Impatiens thomsonii Hook. f. is a member of the family Balsaminaceae and its description is given in the *Flora of British India*. The author has cited three specimens on which the description was based.

Specimen 1: Collected by Thomson from Piti and Kunawar.

Specimen 2: Collected by Strach and Wint from the Kumaon and Garhwal Hills.

Specimen 3: Collected by J D Hooker from Sikkim.

For each specimen, the number, place, and date of collection and the name of the collector are given. Hooker designated specimen 3 as holotype and specimens 1 and 2 as paratypes. If Hooker had not designated specimen 3 as holotype, all three specimens would have been syntypes. One of these syntypes can serve as a lectotype if the holoype is missing. If all three specimens are destroyed for some reason, a fourth specimen (collected by Wallich from Sikkim), which does not find mention in Hooker's description, will be treated as a neotype. Duplicate specimens (if any) collected by Hooker, along with the holotype, are treated as isotypes.

Principle III: Nomenclature of a Taxonomic Group is Based upon Priority of Publication

The rule of priority states, "For any taxon from family to genus inclusive, the correct name is the earliest legitimate one with the same rank, except in cases of limitation of priority by conservation". It also states, "The principle of priority does not apply to names of taxa above the rank of family".

A strict application of this rule may sometimes lead to confusion. Hence, certain well-known and frequently used specific, generic, and family names are conserved in preference to the earlier published but obscure names. There are three Amendments to the ICBN, *Nomina familiarum conservanda*, *Nomina generica conservanda et rejicienda*, and *Nomina specifica conservanda et rejicienda*, which help in conservation of the names. For example, Sterculiaceae Lindl. 1830 is a conserved name and not published earlier. The earlier published name for the same family is Byttneriaceae R. Br. 1814 (Lawrence 1951). The conservation of specific names is restricted to the names of species of major economic importance (Greuter 1981), and was adopted at the International Botanical Congress held at Sydney, Australia, in 1981.

Principle IV: Each Taxonomic Group with a Particular Circumscription, Position, and Rank can Bear Only one Correct Name that is a Validly and Effectively Published Name

According to the ICBN, a scientific name is formally recognized only if it is "validly published". There are four criteria for valid publication.

1. It should be an effective publication. All the names that are published in printed form in scientific journals and are available in botanical institutions with libraries, which are accessible to botanists in general, are effectively published names. Publishing a new name in nursery catalogues, newsprint, seed-exchange lists or email messages does not qualify as an effectively published name. A plant name is not effectively published if it is printed on a label attached to a herbarium specimen, even if the specimens are widely distributed (Jones and Luchsinger 1978).
2. A name is validly published when it is accompanied by a description or a reference to a previously published description of that taxon.
3. From 1 January 1935, names of new taxa of recent plants (with the exception of algae and fossils) must be accompanied by a Latin diagnosis for valid publication. The description itself need not be in Latin, although it is recommended. The description and diagnosis (the distinguishing features of the taxon as mentioned by the author) of new taxa published before 1 January 1935 are treated as valid even if they were in any modern language, including

Japanese, Russian or any other where Roman alphabets are not used.

4. The name of a taxon is not validly published if it is cited merely as a synonym.

Principle V: Scientific Names of Taxonomic Groups are Treated as Latin Regardless of their Derivation

According to this principle, generic names, specific epithets, infraspecific epithets, as well as the names of ranks higher than the genus should all be Latin or Latinized with the addition of prefixes and suffixes, whatever source they might have been taken from.

As mentioned earlier, the scientific names of species are binomials, comprising two parts. The first part of the binomial is a singular noun, the genus name to which the species belongs. The second part can be an adjective modifying the generic name, a noun in apposition or a possessive noun, and is called the specific epithet. Both the generic name and the specific epithet should be in italics when in print and underlined separately when typed or handwritten. The generic name always starts with a capital letter, whereas the specific epithet usually starts with a small letter, except where the use of a capital letter is permissible, for example, *Pinus Roxburghii*.

Most of the specific epithets refer to distinct morphological (for example, *Jacaranda mimosaefolia*: Mimosa-like leaves), ecological (for example, *Anemone sylvestris*: pertaining to woods; *Anemone rivularis*: pertaining to brooks and streams) or chemical (for example, *Cistus ladaniferus*: yielding ladanum, a resinous gum) features (Johnson and Smith 1958).

Some specific epithets refer to the geographic distribution of a species, such as *Menispermum canadense* (Canadian) and *Hamamelis japonica* (Japanese); some are in honour of the person who discovered the plant, such as *Spiraea Aitchinsonii* after Dr Aitchinson and *Oenothera lamarckiana* after Lamarck; while others are simply after a vernacular name: *Psidium guajava* after guava. Often, the specific epithet is made up of two hyphenated words, for example, *Hibiscus rosa-sinensis, Alisma plantago-aquatica*. If the person honoured is a man, the specific epithet ends in "*i*" or "*ii*", for example, *Musa cavendishii* (after William Cavendish, sixth Duke of Devonshire)*, Iris Monnieri* (after Monnier), *Isoetes Panchannani* (after Panchannan Maheshwari), *Abutilon Darwinii* (exception: *Nepenthes Rafflesiana*). If, however, the honoured person is a woman, the specific epithet ends in "*ae*", for example, *Plumbago Larpentae* (after Lady Larpent) and *Cereus Machdonaldiae* (after Ms Mac Donald).

The two parts of a plant name usually belong to the same gender and often have similar endings (Table 3). Usually, the generic ending "*-us*" is masculine, "*-a*" is feminine, and "*-um*" is neuter. By convention, all trees are considered feminine for nomenclatural purposes, with exceptions such as *Quercus rubra* and *Pinus nigra*, where the generic names are both masculine, whereas the specific epithets are feminine.

Table 3 Four different sets of name-endings used in Latin

Masculine	Feminine	Neuter
-us	-a	
Example, *sativus*	Example, *sativa*	Example, *sativum*
-er	-ra	-run
Example, *niger*	Example, *nigra*	Example, *nigrum*
-er	-ris	-re
Examples, *sylvester* and *campester*	Examples, *sylvestris* and *campestris*	Examples, *sylvestre* and *campestre*
-is	-is	-e
Examples, *humilis* and *occidentalis*	Examples, *humilis* and *occidentalis*	Examples, *humile* and *occidentale*

There are four main sets of name-endings in Latin (Table 3). In addition, there can be various sources for Latinized names, as given below.

1. Names from many vernacular languages, for example: *Salmalia* from *Shalmali* (Sanskrit), *Madhuca* from *Madhukaha* (Sanskrit), *Populus* from poplar (English), and *Lychnis* from *lychnos* (Greek). Some names are based on the local names from the areas where they occur, such as *Ginkgo* from the Chinese, *Fatsia* from Japanese, *Ravenala* from Madagascarian, and *Pandanus* from *Padang*, the Malaya name.

2. Some names reflect botanical character, for example, *Leonotis* is a flower having resemblance to a lion's ear, *Callicarpa* (beautiful fruits), and *Liriodendron* (tree with lily-like flowers).

3. Some genera are named in honour of botanists, for example, *Hookera*[1] (for Hooker), *Linnaea* (for Linnaeus), *Puschkinia*[2] (after Russian botanist M Pouschkin). Some are named after famous scientists, for example, *Diervilla*[3] (after M Dierville, a famous French surgeon), *Copernicia* (after Copernicus). Some are named after famous heads of state, for example, *Victoria* (after Queen Victoria), *Washingtonia* (after George Washington).

---

[1] Example taken from Johnson and Smith (1958).
[2] Example taken from Johnson and Smith (1958).
[3] Example taken from Johnson and Smith (1958).

4. Some generic names are of mythological origin, for example, *Narcissus* is named after the Greek god Narcissus. *Dracaena* red resin resembles dragon's blood and *Circaea* refers to Circe, the famous enchantress.
5. Some generic names originate from planets, for example, *Mercurialis* and *Neptunia*.
6. Some generic names are after a place, for example, *Salvadora* after El Salvador and *Heliconia*[4] from Greek Helicon, a hill in Greece.

To complete the botanical/scientific name of a particular plant, the binomial must be followed by the name of the person who identified and described the plant and suggested its name on the basis of the description. For example, *Sesamum indicum* L. means Linnaeus identified, described, and named this particular plant. In all taxonomic work, it is essential to cite the authority of the scientific names. Citation of the author's name is helpful if ever there is confusion due to two plants having the same name. Author's names can be abbreviated or written in full form. A list of abbreviated author's names was prepared by Mabberley (1997). If two persons have named a plant together, their names are joined by "et" or "&", for example, *Antigonon leptopus* Hook. & Arn.

Sometimes a name is proposed by an author but is not validly published, and a second author publishes it validly at a later date and ascribes it to the former author. In such a case, the name of the former author followed by the word "ex" should be inserted before the name of the second author. For example, *Cassia montana* Heyne *ex* Roth. Although Heyne proposed the name, valid publication was undertaken by Roth. When a name proposed, described, and diagnosed by one author is published under the work of another, the two authors' names are linked together by the word "in". For example, *Hygrophila salicifolia* (Vahl). Nees in Wall. and *Euonymus indicus* Heyne *ex* Wall. in Roxb.

The authors' names of higher taxa are usually omitted except in detailed monographic studies. In many floras, scientific publications, and journals, only the species and infraspecific taxa names are usually followed by full authorship.

Scientific names of higher taxa–genera and above are uninominals, meaning that they are represented by a single word. These names are Latinized plural nouns, for example, Umbellales and Apiaceae. The ICBN recognizes seven major ranks (Kingdom, Phylum (or Division), Class, Order, Family, Genus, and Species), but allows the insertion of intermediary ranks by adding the prefix "sub". A higher rank is always inclusive of all lower ranks. Taxa above the rank of genus are not underlined or italicized. Every rank has a specific ending (Table 4).

---

[4] Example taken from Johnson and Smith (1958).

Table 4 Taxonomic ranks and their endings as recognized by the International Code of Botanical Nomenclature

Taxonomic rank	Ending	Taxon as example (*Potentilla glandulosa var. navadensis*)
Kingdom	—	Plantae
Phylum/Division	-phyta	Magnoliophyta
Subphylum/Subdivision	-phytina	Magnoliophytina
Class	-opsida	Magnoliopsida
Subclass	-idae	Rosidae
Order	-ales	Rosales
Suborder	-ineae	—
Family	-aceae	Rosaceae
Subfamily	-oideae	Rosoideae
Tribe	-eae	Potentilleae
Subtribe	-inae	—
Genus	-any	*Potentilla*
Subgenus	-any	—
Section	-any	Drynocallus
Series	-any	—
Species	-any	*Glandulosa*
Subspecies	-any	—
Variety	-any	*navadensis*
Form	-any	—

Principle VI: Rules of Nomenclature are Retroactive Unless Expressly Limited

The principle of Later Homonym emerged from this principle. A name is a later homonym if it is spelt like a name previously and validly published for a taxon of the same rank, based upon a different type specimen. Different genera (of the same family or different families) and different species of the same genus cannot have the same name.

In such instances, the later-formed name or the later homonym is illegitimate and has to be rejected. For example, *Tapienanthus* Boiss ex Benth. 1848 of Labiatae is a later homonym of *Tapienanthus* Herb. 1837 of Amaryllidaceae, and must be rejected. Both *Viburnum fragrans* Bunge 1831 and *Viburnum fragrans* Lois. 1824 belong to the same family Caprifoliaceae, but the type specimens for the two are different. Therefore, the later homonym *V. fragrans* Bunge 1831 should be rejected. To indicate that a plant has a later homonym, the word "non" is used before the author's name of the later homonym and placed after the earlier homonym. For example, "*Viburnum fragrans* Lois. 1824 non Bunge 1831" indicates that this plant has a later homonym. If the plant name itself is a later homonym, the word "nec" is used before the author's name of the earlier homonym and placed after the later homonym, for example, *Viburnum farreri* Stearn 1966 is the new name for *V. fragrans* Bunge 1831 nec Lois. 1824.

The 15th International Botanical Congress was held at Yokohama, Japan, in 1993. The ICBN, called the Tokyo Code, adopted at this Congress was significantly different from the earlier Code (the Berlin Code). Some of the important changes are listed as follows.
- The rules on typification and effective publication had been clarified by creating a logical arrangement of the articles 7–10 and 19–31, respectively.
- The proposals for (1) conservation of species names and (2) rejection of any name that would cause a disadvantageous nomenclatural change were accepted by an overwhelming majority.
- An entirely new concept was incorporated in the Tokyo Code. This concerned the recognition of "Interpretative Type" to serve the requirement of typification when an established name cannot reliably be identified for the purpose of precise application of a name.
- This Code permitted the use of the term "Phylum" as an alternative to "Division".
- An extensive revision of Article 46 clarified the use of the prepositions "ex" and "in" in author citations.
- For valid publication of new taxon of fossil plants on or after 1 January 1996, there must be an accompanying description and diagnosis in Latin or English, (or a reference to such an earlier publication) and not in any language as before.
- The 15th International Botanical Congress proposed that after 1 January 2000 and after approval by the 16th Botanical Congress, new names must be registered.

NOMENCLATURE OF CULTIVATED PLANTS

The nomenclature of cultivated plants, or cultivars, is as important as that of wild plants. Plant breeders develop new plants after selecting desirable features of flowers and fruits and introducing these by various horticultural and floricultural methods. The names of these plants are known as "cultivar names". Most commercially available garden plants are the products of many generations of hybridization maintained and propagated by man. They occur in artificial population only.

The application of cultivar names is covered by the International Code of Nomenclature for Cultivated Plants (ICNCP). The term cultivar, a combination of "cultivated" and "variety", is internationally recognized.

It should not be confused with botanical varieties (usually a geographical race or morphologically different population). Individuals within the same cultivar are usually genetically identical, although genetically diverse plants can still be considered within the same cultivar if they retain the desired characters (Brickell 2004).

The cultivar names can be added to a binomial or a generic name. To make them stand out from the first part of a name, they are (1) not written in italics, (2) start with a capital letter, (3) placed within single or double quotation marks, and (4) preceded by the abbreviation "cv". For example, *Malva sylvestris* cv "Primly Blue" and *Rosa floribunda* cv "Blessings". Since 1 January 1996, however, the use of "cv" has been disallowed (Judd, Campbell, Kellogg, *et al.* 2008).

Cultivar names do not change, even if the binomial name to which it is attached is altered/modified. For example, the cornflower variety "Blue Diadem" accepted will remain so whether the binomial is *Centauria cyanus* or *Cyanus segetum* (earlier name, now changed to *Centauria cyanus*). Cultivars can be subordinate to names of genera, species, hybrids or even vernacular names. For example, *Vanda* "Miss Joaquin", *Camellia japonica* "Purple Dawn", *Bougainvillea spectabilis* "Mary Palmer", *Viburunum bodnantense* "Dawn", and Apple "Cox's Orange Pippin".

New cultivar names since 1 January 1959 must be accompanied by a description written in any modern language. They must not be the same as any botanical or common name of a genus or species, for example, *Camellia*, *Rosa*, Onion or Ginger are not permitted to be used as cultivar names. For many groups of plants such as orchids and roses, there are registration authorities that keep an inventory of all cultivars in the group concerned. Thus, any duplication of cultivar names or any type of misuse of already used names can be prevented. As with specific epithets, cultivar names should not be repeated within a genus.

NOMENCLATURE OF HYBRIDS

Hybrids can be formed between two different genera or species within one genus. The interspecific hybrid between *Erica ciliaris* and *Erica tetralix* has been given the hybrid name *Erica* × *watsonii*, where the multiplication sign (×) denotes its hybrid origin. Some hybrids are not given a separate name. Instead, they are referred to by quoting the parent species linked by the multiplication sign, for example, *Drassera pulchella* × *D. nitidula* and *Verbascum lychnite* × *V. nigrum*.

Intergeneric hybrids are given a new hybrid genus name and the different combinations of species are treated as specific epithets in their own right. For example, the hybrid *Mahonia aquifolium* × *Berberis sargentiana* has been named × *Mahoberberis aquisargentii*. Intergeneric hybrids can also be written as a formula, for example, *Cooperia* × *Zephyranthes* or simply by a formal name preceded by the multiplication sign, as in × *Cooperanthes* and × *Citroncirus* (= *Citrus* × *Poncirus*). The position of the multiplication sign before the formal name denotes that it is an intergeneric hybrid.

The offspring of two varieties of the same species are also known to occur. These are known as "intraspecific" or "intercategory" hybrids. However, no rule has been formulated for their nomenclature so far. There are a few special types of hybrids called "graft hybrids", which are chimeras. Here, the cells/tissues of the parents are mixed physically rather than genetically. Such plants are designated by inserting a plus sign (+) in between the parental genera or species names, for example, +*Laburnocytisus* is a graft hybrid between *Laburnum* + *Cytisus*. A graft hybrid between two species of *Rosa* is indicated as *Rosa webbiana* + *R. floribunda*.

A tri- or tetrageneric hybrid is also possible. Three Orchidaceae genera—*Sophronitis* × *Laelia* × *Cattleya*—were combined to form × *Sophrolaeliocattleya*. Such hybrids may be given a formal name after the person involved in growing or collecting these plants, and the name must have the ending "ara". For example, the name × *Potinara* has been imparted to a tetrageneric hybrid between *Brassavola* × *Cattleya* × *Laelia* × *Sophronitis*.

For both interspecific and intergeneric hybrids, binomials can be used with the correct position of the multiplication (×) sign.

Interspecific Hybrids
- *Calystegia* × *lucaua* is a hybrid of *Calystegia selpium* × *C. sylvatica*
- *Salix* × *capreola* is a hybrid of *Salix aurita* × *S. caprea*

Intergeneric Hybrids
- ×*Gymnaglossum jacksonii* is a hybrid of *Gymnademia conopsea* × *Caeloglossum viridae*
- ×*Agropogon lutosus* is a hybrid of *Agrostis stolonifera* × *Polypogon monspelinesis*.

HARMONIZATION OF NOMENCLATURE

As information on the world's living organisms (biota) is increasing—from bacteria and fungi to plants and animals—there is a rising need for regularized scientific names (Hawksworth 2011). The future endeavour

is, therefore, to bring about harmonization in the nomenclature of all living organisms. Biological organisms are a diversified group, which is why different organisms have been named following different codes of nomenclature. These codes are as follows.
- International Code of Botanical Nomenclature (ICBN) for plants.
- International Code of Zoological Nomenclature (ICZN) for animals.
- International Code of Nomenclature of Bacteria (ICNB), or Bacterial Code (BC) as it is presently known.
- International Code for Nomenclature of Cultivated Plants (ICNCP).
- International Code of Virus Classification and Nomenclature (ICVCN).

With so many different codes, it is rather confusing to apply the correct name. In addition, there are various sets of rules with different conventions for citing valid names and different formats for naming organisms belonging to the same rank. Although priority of publication (of names) is one of the major criteria followed by all codes, methods of determination of correct names may not always be similar.

According to Hawksworth (1995), "Biology as a science is unusual in that the objects of its study can be named according to five different codes as of now". At present, the problems with them are manifold.
- An inordinate amount of time and effort is devoted by the already declining population of trained taxonomists to the purely historical and bibliographical nature of nomenclatural investigation under the terms of existing codes. The expenditure incurred is also very high (McNeill 1996a).
- It is not clear which code should be followed for the nomenclature of "ambiregnal" organisms—whether they are plants, animals or bacteria.
- It is not clear which code should be followed for naming organisms whose current genetic affinity is well-established but whose traditional treatment has been in a different group.
- The development of new knowledge by electronic information retrieval, molecular systematics, and ecology has led to a situation where scientific names are often used without clear taxonomic context. This leads to homonymy (similar names) between plants, animals, and any other groups and, hence, becomes a source of confusion. Over the years, innumerable trials have been conducted to prepare a unified or harmonized code. BioCode and PhyloCode are two such proposals, which emerged following long discussions of over more than two decades.

HISTORICAL ACCOUNT

The advantages of having a unified or harmonized code was realized as early as 1985, during a symposium held at the Third International Congress of Systematic and Evolutionary Biology (ICSEB III) in Brighton, UK. Towards the end of 1985, a Standing Committee on Biological Nomenclature was established at the 22nd International Union of Biological Sciences (22nd IUBS) general assembly in Budapest, Hungary. In 1988, an adhoc group consisting of representatives of the committees in charge of the existing five codes met at Kew, UK. Discussions during the 14th International Botanical Congress in Berlin, Germany, in 1987, and support from IUBS and IAPT (International Association for Plant Taxonomists) led to the 1988 adhoc committee meet. This committee met to consider a common approach towards the protection of names in use at that time.

Two important meetings of this adhoc committee (in 1988, during the 23rd IUBS General Assembly in Canberra, Australia, and in 1989, during ICSEB IV meet at College Park, Maryland, USA) led to a major conference on *"Improving the Stability of Names"* at Kew, UK, in 1991.

At the 24th IUBS General Assembly meet in Amsterdam, the Netherlands, in 1991, a resolution was passed to encourage harmonization between various codes of nomenclature. An important meeting was held at Egham, UK, in March 1994, at the behest of the IUBS, International Union of Microbiological Societies (IUMS), and IAPT, supported by the Linnaean Society of London and the Royal Society of London. An agenda for future action in biological nomenclature was set during this meeting. The establishment of the International Committee on Bionomenclature (ICB) took place during the 25th IUBS General Assembly in Paris, France, later in 1994. The ICB then met in Egham, UK, in 1995, and formulated the first draft of the International Code of Bionomenclature. This document was later presented as the *Draft BioCode: the prospective international rules for the scientific naming of organisms*, at ICSEB V in Budapest, Hungary, in 1996 (McNeill 1996b).

The ICB met again at Egham, UK, in 1997 and after much debate, issued a revised version called the *Draft BioCode (1997)* (Greuter, Hawksworth, Mayo, *et al.* 1998).

From the beginning, the BioCode was regarded as something to deal with "names to be proposed in future" with the existing codes taking care of the names of the past. (All the five currently used codes have been referred to as "Special Codes" while discussing BioCode.) The BioCode and Special Codes were envisaged as operating on parallel lines. Some minor changes have been made in the existing codes to improve harmonization, but an agreed list of names and steps towards compulsory registration of new names are available only in bacteriology. However, with rapid evolution of databases in the present century, production of lists on a group basis has become more practical, and a much needed low-cost system of cataloguing newly proposed names has come up in the disciplines of botany, mycology, and zoology. On the other hand, classification systems of some groups have changed as a result of molecular phylogenetic studies and cladistic studies. With this, the problems of groups potentially being treated under different codes or with the possibility of transfer from one group to another increased. Keeping these developments and changes in mind, the ICB organized a meeting to consider issues such as mandatory registration for new scientific names in 2007 and a workshop on *"Tailoring Biological Nomenclature to User Needs"* in 2009. Both the meetings were held in London. It was decided in the 30th IUBS General Assembly meet in Cape Town, South Africa (in 2009) to renew the preparation of the BioCode. The ICB convened a workshop in Berlin, Germany, in October 2010, to produce an update of the *Draft BioCode (1997)*. Subsequent developments and changes in the five current codes and the possible uses of new technologies were also considered. This document is known as the *Draft BioCode (2011)* (Greuter, Garrity, Hawksworth, *et al.* 2011).

Unlike the *Draft BioCode (1997)*, the present version of BioCode does not aim to replace the current codes. Instead, the effort is to provide an overall common framework for the working of these codes alongside the BioCode.

There are only 35 Articles in the text of the *Draft BioCode (2011)* and wherever required, reference to the Special Codes has been made. New scientific names and nomenclatural acts must be registered with the assistance of recent technological progress. Presently, Latin diagnosis and description is used for plants, and any language for animals. In BioCode 2011, English and Latin descriptions for all organisms have been proposed. Other proposals are the protection of names and their attributes, banishing homonymy for future names, and the use of alternative terminations for suprageneric names in case of ambiregnal organisms. Additional ranks have been introduced, such as "profamily" between family and subfamily, "progenus" between genus and subgenus, and "prospecies" between species and subspecies, so that coordinated status rules can be implemented retroactively without any destabilizing effect on the existing system of names.

BIOCODE (2011)

The BioCode (2011) with 35 Articles has been divided into three parts.

Division I: Principles

The first division deals with the principles, of which there are nine, as follows.

- **Principle I:** The BioCode governs the formation and choice of scientific names of taxa but not the circumscription, position or rank of the taxa themselves. Nothing in this Code may be construed to restrict the freedom of taxonomic action.
- **Principle II:** Scientific nomenclature of organism builds on the Linnaean system of binary names (binominal) for species.
- **Principle III:** The application of names of taxa is determined by means of name-bearing types (hereafter referred to as types), although this principle does not apply to certain names at suprafamilial ranks.
- **Principle IV:** The nomenclature of a taxon is based on priority (precedence by date) of publication, although application of this principle is not mandatory at all ranks.
- **Principle V:** Each taxon in the family group, genus group or species group with a particular circumscription, position, and rank has only one accepted name, except as may be specified in a Special Code.
- **Principle VI:** Scientific names of taxa are treated as Latin, regardless of their derivation.
- **Principle VII:** The name, as applied to a taxon, is not to be changed without sufficient reason, based either on further taxonomic studies or on the necessity of giving up a name that is contrary to the rules of nomenclature.
- **Principle VIII:** In the absence of a relevant rule of where the consequences of rules are doubtful, established custom is followed (see Division III, point 5 of the *Draft BioCode*).
- **Principle IX:** The rules of nomenclature are retroactive, subject to any specified limitations.

Division II: Rules

The second division pertains to rules and is covered under three chapters, as follows.

Chapter I: Taxa and Ranks

The primary ranks of taxa in descending sequence are Kingdom, Phylum, Class, Order, Family, Genus, and Species. Secondary ranks of taxa are allowed and include Domain above Kingdom, Tribe between Family and Genus, Section and Series between Genus and Species, and Variety and Form below Species. If required, an even higher number of ranks can be made by adding prefixes super-, pro- or sub-. Thus, the use of Superfamilies, Progenera, and Subspecies is permitted.

A number of rank groups have also been recognized, such as "supra familial ranks" or all ranks above the family group. However, establishment of new names of infraspecific taxa is strongly discouraged, except where they are traditionally used.

Chapter II: Names

The second chapter pertains to names and has been divided into six sections.

Section 1: Status

Established Names are those published in accordance with the relevant Articles. In this Code, "name" means an established name, whether it is acceptable or unacceptable.

Acceptable Names are the names that are in accordance with the rules and are neither unacceptable nor illegitimate under the relevant Special Code. The name of a genus combined with an epithet is termed a "binomen"; the name of a species combined with a second epithet is a "trinomen". Binomina and trinomina are combinations. It is recommended that scientific names of all ranks preferably be in italics.

Section 2: Establishment

In order to be established, a name must be (1) new, (2) have the form required, and (3) comply with the special provisions of Articles 7–11.

As a rule, on or after a future date to be determined, various categories of organism names, nomenclatural acts, and any name of a new taxon and new combination must be registered with an established entity.

On or after a relevant future date, the name of a new taxon must be accompanied by a Latin or English description or by a bibliographic reference to a previously published description of the taxon.

For a new combination, a bibliographic reference to its basionym, its author, and place of original publication is a must.

The name of a new taxon of the rank of genus or below must be accompanied by its "type" and a specification of the institute where it is conserved, unless it is a published illustration.

The name of a new fossil botanical taxon and a non-fossil algal taxon of species or subordinate rank must be accompanied by an illustration or figure showing diagnostic features, in addition to the description or diagnosis, or by a bibliographic reference to a previously published illustration or figure.

Provision for an illustration or figure showing diagnostic features for all new taxa, particularly zoological fossils as well as ambiregnal and microscopic organisms, has been recommended.

If the corresponding genus or species name is established, the name of a subordinate taxon is also established.

Section 3: Registration

As already mentioned, registration of a name is a must and this is affected by submitting the relevant papers to a registering office designated by the international body. This information should be placed on a global electronic communication network and the corresponding date will be treated as the date of establishment.

Section 4: Typification

Most of the rules for typification are similar to those of the ICBN, particularly for the selection of lectotype and neotype. In the absence of any acceptable designated type for a species or subordinate taxon, a type may be designated, keeping by the rules in the relevant Special Code. The type designations must also be registered.

Section 5: Homonymy

Homonyms are identically spelt names based on different types. A family-group, genus-group or species-group name established on or after a future date (to be determined), unless conserved, is unacceptable if it is a later homonym. The rules followed are similar to those of the ICBN. However, it is recommended that later homonyms acceptable under the relevant Special Code and in current use should not be abandoned but proposed for conservation.

Section 6: Precedence

For precedence or priority, the date of a name is (1) attributed to it in the adopted List of Protected Names, (2) one on which it was validly published under the ICBN or the BC for unlisted names, (3) one that became available under the ICZN or (4) established under the present BioCode.

Names of organisms (except animals and algae) based on a non-fossil type take precedence over names of the same rank based on a fossil type.

"Provisions for conservation" or "limitation to priority" are the same as those in the Special Codes. Once an adopted List of Protected Names is available, entries can be added, modified or removed only by the mechanisms of conservation.

Chapter III: Rank Groups and Their Names

This chapter is divided into four sections and deals with the status of taxa of various ranks.

1. Taxa above the rank of family.
2. Family: group taxa and infrafamilial taxa.
3. Genus: group taxa and infrageneric taxa.
4. Species: group taxa and infraspecific taxa.

For (1) and (2), the names of taxa are nouns in plural starting with a capital letter. The termination of names in each rank has been specified. A name may be published with a Latin termination, but it retains its authorship and date. The name of a genus is a noun in singular and is written with a capital initial letter. It may not have the termination "virus", which is reserved for the names of viral genera.

Chapter IV: Provisions for Special Groups

Rules for naming hybrid plants are included in Chapter IV. Names for hybrids and their progeny are given following the rules of the botanical Code. These are not established names under the BioCode. Distinguishable groups of cultivated plants and fungi whose origin or selection is primarily due to intentional actions of mankind are not covered in this code. Fossil non-algal botanical taxa are "parataxa", for which the rules for nomenclature are provided by the botanical code.

Chapter V: Orthography and Gender of Names

Every established name is deemed to have a single correct orthography. There may be orthographical variants like names with various spellings, compounding, and inflectional forms of a name, although only one type is involved. These variants, including typographical errors, are treated as correctable errors.

A generic name is treated as a noun with masculine, feminine or neuter gender. Gender is established on the basis of classical Latin and Greek grammar. When a new generic name is submitted for registration without indication of gender or the indicated gender is contrary to the Code, the gender is assigned or corrected during registration.

Chapter VI: Authorship of Names

The rules for authorship of names are similar to those of the botanical Code.

Division III: Authority

The third division deals with authority. The BioCode is established under the joint authority of the IUBS and the IUMS, to be exercised through an inter-union—the ICB. The BioCode takes effect on being approved by the ICSEB or any other Congress that may take its place in future, subject to agreement of the members at the IUBS General Assembly or the IUMS Divisional Congress.

The ICB has the power to resolve present or future ambiguity concerning the provisions of the BioCode. The ICB, however, will not interfere with the activities of the nomenclature committees operating under the authority of Special Codes.

CONCLUSION

It can be said that understanding and conserving life on earth is impossible without precise and widespread communication about all kinds of organisms, and this requires a universally accepted mechanism of scientific nomenclature. As Greuter, Garrity, Hawksworth, *et al.* (2011) have pointed out, the *Draft BioCode (2011)* is a precise, coherent, and simple internationally usable system of nomenclature for all organisms, whether eukaryotic or prokaryotic, fossil or non-fossil. In this document, all established names are to be enlisted in the adopted List of Protected Names, and the Special Codes (INCB or BC, ICBN, and ICZN) are to govern names not yet listed. Separate rules have been established for the nomenclature of viruses. The *Draft BioCode (2011)* proposed by the consistent effort of the ICB is, therefore, a welcome avenue towards this realization. This is especially true because the BioCode does not endanger but instead complements the existing nomenclatural system of the Special Codes.

It has been decided by the members of the International Organization for Systematic and Evolutionary Biology (IOSEB) that the ICB will submit a revised BioCode (2011) in early 2012 for approval and further dissemination.

As a matter of fact, biological institutions and societies all over the world should cooperate in promoting the implementation of this coordinated and simplified biological nomenclature in the future.

ACKNOWLEDGEMENTS

The author would like to express her gratitude to Dr Prithipalsingh (retd), Associate Professor, Kirori Mal College, University of Delhi; Dr Aparajita Mohany, Assistant Professor, Gargi College, University of Delhi; and Dr Arindam Bhattacharyya, Scientist, Department of Science and Technology, New Delhi, for providing reference material for this paper.

BIBLIOGRAPHY

Benson L. 1962. *Plant Taxonomy: methods and principles*. New York: Ronald Press

Bhattacharyya B. 2009. *Systematic Botany*, 2nd edn. New Delhi: Narosa

Brickell C D. 2004. *International Code of Nomenclature for Cultivated Plants*, 7th edn. Belgium: International Society for Horticultural Science Gent-Oostakken. [*Acta Horticulturae* 657; *Regnum Vegetabile* 144]

Greuter W. 1981. *XIII International Botanical Congress*. *Taxon* **30**: 904–12

Greuter W. 1988. **International Code of Botanical Nomenclature**. *Regnum Vegetabile* **118**: 1–328

Greuter W, Garrity G, Hawksworth D L, Jahn R, Kirk P M, Knapp S, McNeill J, Michel E, Patterson D J, Pyle R, Tindall B J. 2011. **Draft BioCode (2011): principles and rules regulating the naming of organisms**. *Taxon* **60**: 201–212

Greuter W, Hawksworth D L, Mayo M A, McNeill J, Minelli A, Sneath P H A, Tindall B J, Trechane P, Tubbs P. 1998. **Draft BioCode (1997): the prospective international rules for the scientific names of organisms**. *Taxon* **47**: 127–50

Greuter W, McNeill J, Barrie R, Burdet H M, Demoulin V, Filgueiras T S, Nicolson D H, Silva P C, Skog J E, Trehane P, Turland N J, Hawksworth D L (eds and compilers). 2000. *International Code of Botanical Nomenclature. (St Louis Code)*. Königstein, Germany: Koeltz Scientific Books. [Adopted by the 16th International Botanical Congress St Louis, Missouri, July–August 1999: *Regnum Vegetabile* 138]

Hawksworth D L. 1995. **Steps along the road to a harmonized bionomenclature**. *Taxon* **44**: 447–456

Hawksworth D L. 2011. **Introducing the Draft BioCode (2011)**. *Taxon* **60**: 199–200

Johnson A T and Smith H A. 1958. *Plant Names Simplified: their pronunciation, derivation, and meaning*. London, UK: WH&L Collingridge

Jones S B and Luchsinger A E. 1987. *Plant Systematics*. New York: McGraw Hill

Judd W S, Campbell C S, Kellogg E A, Stevens P F, Donoghue M J. 2008. *Plant Systematics: a phylogenetic approach*, 3rd edn. Sunderland, Massachusetts: Sinauer Associates Inc.

Kelsey H P and Dayton W A (eds). 1942. *Standardized Plant Names*, 2nd edn. Harrisburg, Pennsylvania: J Horace Mc Farland Co. for American Joint Committee on Horticultural Nomenclature.

Knapp S, McNeill J, and Turland N J. 2011. **Changes to publication requirements made at the XVIII International Botanical Congress in Melbourne: What does e-publication mean for you?** *Taxon* **60**(5): 1498–1501

Lawrence G H M. 1951. *Taxonomy of Vascular Plants*. New York: Macmillan.

Mabberley D J. 1997. *The Plant Book: a portable dictionary of the vascular plants*, 2nd edn. Cambridge, UK: Cambridge University Press

McNeill J. 1996a. **The BioCode: integrated biological nomenclature for the 21st century?** In *Proceedings of a Mini-Symposium on Biological nomenclature in the 21st Century*, edited by James L Reveal. [*Mini-Symposium on Biological nomenclature in the 21st Century*, University of Maryland, USA, 4 November 1996]

McNeill J and Turland N J. 2011. **Major Changes to the Code of Nomenclature – Melbourne, July 2011**. *Taxon* **60**(5): 1495–1497

McNeill J. 1996b. **General introduction to the Draft BioCode**. In *Draft BioCode: the prospective international rules for the scientific names of organisms*, pp. 7–18, edited by W Greuter, M A Mayo, J McNeill, A Minelli, P H A Sneath, B J Tindall, P Trehane, and P Tubbs. Paris: International Union of Biological Sciences

McNeill J, Barrie F R, Burdet H M, Demoulin V, Hawksworth D L, Marhold K, Nicolson D H, Prado J, Silva P C, Skog J E, Wiersema J H, Turland NJ (eds). 2006. *International Code of Botanical Nomenclature. (Vienna Code)*. Ruggell, Liechtenstein: ARG Gantner Verlag [Adopted by the 17th International Botanical Congress, Austria, July 2005: *Regnum Vegetabile* 146]

Oren A. 2004. **A proposal for further integration of the Cyanobacteria under the Bacteriological Code**. *International Journal of Systematic and Evolutionary Microbiology* **54**: 1895–1902

Simpson M G. 2010. *Plant Systematics*, 2nd edn. Amsterdam: Elsevier– Academic Press

Smith G F, Figueiredo E, and Moore G. 2011. **English and Latin as alternative languages for validating the names of organisms covered by the International Code of Nomenclature for algae, fungi, and plants: The final chapter?** *Taxon* **60**(5): 1502–1503

Stace C A. 1989. *Plant Taxonomy and Biosystematics*, 2nd edn. London: Edward Arnold

APPENDIX 1

The following major changes have been approved during the Plenary Session of the Nomenclatural Section (18–22 July 2011) of the 18th International Botanical Congress, Melbourne, Australia.

1. Title–name of the code: The name of the code has been changed. The earlier title (International Code of Botanical Nomenclature; abbreviated ICBN) has been replaced with a new name. The code shall henceforth be known as "International Code of Nomenclature for algae, fungi and plants; abbreviated ICN". Amongst the several reasons for this change, the focus was on the suggestions of the mycologists, that the term "botanical" was misleading and could imply that the Code covered the green plants and excluded the fungi and diverse algal lineages. It is also necessary to point out that under the five-kingdom classification, fungi are classified in a separate Kingdom, while the diverse algae are placed in different kingdoms. However, since the names of all algae and the fungi were governed by the rules laid down in the ICBN, this change in the title of the code was accepted. Further, the abbreviation ICN does not compete with the abbreviations of the other codes of biological nomenclature.

2. Effective publication of names: The Melbourne Congress brought about a major shift in the requirement of publication of names. It accepted (by overwhelming majority) to allow "electronic publication" of names from 1 January 2012 as an additional method of effective publication under the code. Earlier, all names had to be published "in printed form in a scientific journal or botanical publication" for making it "effective" under the rules of the code. Thus, now publication is also effected by electronic distribution of material in Portable Document Format (PDF) in an online publication with an International Standard Serial Number (ISSN) or an International Standard Book Number (ISBN).

3. Language of the code: There is a major change in the requirement of Latin as the language of botanical nomenclature. From 1 January 2012, names can be validly published with a description/diagnosis in either English or Latin. If names are published in any other language, then a description/diagnosis shall have to be provided in either English or Latin. This may ultimately lead to completely free botanical nomenclature from Latin.

4. One fungus one name: Mycologists have for a very long time used different names for the asexual and sexual phases of the same fungus. This was due to the fact that very often it was difficult to link the two phases in the life cycle. However, molecular studies have changed this situation. Thus, the asexual phase (the anamorph) and the sexual phase (the telomorph) of one fungus species are increasingly being identified. It was therefore agreed to apply the basic principles of the code and allow the use of "only one correct name" for the fungi (as is the case with all other groups dealt with under the code). This shall require a major exercise to ensure that there is minimal nomenclatural disruption in the proper use of fungal names when the Principle of Priority is applied".

5. One fossil one name: Paleobotanists have used names for plant fossils. Very often, different parts of a plant (leaf/fruit, and so on) found as have been provided separate names because organic connection could not be established. A set of new proposals have been adopted to ensure that "one fossil has only one name".

6. Registration of fungal names: From 1 January 2013, all names of fungi (new to science/new combinations/replacement names) shall have to cite an "identifier issued by a recognized repository" in order to be validly published. Since 2004, whenever a new fungal name is published, the online database Mycobank <www.mycobank.org> registers the description and illustrations. Upon registration, Mycobank issues a unique registration number which can be cited in the publication where the name appears. This requirement shall become mandatory for the valid publication of fungal names.

3

Plants of Delhi: Scientific Names and their Meaning

Neelam Pari Malkani

As accepted by the scientific community, botanical (scientific) names of plant species are binomials comprising a generic name and a specific epithet. The sources of these names may vary, but according to the International Code for Botanical Nomenclature, they are all to be treated as Latin.

Several generic names of plants are derived from their Greek names (for example, *Cestrum*), some from their classical Latin names (for example, *Cicer*), while others are adapted or modified from their vernacular names (for example, *Ceiba*, the South American vernacular name for the silk cotton tree). Quite a large number of names are commemorative, honouring ancient philosophers, explorers, professors, medicine men, and scientists (for example, *Bougainvillea* for Louis Antoine de Bougainville). Some of them refer to a specific morphological feature (for example, *Acacia*, where "Acis" (Greek) means pointed, for the many thorns borne by the plant), or to their useful product (for example, *Olea,* where "Olea" means oil), and yet others to their habitat (for example, *Potamogeton,* where "Potamos" (Greek) means river, and "geiton" (Greek) means neighbour). Some also appear to be constructed in an arbitrary manner.

Generic names are treated as nouns with the suffix indicating gender. Generally, names ending with *-us* or *-on* are masculine; those ending with *-a* or *-is* are feminine; and those ending with *-um* or *-e* are neutral.

The specific epithet may either refer to some character (for example, *Aristida setacea,* where "setacea" means with bristles, referring to bristle-like awns), longevity (for example, *Poa annua,* where "annua" means annual), place of origin or geographical distribution (for example, *Mangifera indica,* where "indica" means from India), habitat (for example,

Veronica agrestis, where "agrestis" means of fields, referring to its common occurrence in fields), or may be commemorative honouring a specific person (for example, *Cassia roxburghii,* where "*roxburghii*" is for William Roxburgh).

Quite often, the names are combinations of two or more words. As such, the meaning refers to the components (for example, *Chenopodium,* where "Cheno-" means goose and "podium" means foot, referring to the shape of the lobed leaves). At times, the meaning of scientific names appears to have no correlation to any feature relating to the plant (for example, *Sida*).

Generally, scientific names appear to be long, difficult to pronounce and remember, and one often wonders about the utility of these names. It is not the purpose of this chapter to explain the advantages of having internationally accepted scientific names. It is sufficient to say that they are here to stay. However, if one tries to understand the meaning of these names, they make a lot more sense. Further, since several names have a common root or prefix or suffix, one can make out the meaning of even those names of plants that appear unfamiliar.

The following is a compilation of names of plants found in Delhi (wild as well as planted), their meanings, and/or derivations. (The language in parenthesis is of the complete generic name or specific epithet or that of its root). Many of these plants are cosmopolitan or pan-tropical. In case of a genus being represented by several species, only one or a few of them have been listed below because the purpose is not to enumerate all species but to give an idea of the variety of names of a single genus and a cross section of the specific epithets.[1]

Acacia auriculiformis	"Acis" (Greek) = pointed, "auriculalatus" (Greek) = lobed like an ear
A. catechu	"catechu" (Indian), vernacular name for betel in Cochin, India
A. farnesiana	"farnesianus" (Latin) = from the Farnese Palace garden of Rome; it has been long cultivated in the Mediterranean region, although a native of South America
A. leucophloea	"leuco" (Greek) = white, "phloea" (Latin) = bark; referring to its white-grey bark
A. nilotica	"nilotica" = from the Nile region
A. Senegal	"senegal" = pertaining to Senegal

---
[1] The language of origin where not mentioned is Latin.

Acalypha indica	"Akalephes" (Greek) = nettle, used in ancient times for a kind of nettle and applied to this plant by Linnaeus due to its nettle-like leaves; "indica" = pertaining to India
Achyranthus aspera	"Achyranthus" (Greek) = chaff-flower; "aspera" (Latin) = rough; referring to the surface of its leaves
Adansonia digitata	"Adansonia", after Michael Adanson (1727–1806), a French, West African botanist; "digitata" (Latin) = like fingers on the hand; referring to its palmated compound leaves
Adhatoda vasica	"Adhatoda" (Tamil) = plant shunned by herbivores, "vasic" (Latin) = duct
Adina cordifolia = Anthocephalos cadamba	"Adinos" (Greek) = crowded; referring to flowers being clustered in a head; "cordifolia" (Greek); referring to the plant's heart-shaped leaves
Aegle marmelos	"Aegle" (Greek), after the name of the most beautiful daughter of Zeus, a famous king in Greek mythology; "marmelos" (Portuguese) is the vernacular name for marmalade in Portugal; referring to the sweet pulp
Agave Mexicana	"Agauos" (Greek) = admirable, noble; referring to its appearance; "mexicana" pertains to Mexico
Ageratum conyzoides	"Ageratos" (Greek) = not growing old; referring to retention of colour by its flowers for a long time; "conyza", a name used by Theophrastus, Greek philosopher; "-oides" (Latin) = resembling
Ailanthus excels	"Ai" (Latin) = eternally, always; "Ailanthus", from a Molluccan name which means reaching heaven; referring to its very tall nature; "excelsa" (Latin) = tall; referring to its magnificent height and appearance
Albizia amara	"Albizia", after Filippo degli Albizzi, an Italian naturalist; "amarus" (Latin) = bitter
Aleurites moluccana	"Aleur" (Greek) = mealy or wheaten flour; referring to the undersurface of leaves; "moluccana", from Moluccas, Indonesia

Alhagi pseudo-alhagi	"Alhagi" (Arabic) = pilgrim; "pseudo" (Greek) = false
Aloe barbadensis = Aloe vera	"Aloeh" (Arabic) = bitter; referring to the juice; "barbadensis" = pertaining to Barbados
Alstonia macrophylla	"Alstonia", after Professor Charles Alston (1716–60) of Edinburgh; "macro" (Greek) = large, "phyla" (Greek) = leaves
A. scholaris	"scholaris" (Latin) = of the school; referring to the wood used earlier to make wooden slates for school children
Alternanthera pungens	"Alternus" (Latin) = alternate; "anther" (Latin) = anthers; referring to alternate fertile and sterile stamens; "pungens" (Latin) = pricking
A. sessilis	"sessilis" (Latin) = without stalk; referring to sessile leaves
Amaranthus gracilis	"Amarantos" (Greek) = unfading; referring to the long-lasting flowers; "gracilis" (Latin) = slender, graceful
Ammania baccifera	"Ammania", after Paul Ammann (1634–91), a German botanist and professor at Leipzig; "baccifera" (Latin) = bearing berries
Anagallis arvensis	"Ana" (Latin) = again; "agallein" (Latin) = to delight in; referring to the delight one gets in seeing the flowers open every time they get the rays of the sun; "arvensis" (Latin) = of the fields, since the plant is commonly found growing in fields
Annona squamosa	"Annona", a Haitian name; "squamosa" (Latin) = scale-like; referring to the outer appearance of its fruits
Anogeissus acuminata	"Anogeissus" (Latin) = towards the top; referring to its top-tiled fruiting heads; "acuminata" (Latin) refers to the pointed apex of leaves
A. pendula	"pendula" (Latin) refers to its drooping/pendulous branches
Antirrhinium orontium	"Anti-" (Greek) = like; "rhinon" (Greek) = nose; referring to the nose-like appearance of the flower

Aponogeton natans	"Aponos" (Latin) = the healing springs at Aquuae Aponi, Italy; "geiton" (Greek) = neighbour; the name being applied to this plant because of its aquatic habitat; "natans" (Latin) = floating
Aristida setacea	"Arista" (Latin) = awn; "setacea" (Latin) = with bristles; referring to its bristle-like awns
Artemisia scoparia	"Artemisia", ancient Greek name for wormwood; also refers to the Greek goddess Artemis who was healed by this herb; "scoparius" (Latin) = broom-like
Artocarpus heterophyllus	"Artopta" (Greek) = baker; "carpos" (Greek) = fruit, literally meaning bread fruit; "hetero" (Greek) = different; "phyllus" (Greek) = leaf
Arundo donax	"Arundo" (Latin) is the old Latin name for reed; "donax" (Greek) is the old Greek name for reed
Asparagus racemosus	"Asparagus", an ancient Greek name "aspharagos", originating from the Persian word "asparag", which means sprout, stalk or shoot; "racemosus" (Latin) refers to the racemose inflorescence
Asphodelus tenuifolius	"Asphodelus", a Greek name; "tenuis" (Latin) = slender; "folius" (Latin) = leaved; referring to the slender leaves of the plant
Averrhoa carambola	"Averrhoa", after Ibn Rushd Averrhoes, a 12th century Arabian physician who translated Aristotle's work
Azadirachta indica	"Azadirachta", after the Persian name "Azad dirakht"; "indica" = pertaining to India
Balanites roxburghii	"Balanus" (Greek) = acorn; "roxburghii", after William Roxburgh, a Scottish botanist (1751–1815)
Basella rubra	"Basella" (Malayalam), vernacular name for spinach in Malabar region of India, where the language spoken is Malayalam; "rubra" (Latin) = red; referring to pink flowers
Bauhinia purpurea	"Bauhinia", after the botanists Casper Bauhin (1550–1624) and his brother, John Bauhin;

	referring to the two-lobed leaves; "purpurea" (Latin); referring to its purple flowers
B. racemosa	"racemosa" (Latin) = racemose; referring to its inflorescence
B. variegata	"variegata" (Latin) = variegated; referring to its variegated petals, particularly the central, largest one
Beta vulgaris	"Beta", its ancient Latin name; "vulgaris" (Latin) = common
Bidens biternata	"Bidens", from "bis" (Latin), meaning two; "dens" (Latin) = tooth; referring to the two bristles on the plant's achenes; "biternata" (Latin) = two sets of three
Bischofia javanica	"Bischofia", after Gottlieb Wilhelm Bischoff, a 19th century German professor of botany at Heidelberg; "javanica" = from Java
Blumea lacera	"Blumea", after Karl L Blume, a German botanist; "lacera" (Latin) = shredded; referring to the irregular margin of its leaves
Boerhavia diffusa	"Boerhavia", after the Dutch botanist Hermann Boerhaave (1668–1738); "diffusa" (Latin) = spreading; referring to the plant's trait
Bombax ceiba	"Bombax" (Greek) = silkworm; "ceiba", South American vernacular name for silk cotton tree
Bougainvillea spectabilis	"Bougainvillea", after Louis Antoine de Bougainville (1729–1811), noted mathematician and scientist; "spectabilis" (Latin) = admirable, good-looking
B. glabra	"glabra" (Latin) = without hair
Brachiaria ramosa	"Brachiatus" (Latin) = branched at right angles; "ramosa" (Latin) = branched
Brassica campestris	"Brassica" (Latin) = cabbage-like plant; "campestris" (Latin) = of pastures
B. juncea	"junceus" (Latin) = rush-like
B. rapa	"rapa" (Latin) = turnip
Bridelia retusa	"Bridelia", after Samuel Elisée Bridel-Brideri, a Swiss-German bryologist; "retusa" (Latin); referring to the retuse leaf apex

Broussonetia papyrifera	"Broussonetia", after French naturalist T N V Broussonet (1761–1807); "papyrifera" (Latin) = paper-bearing; referring to the thin, papery bark used in Japan for almost 1500 years to produce high-quality paper
Butea monosperma	"Butea", after the Earl of Bute, John Stuart (1713–92); "monosperma" (Greek) = one-seeded fruit
Cajanus cajan	From "Katjan", the Malay name for pigeon pea
Callistemon lanceolatus	"Calli" (Greek) = beautiful; "stemon" (Greek) = stamen; "lanceolatus" (Latin); referring to the shape of leaves
Cannabis sativa	"Cannabis", name given by Greek physician, pharmacologist, and botanist Dioscorides (c. 40–90 AD) to this plant (hemp); "sativus" (Latin) = cultivated, not sown
Capparis decidua	"Capparis", from the common name "caper"; "decidua" (Latin) = not persisting, falling off quite early; referring to its shedding of leaves during the non-seasonal period
Capsicum annuum	"Kapsimo" (Greek) = to bite; "annuum" (Latin) = annual
Cardiospermum helicacabum	"Cardios" (Greek) = heart; "spermum" (Greek) = seeds; referring to the three-sided slightly winged fruits
Carissa congesta = C. carandas	From "Corissa", the Sanskrit name for the species; "carundus", from the vernacular name "caraunda", in turn from "kurundum" (Tamil); referring to the ruby red rocks whose colour resembles that of its fruits
Carthamus tinctorius	"Carthamus" (Hebrew) = painted-one; "tinctorius" (Latin) = used in dyeing
Caryota urens	"Caryo" (Greek) = nut; "uro" (Greek) = burn; referring to the fruits, which contain sharp crystals under their skin that can cause severe irritation and chemical burns
Cassia fistula	"Cassia", name used by Dioscorides for this medicinal plant; "fistula" (Latin) refers to a hollow, pipe-like fruit

C. grandis	"grandis" (Latin) = showy, magnificent
C. javanica	"javanica" = from Java
C. occidentalis	"occidentalis" (Latin) = from the West
C. roxburghii	"roxburghii", after William Roxburgh
C. siamea = *Senna siamea*	"siamea" = from Siam
Casuarina equisetifolia	"Kesuari" (Malay) or "cassowary"; referring to the resemblance of the branches to the feathers of the cassowary bird; "equisetifolia" (Latin) = branches and scale leaves resembling those of the *Equisetum*.
Catharanthus pusilus	"Catharanthus" (Greek) = pure flower; "pusillus" (Greek) = slender, weak
Ceiba insignis	"Ceiba", from a South American vernacular name for silk cotton tree; "insignis" (Latin) = remarkable
C. pentandra	"pentandra" (Latin) = with five stamens
C. speciosa	"specre" (Latin) = to look; referring to the showy, handsome flowers
Celtis tetrandra	"Celtis" (Greek), name for a tree with a sweet fruit; "tetrandra" (Latin) = with four stamens
Centella asiatica	"Kenteo" (Greek) = to pierce or prick; "asiatica", after the continent Asia
Cestrum nocturnum	"Cestrum" (Greek), an ancient Greek name; "nocturnum" (Latin) = night; referring to its flowers opening at night
Chenopodium album	"Cheno" (Latin) = goose; "podium" (Greek) = foot; referring to the shape of the leaves, which are lobed; "album" (Latin) = white; referring to the milky white appearance of its leaves
C. ambrosioides	"ambrosia" (Latin) = elixir of life; "-oides" (Latin) = resembling; referring to the beverage made from the plant
C. murale	"murale" (Latin) = growing on walls
Chloris barbata	"Chloris", the Greek goddess of flowers; "barbata" (Latin) = with tufts of hair, bearded; referring to its glumes, which are bearded on the margins

Chrysalidocarpus lutescens = *Dypsis lutescens*	"Chrysalidocarpus" (Greek) = golden fruit; "lutescens" (Latin) = turning yellow
Chrysophyllum oliviforme	"Chryso" (Greek) = golden; "phylum" (Greek) = leaf; "oliviforme" (Latin) = shaped like an olive
Cicer arietinum	Cicer", classic Latin name for chick pea; "arietinum" (Latin) = like a ram's head
Cichorium intybus	"Cichorium", name used by Theophrastus, Greek philosopher (c. 371–287 BC); "intybus", classic Latin name for wild chicory
Clerodendrum phlomidis	"Kleros" (Greek) = chance or fate; "dendron" (Greek) = tree; "phlomis" (Greek) = flame
Clitoria ternatea	"Clitoria" (Greek); referring to the shape of the plant's flowers; "ternatea" (Latin) = parts in three
Coccinea cordifolia	"Coccinea" (Latin) = scarlet red; "cordifolia" (Latin) = heart-shaped leaves
Commelina benghalensis	"Commelina", after the two Dutch botanists, Jan Commelijn and his nephew, Caspar (1667–1731), each representing one of the showy petals of *Commelina communis;* "benghalensis" = from Bengal, India
Convolvulus arvensis	"Convolvulus" (Latin) = interwoven; referring to the twining habit; "arvensis" (Latin) = of the fields; referring to the plant generally occurring as a common weed in cultivated fields
Corchorus aestuans	"Corchorus", ancient Greek name for jute; "aestuans" (Latin) = glowing
C. capsularis	"capsularis" (Latin) = producing capsules
C. olitorius	"olitorius" (Latin) = of gardens
Cordia dichotoma	"Cordatus" (Greek) = heart-shaped; referring to the plant's heart-shaped leaves
Coriandrum sativum	"Coriandrum" (Greek), name used by Pliny, a Roman author, naturalist (23–79 AD), derived from "koros" (Greek) = a bug; referring to the foetid smell of the leaves; "sativum" (Latin) = planted, not wild

Coronopus didymus	"Coronopus", name used by Theophrastus, a Greek philosopher, for crow-foot; referring to the plant's leaf shape; "didymus" (Greek) = twinned or double; referring to the two stamens of the plant
Cotula hemispherica	"Cotula" (Greek); referring to a small cup, specifically to the broad saucer-shaped heads; "hemispherica" (Greek) = half-sphere
Crataeva adansonii	"Crateva", after the ancient Greek botanist Crateva
Crossandra infundibuliformis	"Crossandra" (Greek) = fringed anthers; "infundibuliformis" (Latin) = funnel-shaped
Crotalaria juncea	"Krotalon" (Greek) = rattle or clapper; referring to the rattling seeds in the inflated pods; "junceus" (Latin) = rush-like
Cuscuta hyaline	"Cuscuta", medieval name for dodder; "hyaline" (Latin) = nearly transparent
Cyanotis axillaris	"Cyanotis" (Latin) = blue-ear; referring to the plant's blue flowers; "axillaris" (Latin) = in axils; referring to flowers born in axillary fascicles
Cynodon dactylon	"Cyno" (Greek) = dog; "don" (Greek) = tooth; "dactylos" (Greek) = finger-like; referring to the form of its inflorescence
Cyperus compressa	"Cyperus", Greek name for several species of sedges; "compressa" (Latin) = flattened sideways; referring to the spikelets being compressed laterally
C. rotundus	"rotundus" (Latin) = rounded; referring to the umbel of condensed spikes
Dactyloctenium aegyptium	"Daktylos" (Greek) = finger; "ktenion" (Greek) = little comb; referring to the arrangement of the spikelets; "aegyptium" (Latin), presumably Egyptian
Dalbergia sissoo	"Dalbergia", after the Swedish brothers Nils (botanist) and Carl Dalberg who sent Linnaeus specimens from Surinam; "sissoo", from "sisham" (Hindi); referring to an east Indian tree whose leaves are used for fodder

Datura innoxia	"Datura", an ancient Indian vernacular name for this plant; "innoxius" (Latin) = harmless; in particular, leaves that are useful in medicine
Delonix regia	"Delonix" (Latin) = conspicuous claw (present on the petals); "regia" (Latin) = royal, splendid; referring to the plant's look in full bloom
Dichanthium annulatum	"Dicha" (Latin) = in two or bifid; "anthos" (Greek) = flower; referring to flowers in bifid spikes; "annulatum" (Latin) = ring-shaped
Dichrostachys cinerea	"Dichro" (Latin) = two; "stachys" (Greek) = spike, probably referring to the two kinds of flowers, sterile pink ones and fertile yellow ones on the spike; "cinerea" (Latin) = ash grey; referring to its ash grey bark
Digitaria adscendens	"Digitaria" (Latin) = fingered; referring to its spikes arranged like fingers on a hand; "adscendens" (Latin) = curving up from a prostrate base, half-erect; referring to its habit
Diospyros cordifolia	"Diospyros" (Latin) = divine fruit; "cordifolia" (Latin) = heart-shaped leaves
D. malabarica	"malabarica" = from Malabar
Duranta repens	"Dura" (Latin) = hard; "repens" (Latin) = creeping
Echinochloa colonum	"Echinatus" (Greek) = prickly; "colonum" (Latin) = humped
Eclipta prostrata	"Ekleipo" (Greek) = deficient; referring to the absence of a pappus; "prostrata" (Latin); referring to the plant's prostrate habit
Ehretia acuminata	"Ehretia", after the botanical artist G D Ehret (1710–70); "acuminata" (Latin) = with a long narrow tip; referring to the pointed apex of the plant's leaves
E. laevis	"laevis" (Latin) = polished, not rough
Eichhornia crassipes	"Eichhornia", after J A F Eichhorn (1779–1856), Prussian Minister of Education and Public Welfare; "crassi" (Latin) = thick; referring to its thick, spongy leaves

Eleusine compressa	"Eleusine", from Eleusis, an ancient city in Greece; "compressa" (Latin) = flattened sideways
Elytraria acaulis	"Elytri" (Latin) = covering; "acaulis" (Greek) = without stem; referring to the apparent absence of a stem, which is condensed near the base
Emblica officinalis	"Emblica", an old generic name for this plant; "officinalis" (Latin) = of the apothecaries, sold in shops; referring to the plant's several medicinal uses
Eragrostis tenella	"Eros" (Greek) = love; "grostis" (Greek) = grass; "tenella" (Latin) = delicate; referring to the inflorescence
Eranthemum nervosum	"Eros" (Greek) = love; "anthos" (Greek) = flower; referring to the beautiful flowers; "nervosum" (Latin) = with prominent nerves or veins
Erigeron bonariensis	"Eri" (Greek) = early; "geron" (Greek) = old man, meaning "old man in the spring"; referring to the fluffy, white fruiting heads and the early flowering and fruiting of many species; "bonariensis" = from Buenos Aires, Argentina
Erythrina variegata	"Erythros" (Greek) = red; referring to its bright red flowers; "variegata" (Latin); referring to the leaflets of the plant, which are sometimes variegated
Eucalyptus camaldulensis	"Eu" (Greek) = true, good, well; "calyptus" (Greek) = covering; referring to its floral parts being well covered at first by the cap-like perianth; "camaldulensis" = from the Camalduli gardens near Naples
Euphorbia hirta	"Euphorbia", after Euphorbus who used the plant's latex for medicinal purposes; "hirta" (Latin) = shaggy, hairy
E. neriifolia	"nerium" (Greek), ancient Greek name for oleander; "neriifolia" (Greek); referring to its nerium-like leaves
E. prostrata	"prostrata" (Latin); referring to the prostrate-spreading habit of the plant

E. pulcherrima	"pulcherrima" (Latin) = most beautiful; referring to its pretty, red bracts
Ficus amplissima	"Ficus", ancient Latin name for fig; "amplissimus" (Latin) = very large, the biggest
F. benghalensis	"benghalensis" = from Bengal, India
F. benjamina	"benjamina", from Indian vernacular ben-yen or bania (Hindu merchants who assembled under the plant's shade for trade or worship)
F. elastica	"elastica" (Latin); referring to the latex that exudes from the plant when cut and solidifies to give elastic rubber
F. lyrata	"lyrata" (Latin) = shaped like the leaves of the lyre plant
F. microcarpa	"microcarpa" (Greek) = small fruits
F. palmata	"palmata" (Latin); referring to the three to five palmately lobed leaves
F. racemosa	"racemosa" (Latin); referring to the arrangement of the inflorescences (syconia) in a racemose manner
F. religiosa	"religiosa" (Latin); referring to the tree being considered sacred because Gautama Buddha attained enlightenment under this tree
F. virens	"virens" (Latin) = green, vigorous
Fimbristylis dichotoma	"Fimbriae" (Latin) = shreds, fringe; "stilus" (Latin) = style; referring to the ciliate style
Flacourtia indica	"Flacourt", after Etienne de Flacourt (1607–61) of the French East India Company; "indica" = pertaining to India
Foeniculum vulgare	"Foeniculum" (Latin) = fennel; "vulgare" (Latin) = common
Fumaria indica	"Fumaria" (Greek) = smoke; referring to the effect of the plant's juice, which is similar to that of smoke; "indica" pertains to India
Gliricidia sepium	"Gliricidia" (Latin) = mouse killer; referring to the plant's poisonous seeds and bark; "sepium" (Greek) = growing in hedges
Gmelina arborea	"Gmelina", after German botanist Johann Georg Gmelin (1709–55); "arborea" (Latin) = tree-like

G. asiatica	"asiatica" = from Asia
Gnaphalium luteo-album	"Gnaphalon" (Greek) = lock of wool; referring to the woolly surface of the entire plant; "luteo-album" (Greek) = white, the hairs give the plant a white appearance
Gomphrena celosioides	"Gomphrena", an ancient Latin name; "celosia" (Greek) = burnt; "-oides" (Latin) = like; referring to the burnt, dry look of the flowers
Gossypium herbaceum	"Gossypium", name used by the Roman naturalist and philosopher Pliny for this plant; "herbaceum" (Latin); referring to the herbaceous habit of the plant
Grevillea robusta	"Grevillea", after Charles F Greville FRS (1749–1809) who was founder member of the Royal Horticultural Society; "robusta" (Latin) = strong
Gynocardia odorata	"Gynocardia" (Greek); referring to the plant's large fruits; "odoratus" (Latin); referring to the fragrant yellow flowers borne in clusters directly on the stem
Haplophragma adenophyllum	"Haplo" (Greek) = simple; "phragma" (Greek) = fence; "adeno" (Greek) = gland; "phylum" (Greek) = leaf
Helianthus tuberosus	"Helios" (Greek) = sun; "anthos" (Greek) = flowers, especially those turning in the direction of the sun; "tuberosus" (Latin) = bearing tubers
Heliotropium supinum	"Helios" (Greek) = sun, "tropium" (Latin) = turning; referring to turning of flowers towards the sun; "supinum" (Latin) = lying flat; referring to its posture
Hibiscus tiliaceus	"Hibiscus", an old Greek name for mallow; "tiliaceus" (Latin) = lime-like or resembling tilia
Hydrilla verticillata	"Hydrilla", may be derived from the Greek word "hydra", which means a water serpent; referring to its aquatic habitat; "verticillatus" (Latin) = arranged in whorls; referring to arrangement of its leaves

Imperata cylindrica	"Imperata", after Ferrante Imperato (1550–1625), the Neapolitan pharmacist who had one of the earliest collections of natural history specimens in Italy (possibly in Europe); "cylindrica" (Latin) = cylindrical
Indigofera enneaphyla	"Indigofera" (Latin) = bearer of indigo (dye); "ennea" (Greek) = nine; "phyla" (Greek) = leaves; referring to its bearing, usually nine pairs of leaflets
I. tinctoria	"tinctoria" (Latin) = used for dyeing
Ipomoea carnea	"Ipomoea" (Greek) = resembling a worm; referring to the stems that often coil and twine like worms; "carnea" (Latin) = flesh-coloured, its flowers being pink in colour
I. pilosa	"pilo" (Greek) = covered with soft, long hair; referring to leaves, sepals, capsules, and seeds being covered with soft, long hairs
Ixora coccinea	"Ixora", after a Malabar deity *Isvara*; "coccinea" (Latin) = crimson; referring to its deep red flowers
Jacaranda mimosifolia	"Jacaranda", taken from the Tupi–Guarani name "jakaranda"; "mimosa" (Greek) = imitator; referring to sensitivity of the leaves; "folia" (Latin) = leaves; "mimosifolia" = mimosa-like leaves
Jasminum grandiflorum	"Jasminum", from the Persian name Yasmin; "grandiflora" (Latin) = large, showy flowers
J. officinale	"officinale" (Latin) = used in medicine
Jatropha curcas	"Jatropha" (Latin) = physician's food; "curcas", ancient Latin name for Jatropha
Juncus bufonius	"Juncus" (Latin), classical Latin name for the rush; "bufonius" (Latin) = of the toad; referring to this plant inhabiting damp places like the toad
Jussiaea perennis	"Jussiaea", after the French naturalist Bernard de Jussieu (1699–1777); "perennis" (Latin) = perennial
Justicia diffusa	"Justicia", after James Justice (1698–1763), a Scottish botanist and horticulturist; "diffusa" (Latin) = spreading; referring to the plant's trait

J. simplex	"simplex" (Latin) = simple, undivided; referring to plants of this species being erect and relatively less branched
Kickxia ramosissima	"Kickxia", after J J Kickx, a Belgian botanist who specialized in cryptogams; "ramosus" (Latin) = branched; "-issimus" (Latin) = much, greatly (suffix used as superlative); referring to the much branched nature of the plant
Kigelia pinnata	"Kigelia", the native Mozambique name for this tree, also called "sausage tree"; "pinnata" (Latin); referring to its distinct, large, pinnate compound leaves
Kochia indica	"Kochia", after Wilhelm Daniel Joseph Koch (1771–1849), a German professor of botany; "indica" = pertaining to India
Lagerstroemia floribunda	"Lagerstroemia", after Magnus von Lagerstrom of Goteborg, a friend of Linnaeus; "floribunda" (Latin) = bearing many flowers
L. indica	"indica" = pertaining to India
L. microcarpa	"microcarpa" (Greek) = with small fruits
L. tomentosa	"tomentosa" (Latin) = with thick, matted hair
Lathyrus aphaca	"Lathyros", old Greek name for pea; "aphaca" (Greek) = name used by Pliny, the Roman naturalist (23–79 AD) for pea-like plant
L. sativa	"sativa" = cultivated, planted
Launaea nudicaulis	"Launaea", after Jean Claude Mien Mordant de Launay (1750–1816), a French lawyer; "nudicaulis" (Latin) = naked stem
Lawsonia inermis	"Lawsonia", after Dr Isaac Lawson, a botanical traveller; "inermis" (Latin) = without spines, unarmed
Lemna paucicostata	Lemna", name used by Theophrastus for a water plant; "paucicostata" (Latin) = few or sparsely nerved/veined
Lens culinaris	"Lens" (Latin); referring to the shape of the seeds; "culinaris" (Latin) = pertaining to food or kitchen; referring to the plant's extensive use as food
Lepidagathis cristata	"Lepidos" (Greek) = scaly; "agathis" (Latin) = ball of twine; referring to the appearance of

	the inflorescence; "cristata" (Latin) = tassle-like at the tips
Lepidium sativum	"Lepidium" (Greek) = scale-like; referring to the appearance of the plant's fruits; "sativum" (Latin) = cultivated; referring to its occurrence in cultivated fields
Leptadaenia pyrotechnica	"Lepto" (Greek) = thin, slender, weak; referring to the plant's thin twining stems; "pyro" (Greek) = fire; "technica" (Latin) = special, technical
Leucaena leucocephala	"Leuco" (Greek) = white; "cephalos" (Greek) = head; referring to the white flowers clustered in heads
Leucas aspera	"Leuco" (Greek) = white; referring to the white flowers; "aspera" (Latin) = rough; referring to the surface of the leaves
L. cephalotes	"cephalotes" (Greek) = having a small head-like appearance; referring to the clusters of the plant's flowers in terminal heads
Limonia acidissima	"Limon", Persian name for citrus fruits; "acidissima" (Latin) = very acidic or sour
Linum usitatissimum	"Linum", classical Greek name for flax, also from "linon", the name used by Theophrastus, a Greek philosopher; "usitatissimum" (Greek) = most used
Liquidambar formosana	"Liquid-amber" (Latin), the resin from the bark of sweet gum; "formosana" = from Formosa (now Taiwan)
Livistona chinensis	"Livistona", after Patrick Murray, Lord Livingstone, whose garden formed the nucleus of the royal botanic garden, Edinburgh; "chinensis" (Latin) = from China
L. rotundifolia	"rotundifolia" (Latin) = round leaves
Lolium temulentum	"Lolium" (Latin) = name for a kind of grass; "temulentum" (Latin) = intoxicating (the grains of the plant are supposedly poisonous)
Luffa acutangula	"Luffa", from the Arabic name "louf"; "acutangula" (Latin) = sharply angled; referring to the plant's ridged (angular) fruits

L. cylindrica	"cylindrica" (Latin) = cylindrical; referring to the plant's unangled, cylindrical fruits
Lycium europium	"Lycium", ancient Greek name for a tree from Lycia; "europium" = from Europe
Lycopersicon esculentum	"Lyco" (Greek) = wolf; "persica" (Latin), ancient name for peach (literally meaning wolf-peach, used for tomato); "esculentum" (Latin) = edible
Maerua arenaria	"Maerua", taken from the word "Meru", which is an Arabic name; "arenaria" (Latin) = sand dweller
Magnolia grandiflora	"Magnolia", after Pierre Magnol (1638–1715), director of the Montpelier Botanic Garden; "grandiflora" (Latin) = big flowers; referring to large showy flowers
Mallotus philippensis	"Mallotus" (Latin) = woolly; referring to the powdery substance covering the fruits and hairy leaves; "philippensis" = from the Philippines
Malva parviflora	"Malva" (Latin) = soft; "parvi" (Latin) = small; "flora" (Latin) = flowered
Mangifera indica	"Mangifera", from the Tamil word "mangai", which changed to "mango" in English; "indica" = pertaining to India
Mazus japonicas	"Mazus" (Greek) = nipple; referring to the shape of the corolla; "japonicas" = from Japan
Medicago denticulate	"Medicago", from a Persian name for a grass; "denticulata" (Latin) = with very small teeth; referring to the plant's leaf margins
M. sativa	"sativa" (Latin) = sown or cultivated; referring to the plant's common occurrence in cultivated fields
Melaleuca bracteata	"Mela" (Greek) = black; "leuca" (Greek) = white; referring to the black and white colours on the bark; "bracteata" (Latin) = with bracts
Melia azadirach	"Melia", from the Greek name of the ash tree whose leaves resemble those of the *Azadirchta* plant ("Azadirachta", after the Persian name "Azad dirakht")
Melilotus alba	"Melilotus" (Greek) = honey-lotus, name given by Theophrastus due to its attraction to

Plants of Delhi: Scientific Names and their Meaning

	honey bees; "alba" (Latin) = white; referring to the plant's white flowers
M. indica	"indica" = pertaining to India
Mentha spicata	"Mentha", after the unfortunate Greek nymph, Mentha, who was turned into a mint plant; "spicata" (Latin) = in spikes; referring to the whorls of flowers being arranged in a spike
Millingtonia hortensis	"Millingtonia", after Sir Thomas Millington (1628–1704); "hortensis" (Latin) = of gardens
Mimosa pudica	"Mimosa" (Greek) = mimic; referring to the sensitivity of the plant's leaves; "pudica" (Latin) = retiring, modest; referring to its leaflets shying away or folding up when touched
Mitragyna parvilora	"Mitra" (Latin) = turban; "gyna" (Greek) = ovary, meaning turban-shaped ovary; "parvi" (Latin) = small; "flora" (Latin) = flowers; referring to the plant's small, clustered flowers
Mollugo cerviana	"Mollugo" (Latin) = soft, a name cited by Pliny, a Roman naturalist; "cerviana" (Latin) = stag-coloured
Moringa oleifera	"Moringa", from an Indian vernacular name for the fruit; "oleifera" (Latin) = oil yielding
Morus alba	"Morus", ancient Latin name for mulberry; "alba" (Latin) = white
Nasturtium officinale	"Nasturtium" (Latin) = nose twist; referring to the offensive mustard-oil smell; "officinale" = of the apothecaries, sold in shops
Nelumbo nucifera	"Nelumbo", from a Sinhalese name; "nucifera" (Latin) = nut-bearing
Neolamarckia cadamba	"Neo" (Latin) = new; "lamarckia", after the French evolutionary biologist Jean Baptiste Antoine Pierre Monnet de Lamarck (1744–1829); cadamba", from the vernacular name "kadamba"
Nepeta hindostana	"Nepi" = from Nepi, Italy; "hindostana" = pertaining to Hindustan
Nicotiana plumbaginifolia	"Nicotiana", after Jean Nicot (1530–1600), French ambassador to Portugal, also supposedly responsible for introducing tobacco to France in the 1560s; "plumbaginifolia" (Latin) = leaves

	like *Plumbago*, a plant whose flowers have the colour of "plumbum" (Greek), meaning lead
Nyctanthes arbor-tristis	"Nyctanthes" (Latin) = night flower; "arbor-tristis" (Latin) = sad tree, so called because flowers bloom at night and fall off with the first rays of the sun
Nymphaea cristatum	"Nymphaea", after Nymphe, a Greek mythological freshwater nymph; "cristatum" (Latin) = tassle-like at the tips, crested
Ochna obtusata	"Ochna", an ancient Greek name used by Homer for wild pear; "obtusata" (Latin) = blunt, obtuse
Ocimum americanum	"Ocimum" (Greek) = aromatic plant; "americanum" = pertaining to America
O. sanctum	"sanctum" (Latin) = holy, sacred; this plant is considered holy by Hindus
Oldenlandia aspera	"Oldenlandia", after Danish botanist Henrik Bernard Oldenland (1663–1699); "aspera" (Latin) = rough; referring to its scabrid branches
O. corymbosa	"corymbosa" (Latin) = corymb-like inflorescence
Olea europaea	"Olea" (Latin) = oil; "europaea", from Europe, since it is the major oil-yielding plant of Europe
Orobanche aegyptiaca	"Orobanche" (Greek) = legume-strangler, since one species parasitizes legumes; "aegyptiaca" = pertaining to Egypt
Oryza sativa	"Oryza", from the Arabic name "Eruz"; "sativa" (Latin) = sown, cultivated
Oxalis corniculata	"Oxalis" (Greek) = acid-salt; referring to the acidic (sour) taste; "corniculatus" (Latin) = having small horn or spur-like appendages
Parkinsonia aculeate	"Parkinsonia", after the author of *Paradisi in Sole*, John Parkinson (1569–1629); "aculeata" (Latin) = having prickles
Paspalidium flavidum	"Paspalum" (Greek) = millet grass; "idium" (Latin) = resembling millet grass; "flavidum" (Latin) = yellowish

Paspalum distichum	"Paspalum" (Greek) = millet grass; "distichum" (Greek) = in two opposed ranks; referring to the arrangement of the plant's spikelets in two rows
Peltophorum pterocarpum	"Pelto" (Latin) = shield; "phorum" (Greek) = bearing or carrying; referring to the shape of the stigma; "ptero" (Greek) = winged; "carpos" (Greek) = fruit; referring to the short wings along the edges of the fruit
Pennisetum typhoides	"Penna" (Latin) = feathery; "setum" (Latin) = bristle; referring to the persistent involucres of numerous ciliate bristles; "typhoides" (Latin) = resembling Typha (cattail), a name used by Theophrastus, a Greek philosopher and referring to the plant's long, linear leaves
Pentas lanceolata	"Pentas" (Greek) = five-fold; "lanceolata" (Latin) = lanceolate; referring to the shape of the plant's leaves
Pergularia daemia	"Pergularia" = arbour; referring to the plant's twining habit
Peristrophe bicalyculata	"Peri" (Latin) = around; "strophio" (Greek) = turned over; "bi-" (Latin) = two; "calyculata" (Latin) = with a small calyx; referring to its small bi-lipped calyx
Phalaris minor	"Phalaris", name used by the Greek philosopher Dioscorides for plume-like grass; "minor" (Latin)= smaller
Phaseolus aureus	"Phaseolus", name used by the Greek philosopher Dioscorides for a kind of bean; "aureus" (Latin) = golden; referring to the colour of seeds after removal of the seed coat
P. vulgaris	"vulgare" (Latin) = common
Phoenix sylvestris	"Phoenix" = for Phoenix, son of the Greek king Helias, who accompanied Achilles to Troy; "sylvestris" (Latin) = of the forest
Phragmites maxima	"Phragmites" (Greek) = hedge dweller; referring to its hedge-forming ability; "maxima" (Latin) = largest; referring to the plant's height
Phyla nodiflora	"Phyle" (Greek) = tribe, probably referring to the flowers that are tightly

	clustered in heads; "nodiflora" (Greek) = with flowers borne at the nodes
Pithhecellobium dulce	"Pithece" (Latin) = monkey; "lobium" (Latin) = podded; referring to the shape of its pods; "dulce" (Latin) = sweet tasting
Platanus orientalis	"Platanus" (Greek), an ancient Greek name for the long-lived plane tree; "orientalis" refers to the Old World
Pluchea lanceolata	"Pluchea", after French naturalist Noël-Antoine Pluche (1688–1761); "lanceolata" (Latin) refers to the shape of its leaves
Plumbago zeylanica	"Plumbum" (Latin) = lead; "-ago" (Latin) = resemblance, name used by Pliny, a Roman naturalist, to refer to the colour of the flowers; "zeylanicum" = pertaining to Ceylon (now Sri Lanka)
Plumeria obtuse	"Plumeria", after Charles Plumier (1664–1706), a French botanist; "obtuse" = obtuse; referring to the shape of the tips of the plant's leaves
P. rubra	"rubra" (Latin)= red; referring to the plant's red flowers
Poa annua	"Poa" (Greek) = pasturage, Greek name for a fodder grass; "annua" (Latin) = annual
Polyalthia longifolia	"longifolia" (Latin); referring to the plant's slim, long leaves
Polycarpaea corymbosa	"Polycarpon" (Greek) = many-fruited, a name used by the ancient Greek physician Hippocrates; "corymbosa" (Latin) = flowers in corymbs
Polygonum glabrum	"Polys" (Greek) = many; "gony" (Greek) = knee or joint; referring to the thickened joints on the stem; "glabrum" (Latin) = without hair
P. plebeium	"plebeium" (Latin) = common, inferior; referring to the plant's occurrence as a weed
Polypogon monspeliensis	"Polys" (Greek) = many; "pogon" (Greek) = bearded; referring to the dense covering of hair on the spikes; "monspeliensis" = from Montpellier, France
Populus deltoids	"deltoids" (Greek); referring to the plant's delta-shaped/triangular leaves

Portulaca grandiflora	"Portulaca" (Latin) = milk carrier; "grandiflora" (Latin) = beautiful, showy flowers
P. oleracea	"oleracea" (Latin) = of cultivation; referring to the plant's occurrence as a weed in cultivated fields
Potamogeton crispus	"Potamos" (Greek) = river; "geiton" (Greek) = neighbour; referring to its habitat; "crispus" (Latin) = with a wavy or curled margin; referring to the margins of the plant's leaves
Potentilla supine	"Potentilla" (Greek) = powerful; referring to the plant's use as a medicine; "supina" (Latin) = lying flat; referring to the plant's prostrate habit
Prosopis spicigera	"Prosopis" (Greek) = obscure; "spicigera" refers to the plant's several spines
Prunus dulcis	"Prunus", ancient Latin name for plum; "dulcis" (Latin) = sweet
P. persica	"persica" (Latin) = reached by way of Persia
Psidium guajava	"Psidium", the plant's Greek name; "guajava", the Spanish name of the plant
Psoralea corylifolia	"Psoralea" (Latin) = manged; referring to the plant's vegetative parts being marked with dots (glands); "corulus" (Latin) = helmet, "folia" (Latin) = leaves; referring to the shape of the plant's leaves
Pterospermum acerifolium	"Pteros" (Greek) = wing; "spermum" (Greek) = fruit; referring to the plant's winged fruits; "acerifolium" (Latin) refers to the plant's acer-like leaves
Pterygota alata	"Pterygota" (Greek) = winged; "alata" = wing-like; referring to its winged fruits
Pulicaria angustifolia	"Pulicaria" (Latin) = fleabane, name for a plant that wards off fleas; "angustifolia" (Latin) = narrow-leaved
Punica granatum	"Punica" (Latin), name used by Pliny, a Roman philosopher; "granatum" (Latin) = many-seeded

Pupalia lappacea	"lappacea" (Latin) = bud-like; referring to the slightly rounded bud-like fruits covered with stiff hair
Putranjiva roxburghii	"Putra-jeev" (Sanskrit); referring to the ancient belief that the plant's strong stone-like seeds (which when crushed have several medicinal properties) ward off evil. Hence, the seeds of this plant were strung into necklaces for little children; "roxburghii", after William Roxburgh
Pyrus pyrifolia	"Pyrus", ancient Latin name for pear; "pyri" (Latin) = pear-shaped, "folia" (Latin) = leaves
Quercus leucotrichophora	"Quercus", ancient Latin name for oak; "leuco" (Greek) = white, "tricho" (Greek) = hair, "phora" (Greek) = to carry or bearing
Roystonea regia	"Roystonea", after American soldier General Roy Stone; "regia" (Latin) = royal, splendid; referring to the plant's magnificent appearance
Rhus mysorensis	"Rhous", ancient Greek name for sumac, which refers to the various shrubs or small trees of the genus *Rhus*; "mysorensis" = from Mysore, India
Rhyncosia minima	"Rhyncosia" (Latin) = beak; referring to the beak-like keel petals; "minima" (Latin) = smallest, least
Rorripa indica	"Rorripa", an old Saxon name; "indica" = pertaining to India
Rumex dentatus	"Rumex", name used by Pliny, a Roman naturalist, for sorrel; "dentatus" (Latin) = having teeth; referring to the teeth-like segments of the innerperianth
Russelia coccinea	"Russelia", after Dr Alexander Russell, who authored *Natural History of Aleppo* (1775); "coccinea" (Latin) = crimson; referring to the colour of the flowers
Saccharum officinarum	"Sakcharon" (Latin) = sugar; there are similar words in Malay and Sanskrit for sugar or "the juice made from sugarcane"; "officinarum" (Latin) = of the apothecaries or pharmacists, sold in shops

Salix tetrasperma — Latin name for willow

Salvia aegyptiaca — "Salvia" (Latin) = healer, old Latin name for sage, a plant with medicinal properties; "aegyptiaca" = pertaining to Egypt

Saraca indica — "Saraca", from an Indian vernacular name; "indica" = pertaining to India

Scirpus affines — "Scirpus", an old Latin name for a rush-like plant; "affines" (Latin) = similar to or resembling

S. tuberosus — "tuberosus" (Latin) = swollen or tuberous

Sesamum indicum — "Sesamum" (Greek), name used by Hippocrates for this plant; "indicum" = pertaining to India

Setaria glauca — "Setaria" (Latin) = bristly; referring to the hairs subtending the spikelets; "glauca" (Latin) = with a white or greyish colour

S. verticillata — "verticillata" (Latin) = arranged in whorls

Sida cordifolia — "Sida", from a Greek name for water lily; "cordatus" (Greek) = heart-shaped, "folia" (Latin) = leaf; referring to the shape of the plant's leaves

Solanum melongena — "Solanum" (Latin) = comforter; "melongena" = apple bearer

S. nigrum — "nigrum" (Latin) = black; referring to the plant's black fruits

S. tuberosum — "tuberosum" (Latin) = tuberous, bearing tubers

Sonchus arvensis — "Sonchus" (Greek) = thistle; "arvensis" (Latin) = cultivated, since the plant is commonly found as a weed in cultivated fields

S. oleraceus — "oleraceus" (Latin) = of cultivation, since the plant is commonly found as a weed in cultivated fields

Spathodea campanulata — "Spathodea" (Latin) = spathe-like; referring to the calyx; "campanulata" (Latin) = bell-shaped; referring to the shape of the plant's flowers

Spergula arvensis — "Spergula" (Latin) = scatterer; referring to the plant's dispersal of seeds; "arvensis" (Latin) = cultivated, since the plant is commonly found as a weed in cultivated fields

Sporobolus diander	"Spora" or "sporos" (Greek) = seed or spore; "bolis" (Greek) = casting; referring to the dispersion of seeds
Stellaria media	"Stellaria" (Latin) = star; referring to the shape of the flower; "media" (Latin) = middle-sized
Sterculia asper	"Sterculia" (Latin) = dung; referring to foul-smelling flowers of some species; "asper" (Latin) = rough; referring to the rough surface of the plant's leaves
S. foetida	"foetida" (Latin) = foul-smelling; referring to the foul-smelling flowers
Striga euphraoides	"Striga" (Latin) = swathe, straight; "euphrasia" (Latin) = good cheer; "-oides" (Latin) = like
Suaeda fruticosa	"Suaeda", taken from an Arabic name for this plant; "fruiticulosus" (Latin) = shrubby, dwarf
S. maritima	"maritimus" (Latin) = growing by the sea
Syzigium cumini	"Suzugos" (Greek) = paired; referring to the plant's paired leaves
S. nervosum	"nervosum" (Latin) = nerved; referring to the 8–12 pairs of lateral nerves in the plant's leaves
Tabebuia aurea	"Tabebuia", an ancient Brazilian vernacular name; "aurea" (Latin) = golden; referring to the plant's golden-coloured flowers
Tabernaemontana divaricate	"Tabernaemontana", Latinized form of the name of J T Bergzabern, a physician and herbalist of Heidelberg; "divaricata" (Latin); referring to the branching of the plant's inflorescence
Tamarindus indica	"Tamar-i-Hind" (Persian) = date of India; "indica" = pertaining to India
Tamarix aphylla	"Tamarix", ancient Latin name for the Spanish area of the river Tambo; "aphylla" (Greek) = without leaves; referring to the plant's stem appearing to be leafless (only scale leaves, pressed close to the stem are present)
Tecoma castanifolia	"Tecoma", the Mexican name of the plant; "castanea", the ancient Latin name for chestnut; "folia" (Latin) = leaves; referring to chestnut brown leaves of this plant

T. stans	"stans" = upright; referring to the plant's erect habit
Tecomella undulata	"Tecoma", the Mexican name of the plant; "-ella" (Latin) = small, diminutive; "undulatus" (Latin) = wavy; referring to the plant's wavy leaf margins
Tectona grandis	Originally from "Tekka" (Malayalam), this word was translated to "teca" in Portuguese, meaning carpenter; referring to the wood being used by carpenters; "grandis" (Latin) = large; referring to the plant's huge size
Tephrosia purpurea	"Tephrosia" (Greek) = ashen; referring to the leaf colour; "purpurea" (Latin) = reddish-purple
Terminalia arjuna	"Terminalia" (Latin); referring to the leaves being frequently crowded near the terminal part of the branches; "arjuna", after Arjuna, a Pandava in the Hindu epic *Mahabharata*
T. bellirica	"bellis" (Latin) = pretty
T. myriocarpa	"myrio-" (Greek) = numerous; "carpos" (Greek) = fruits; referring to the several fruits clustered in slender spikes, which in turn form large branching clusters
Thespesia populnea	"Thespesia" (Greek) = divine, so called because Captain Cook, who discovered the plant in Tahiti, found it planted in the vicinity of temples; "populnea" (Latin); referring to the plant's *Populus*-like leaves, *Populus* being the ancient name for the "tree of people"
Thevetia peruviana	"Thevetia", after Andre Thevet (1502–92), a Frenchman who travelled to Brazil; "peruviana", from Peru
Thysanolaena maxima	"Thysano" (Latin) = fringed; "maxima" (Latin) = largest; referring to the plant's large inflorescence
Toona ciliate	"Toona", modified from "Doon", short for Dehradun, India; "ciliata" (Latin) = fringed with hairs
Trapa bispinosa	"Trapa", taken from the word "calcitrapa" (Latin), a spiked weapon used in battle to maim the

	hooves of cavalry horses; "bi-spinosa" (Latin); referring to the fruits bearing two spikes
Trianthema portulacastrum	"Treis" (Greek) = three; "anthema" (Greek) = flowered; "portulacastrum" (Latin) = somewhat resembling "Portulaca"; referring to the plant's leaves
Tribulus terrestris	"Treis" (Greek) = three; "bullatus" (Latin) = bumpy surface, probably referring to the rough surface of the plant's fruits; "terrestris" (Latin) = growing on the ground; referring to its spreading on the ground
Tridax procumbens	"Tridax" (Latin) = three-cornered; referring to the plant's fruits; "procumbens" (Latin) = prostrate; referring to its trait
Trifolium alexandrianum	"Treis" (Greek) = three; "folium" (Latin) = leaf; referring to its three leaflets; "alexandrianum" = pertaining to Alexandria
Trigonella corniculata	"Trigonella", originates from the word "trigonum", meaning a triangle or three corners; referring to the corolla of one of the species; "corniculata" (Latin) = having a small spur or horn-like appendage
T. foenum-graecum	"foenum-graecum" (Latin) = Greek-hay; referring to the plant being used as fodder
Triticum aestivum	"Triticum" (Latin) = classical name for wheat; "aestivum" (Latin) = developing in summer
Typha angustata	"Typha" (Greek), name used by Theophrastus, a Greek philosopher; "angusti" (Latin) = narrow; referring to its long, narrow leaves
Urena lobata	"Urena" (Malayalam), the Malabar name of the plant; "lobata" (Latin) = lobed; referring to its lobed leaves
Urginea indica	"Urginea", the area of the tribe Beni Urgin, Algeria; "indica" = pertaining to India
Vaccaria pyramidata	"Vacca" (Latin) = cow; referring to the plant's use as fodder or its prevalence in pastures; "pyramidata" (Latin) = conical
Vallisneria spiralis	"Vallisneria", after Antonio Vallineri de Vallisnera (1661–71); "spiralis" (Latin)

	= spiral; referring to the coiled stalks of the female flowers
Verbena bipinnatifida	"Verbena" (Latin) = leafy twigs, used in wreaths for rituals; "bipinnatifida" (Latin) = twice pinnate
Vernonia cinerea	"Vernonia", after the late 17th century English botanist, William Vernon; "cinerea" (Latin) = ash-grey
Veronica agrestis	"Veronica", after St Veronica; "agrestis" (Latin) = of fields
Vicia faba	"Vicia", a classical Latin name for this genus; "faba", an old Latin name for broad bean
Vigna sinensis	"Vigna", after Dominico Vigna, a 17th century professor of Botany; "sinensis" (Latin) = from China
Washingtonia filifera	"Washingtonia", after George Washington (1732–99), the American President; "filifera" (Latin) = bearing threads or filaments; referring to the cottony threads present on the leaf margins
Ziziphus mauritiana	"Ziziphon" (Greek), an edible plum-like fruit; "mauritiana" = from Mauritius

BIBLIOGRAPHY

Gandhi M. 1989. ***Brahma's Hair: the mythology of Indian plants***. New Delhi: Rupa and Co.

Gledhill D. 2002. ***The Names of Plants***. Cambridge, UK: Cambridge University Press

Krishen P. 2006. ***Trees of Delhi: a field guide***. Noida, India: Dorling Kindersley

Maheshwari J K. 1963. ***The Flora of Delhi.*** New Delhi, India: Council of Scientific and Industrial Research

Stearn W T. 1973. ***Botanical Latin***. Newton Abbot, UK: Devon David and Charles

WEBSITE

<www.calflora.net/botanical names>

4

Species and Speciation

M A Khalid

Before discussing the interesting subject of species and speciation, I would like to share my memorable moments with Dr Prithipalsingh, my dear friend and colleague for more than a decade. Our association began when he was invited to be a subject specialist in the World Bank-aided forestry projects of TERI in Uttar Pradesh and Uttarakhand. This relationship grew further when we spent many weeks together in the forests of Shivalik Hills in Punjab conducting biodiversity surveys. I learnt a lot from him during those days and was extremely impressed with his plant-identification skill, which is becoming rare; I am yet to meet a plant taxonomist of his stature. There is indeed a decline in the number of professional taxonomists in both botany and zoology.

SPECIES AND SPECIATION

Various scientists have conducted exhaustive studies on plant species (Davis 1995; Baum and Donoghue 1995; Doyle 1995; Luckow 1995; McDade 1995; Olmstead 1995), and almost every biologist has faced challenges in carrying out taxonomic classification and studying evolutionary aspects, especially the process of speciation in nature. This is because species is the ultimate unit for identifying a plant or animal group as well as for assigning its right place in the respective classification. Also, species are the basic blocks that bring in a variety of existential or biological diversity (called biodiversity) in ecosystems.

Our planet earth is inhabited by millions of species, many of which have a unique role to play. Some of them are vectors of diseases, while others are important constituents of the food web. Biologists have recognized only 2–3 million species till date, which have been named, but probably 2–10 times the known species are yet to be identified. Some studies suggest that the species existing today may be only 1% of all the species that might have existed; the remaining 99% might have become extinct during the evolutionary process. This leads to the idea that species are products of the evolutionary process. Figure 1 shows the discovered and undiscovered species.[1]

New species evolve from pre-existing species as a result of adaptation to local environmental conditions. Charles Darwin first proposed this concept, and it is now uniformly accepted by all biologists. Although every biologist accepts the evolutionary origin of species in nature, there is no uniformity in defining species. In science, it is generally believed that "the more one tries to precisely define a concept, the more difficult it becomes". There has always been a great difficulty in defining "what a species is". Numerous definitions have been put forward over the years, but no single definition fits all species. Charles Darwin remarked that although every biologist knows approximately what a species is, there is no other taxon that has been subjected to numerous controversies regarding its definition. Darwin's theory of evolution, which leads to

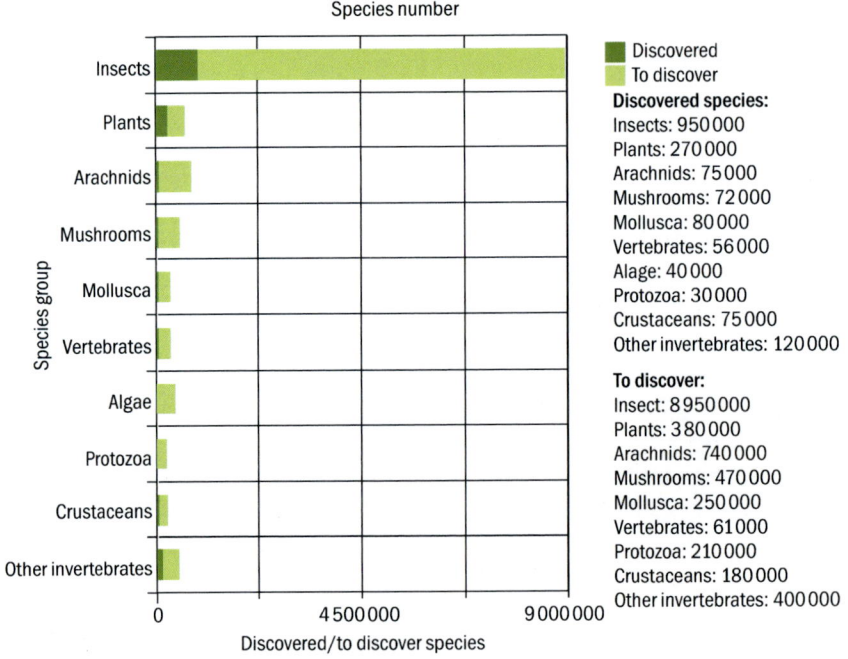

Figure 1   Bar chart showing discovered and undiscovered species

Source   <http://en.wikipedia.org/wiki/Species>

---

[1] Details available at <www.lemonde.fr/archives/article/2006/06/27/protection-de-la-biodiversite-un-inventaire-difficile_788741_0.html>

the origin of species, is also based on various notions like variation (in terms of numerous traits and adaptations that differentiate species from each other) and natural selection (explaining the relationship between variation and evolution of species and geographical isolation).

Some of the scientific arguments given by Darwin are (1) hereditary variation (which is the cause of species development), (2) struggle for existence (which defines advantageous variation), (3) natural selection (which is the mechanism of evolution), and (4) geographical isolation (which basically influences the divergence of species).

SPECIES CONCEPT

Species concept has been an interesting and favourite topic of various national/international seminars for the last two centuries. Various proposed concepts no longer exist, but attempts are being made to explain species corroborated with advancement of scientific knowledge. Attempts to define species led to many interesting references by various authors, which are as follows.

- A species is a creation of God.
- A species is a name in biology.
- A species is the lowest unit of biological classification.
- A species is the fundamental category of biological classification.
- A species is the smallest group of organisms that can be consistently distinguished by their morphology.
- A species is an assemblage of individuals with morphological features in common and separable from other such assemblages by correlated morphological discontinuities in a number of features.
- A species refers to the smallest natural population permanently separated from each other by a distinct discontinuity in the series of biotypes.
- A species is a community of cross-fertilizing individuals linked together by bonds of mating and isolated from other species by barriers to mating.
- A species is a group of inter-fertile populations that are reproductively isolated from other populations.
- A species is a group of similar-looking individuals that can produce fertile offspring in the natural environment.

Referencing to the level of species is the common practice adopted in most biological studies where the typical criteria include species richness, number of endemic species, and number/presence of endangered species in a given area. Various evidences to understand the species concept are given in Figure 2. Biologists use data on phylogenetic, genotypic, morphological, and ecological evidences to understand the species concept.

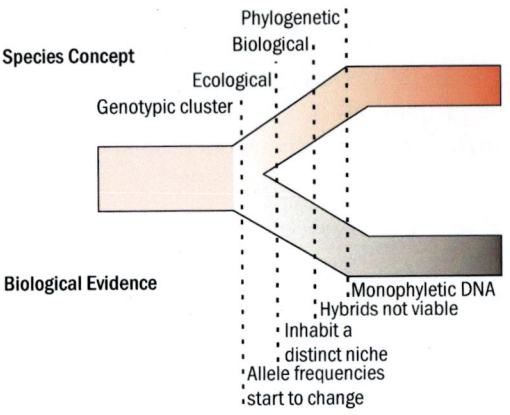

Figure 2 Various evidences to understand species concept

Source <theatavism.blogspot.com>

There are six major kingdoms. Kingdom Eubacteria and Kingdom Archaebacteria originated initially, which later culminated into Kingdom Protista. From this, three branches originated: Kingdom Fungi (Fungi), Kingdom Plantae (Plants), and Kingdom Animalia (Animals).

Spatial variability and distribution of species have been influenced by various factors like continental drifts, which in turn have influenced the biographic distribution of both plants and animals. Figure 3 shows a map of global vegetation types.

Morphologically, species are the smallest groups that are consistently and persistently distinct and distinguishable by ordinary means (Cronquist 1968). Since long, species have been treated as fundamental units of biology (Hull 1977) and have been recently considered in biodiversity conservation (Sites and Crandall 1997). Species should be identified accurately as they are vital pieces of research both in biology and biodiversity conservation. In Conservation Biology, biodiversity has been assessed at the level of species richness, number of endemic species, and presence of endangered species in given areas (Myers, Mittermeier, Mittermeier, *et al.* 2000). In biological systematics, species are mostly used as terminal taxa in the reconstruction of phylogenetic trees, whereas the methods by which they are delimited and identified receive scant attention (Wiens and Penkrot 2002).

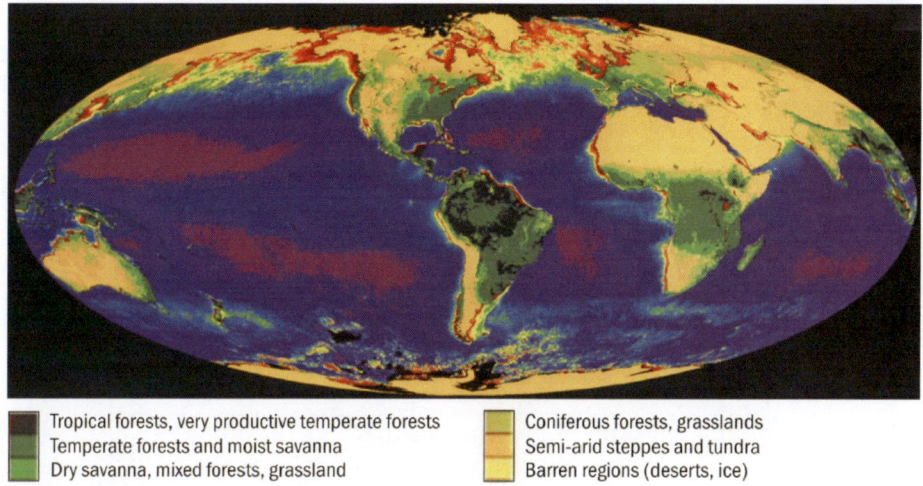

- Tropical forests, very productive temperate forests
- Temperate forests and moist savanna
- Dry savanna, mixed forests, grassland
- Coniferous forests, grasslands
- Semi-arid steppes and tundra
- Barren regions (deserts, ice)

Figure 3 Map of global vegetation types

Source <http://upload.wikimedia.org/wikipedia/commons/c/c3/Global_Vegetation_map_-_GPN-2003-00029.jpg>

Species have evolved over time as can be seen from records available for the last 650 million years. Figure 4 shows that there has been evolutions and extinctions during various periods.

Morphological Species Concept

The most commonly used species concept in biology is based on the morphological distinctness of species. It is very practical because it helps taxonomists to identify and name species on the basis of morphological appearance. It is also called the classical species concept because it has probably been used since people first began to classify organisms. Thus new species are usually named on the basis of the morphological species concept. A proper understanding of the morphology of individuals constituting a species becomes a prerequisite for distinguishing a new species from a previously defined one. Virtually, all names of plant species used in botany are based on the morphological species concept. Figure 5 shows the evolutionary classification of the plant kingdom.

Any botanist can distinguish a rice plant from a rose plant, or oak from oat, or for that matter *Spirogyra* from *Spirodela*. With more effort, a taxonomist can learn to distinguish many species of the same genus. For example, Randhawa (1959) described 289 species of *Spirogyra* in the monograph of the Family Zygnemaceae. Similarly, a specialist in genus *Pinus* is able to identify the 100 odd species described, or an oak specialist knows the 600 species of genus *Quercus*. This is also

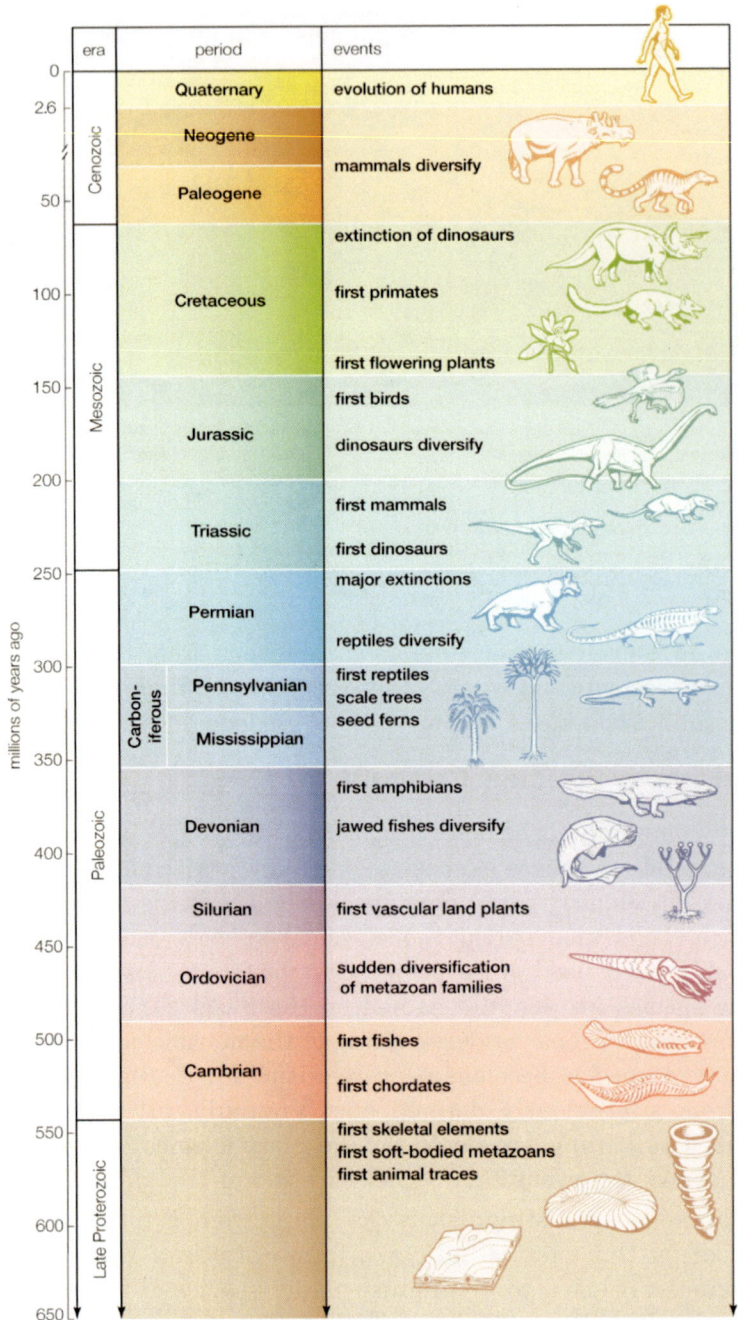

Figure 4 Geological timescale showing evolution of species during the last 650 million years

Source <www.britannica.com/bps/media-view/1650/1/0/0>

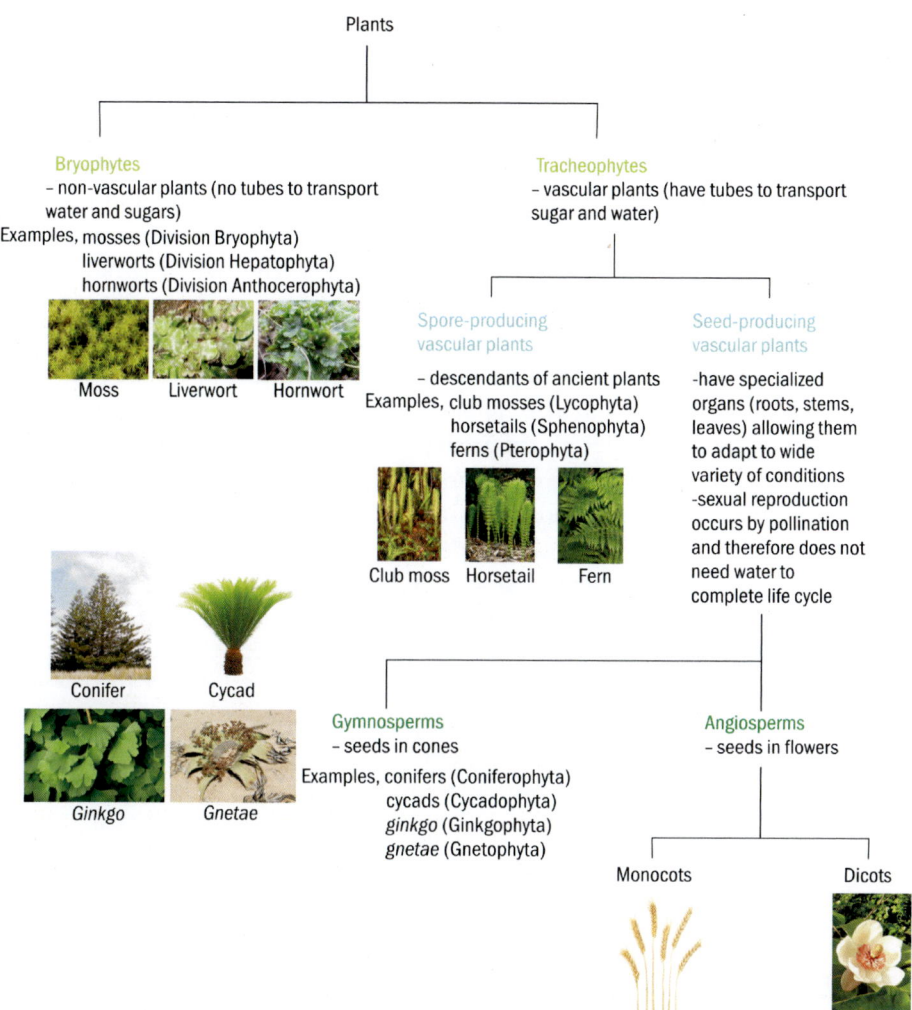

Figure 5 Evolutionary classification of the plant kingdom

Source <http://11biology.blogpost.com/2010/03/kingdom-plantae.html>

called the taxonomic species concept because it is commonly used by a majority of taxonomists. Thus it appears that the morphological species concept allows taxonomists to recognize different species on the basis of morphological differences among species. Therefore, since most species are known only by their morphology, this concept is widely used in biology. During the post-Darwinian period, the morphological species concept became more refined and it accepted the facts of variation. When species descend from a common ancestor and gradually diverge in their morphology to acquire unique adaptation, it is called gradualism;

whereas, if there is very little change in the morphology of offspring during the rest of its existence, it is called punctuated equilibrium, as shown in Figure 6.

In addition to morphology, geographical distribution of populations constituting a species began to be taken into account for a better definition. Therefore, from a purely morphological definition, the emphasis changed to a morphological–geographical approach. This helped in a better understanding of the concept of species. However, the use of this concept requires a proper understanding of the extent of morphological variation within species as well as proper information about their distribution. This usually requires an in-depth analysis based on many years of exposure to the vast diversity of nature. Therefore, it is based on the experience of the taxonomist. The real difficulty arises here. Taxonomists often disagree on which plants are of same species or which are different. For example, if a taxonomist recognizes a species as a group of individuals growing together in a natural population on the basis of morphological characters, another taxonomist may not agree on this recognition. Thus in the single-genus *Opuntia,* the number of species described by different taxonomists varies from 400 to 1000 depending on the judgement of individual taxonomists. This has made the application of this concept subjective. A species, therefore, cannot be precisely defined. At the same time, botanists also come across population of individuals that really cannot be strictly segregated on a morphological basis because of overlap

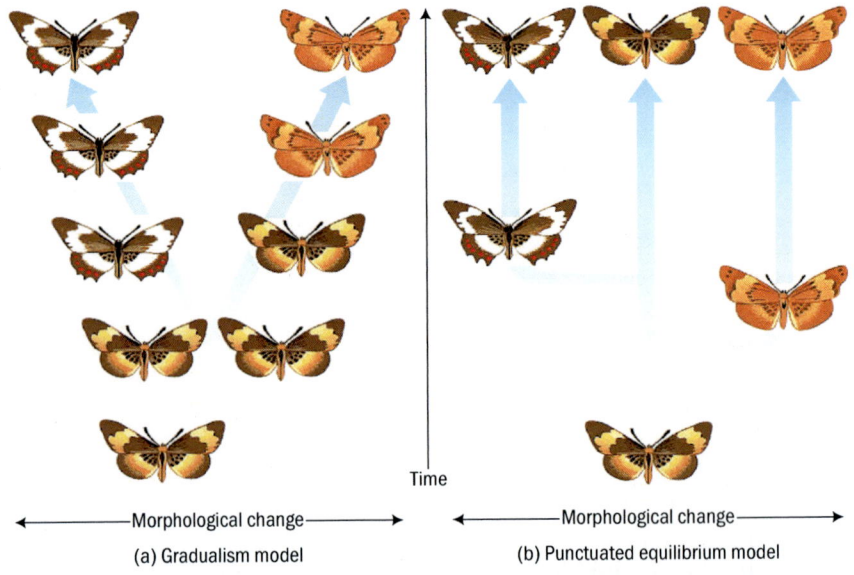

Figure 6   Various models showing morphological changes in butterflies

Source   <www.pearsoned.co.uk/imprints/BenjaminCummings>

of characters. Thus the lines between species are not clearly demarcated, and in such instances, different taxonomists give different judgements. Therefore, the morphological species concept does not completely help in defining species, especially when there is overlap of characters.

Biological Species Concept

In contrast to a purely morphology-based concept of species, the biological species concept uses reproductive biology to define species. This concept developed along with the study of the genetic basis of variation, reproductive mechanisms, and population structure. This concept has two components, and these are frequently stressed in the different definitions of the species concept. These are (1) interbreeding between component members and (2) reproductive isolation from other populations. A good example of using these components for defining species was provided by Grant (1957): "A species is a community of cross-fertilizing individuals linked together by bonds of mating and isolated from other species by barriers to mating." According to this definition, a species is a group of inter-fertile populations that are reproductively isolated from other such groups. Interestingly, this concept is very useful in zoology but is difficult to be applied in botany. For example, morphologically different species that are geographically isolated may hybridize readily when they are grown together. Are they really separate species? Similarly, identical species that are geographically isolated may not hybridize readily when they are grown together. Can they be accepted as the same species? These and other observations require a better understanding of the application of the biological species concept. Reproductive isolation may be interpreted in a strict sense as internal genetic or genetic–physiological mechanisms. It is often found that species delimited by orthodox taxonomic methods may not coincide with species defined in terms of sterility barriers. A practical difficulty arises from the need to test for sterility or fertility by conducting experiments. The fertility–sterility test is not an all-or-nothing criterion, because every degree of inter-fertility can occur between populations. This test becomes largely of theoretical value in allopatric populations. Further, if reproductive isolation is considered in a wider sense so as to include spatial, ecological, ethological, and other factors, it becomes even more difficult to apply the biological species concept.

It is also important to know that the necessary cytogenetic and experimental knowledge is only available for a minute fraction of the world's flora. In spite of these observations, the proponents of the biological species concept suggest that (1) it is objective (unlike the morphological–geographical species concept, which is subjective) and

(2) it does not require much experience to apply this concept because it can be easily experimented with by performing a few hybridization tests (unlike the other concept, which requires an in-depth analysis based on many years of exposure to the vast diversity of nature). The biological species concept is, however, evolutionarily important because it requires reproductive isolation. This concept emphasizes a fundamental assumption about speciation: a population is a distinct species only if it does not mix genes with other populations. Speciation can begin only when reproductive isolation occurs. Table 1 summarizes this aspect of understanding species.

Table 1	The biological species concept	
Individuals are	Not reproductively isolated	Reproductively isolated
Identical in morphology and sympatric	Same population	Sibling species
Identical in morphology and allopatric	Same subspecies	Sibling species
Different in morphology and sympatric	Variant individuals of the same population	Separate species
Different in morphology and allopatric	Separate subspecies	Separate species

Source The College of Science, SIUC (2007)

Genetic Species Concept

This is a relatively recent concept based on a quantifiable comparison of genetic data. Accordingly, the genetic distance from its nearest relatives determines the uniqueness of a species. Thus the genetic species concept is like the morphological species concept except that the recognition is based on genetic data instead of morphological data. The main drawback of measuring the genetic distance is that the distance is based on only a small part of the genome. For example, allele frequencies can be obtained by electrophoresis for only about 30 different enzymes; several thousand other enzymes cannot be used to measure genetic distance because we do not know how to detect them. Although recent advances have been made in calculating the genetic distance by comparing DNA fragments (RFLPs) and gene sequences, the data are limited to a small fraction of the genome. However, the information on gene distance has become important for evaluating endangered species because "genetic distance is being considered a tool for conservation". The following "species tree" (Figure 7) shows the evolution of species, while Figure 8 shows the phylogenetic tree of life.

Evolutionary Species Concept

This is another recent idea of understanding species. None of the species concepts refer directly to evolution. Therefore, evolutionary biologists have proposed that the ancestral–descendent sequence of populations should be called an evolutionary species. This concept is based on distinguishing morphological species over geological time. Thus a sequence of paleontological species along one evolutionary line is considered to be an evolutionary species.

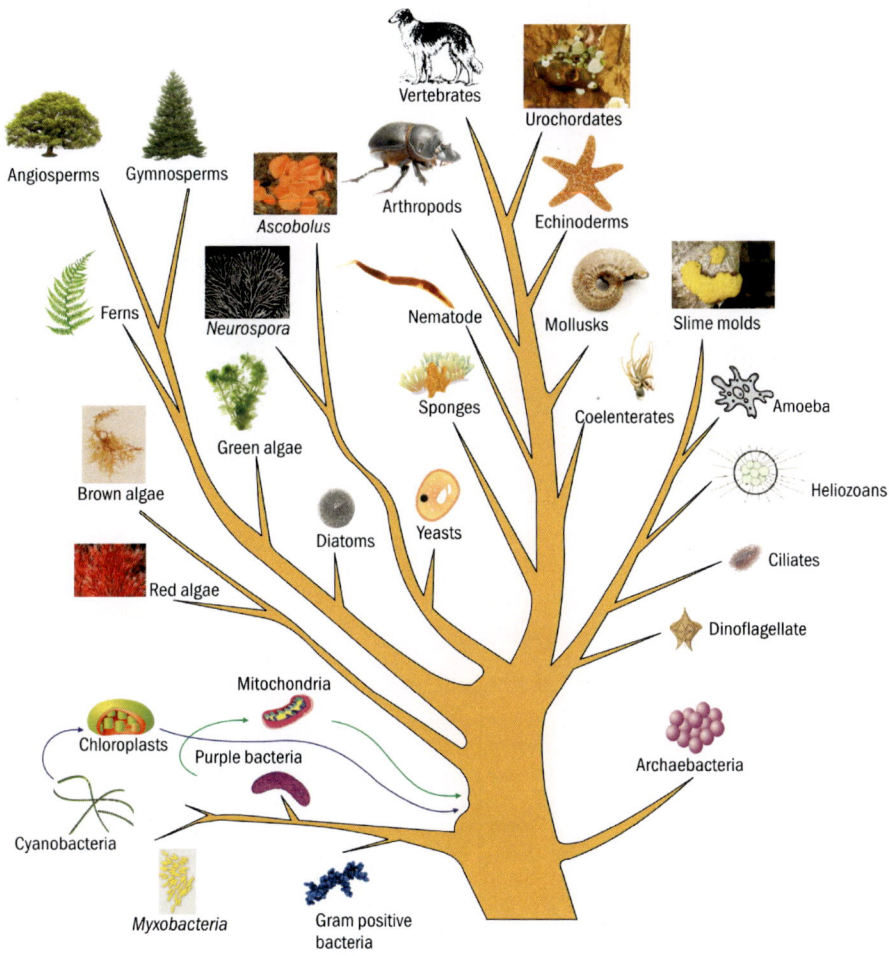

Figure 7   Phylogenetic tree of species evolution

Source   <www.nbii.gov/portal/server.pt/community/taxonomy%2C_phylogenetics___systematics/404/phylogenetic_trees/1617>

Figure 9 depicts the time line of species that evolved in different periods and eras starting from the Cambrian period of Paleozoic Era till the Cenozoic Era. The figure shows seed ferns originating in the Mississippian period and mammals appearing in the late Tertiary period.

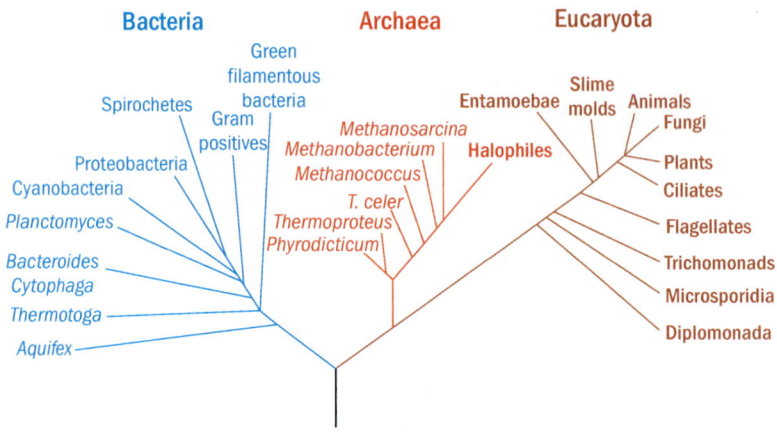

Figure 8 Phylogenetic tree of life

Source <http://nai.arc.nasa.gov/library/images/news_articles/big_274_3.jpg>

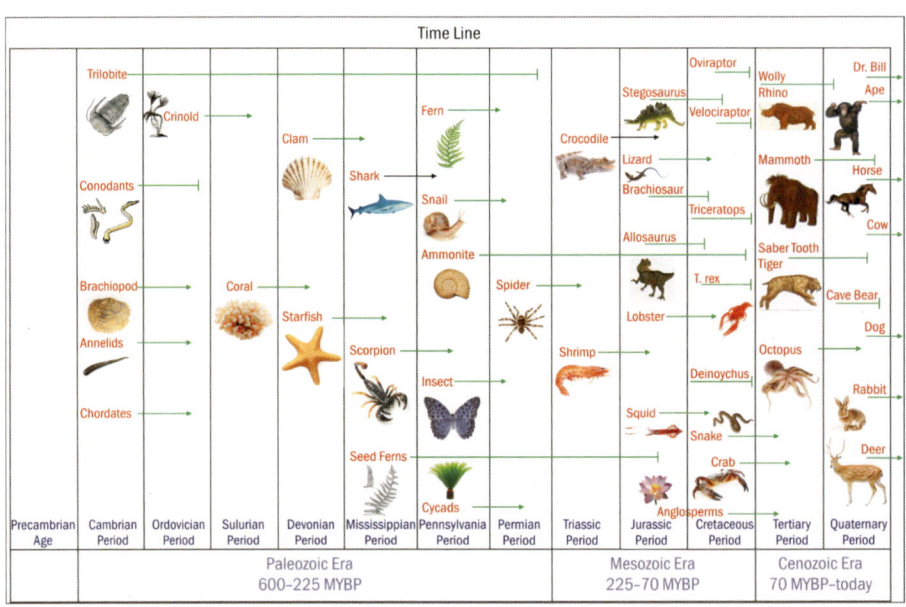

Figure 9 Species and phyla time line

Source <http://cas.bellarmine.edu/tietijen/Evolution/PhylaTimeLine.htm>

A major factor that emerges is the need to be able to know what constitutes this unit called species. What is the best manner for determining a species? Is the morphological distinctness of the unit important because this is the most easily observed parameter? Information about breeding behaviour, cross-ability or genetic distance is always welcome since this helps in confirming the hypothesis based on morphology. Unfortunately, information on these modern approaches for understanding species is meagre. Therefore, it is applied to a very few species.

For a vast majority of species recognized today, morphological distinctness is basically the only means of identification. We may conclude this discussion by accepting the comments made by Davis and Heywood (1963): "We may regard species as morphologically definable units, made up of groups of individuals (populations) which, it is assumed, are usually interbreeding, the containers and expression of one or more gene pools".

Population Divergence and Speciation

Having understood a species, it would be interesting to know that for the formation of a new species, the gene pool of a population must diverge from that of the parent (ancestral) population. Individuals in a diverging population become so different that they usually look different from the ancestral population. Over a period of time, these different individuals exploit different habitats, respond to different environments, and establish themselves as new species. Various steps recognized for this process are as follows.

- Partial reproductive isolation
- Changes in gene frequencies
- Genetic drift
- Mutations in subsequent generations
- Isolation
- Random accumulation of different mutations
- Environmental variation
- Morphological competitiveness
- Separation of the gene pool of a population from other populations

Natural speciation takes place over course of evolution (Figure 10). Speciation involves reproductive isolation of a gene pool and results from genetic divergence of populations. Reproductively isolated populations continue to evolve through mutation, genetic drift, and selection. Therefore, physical or temporal separation of populations allows genetic divergence to increase.

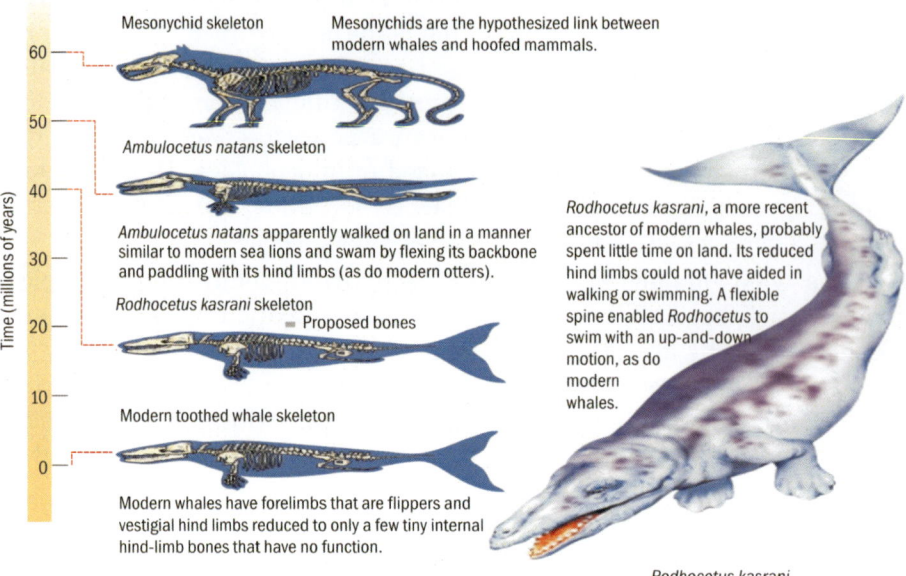

Figure 10 Separation of species from the same ancestor during course of time

Most biologists recognize three modes of speciation, each based on the geographical relationship of a new species to its ancestral species (Figure 11). These are as follows.
1. **Allopatric speciation** (meaning different homeland)—genetic divergence of two populations that are geographically separated
2. **Parapatric speciation** (meaning parallel native land)—populations have separate ranges that meet along a common border. This could be due to a discontinuity in some important environmental features such as soil changes.
3. **Sympatric speciation** (meaning same homeland)—populations that overlap geographically, may have mechanisms that bring about reproductive isolation within the same environment
4. **Peripatric speciation** (meaning around/surrounding homeland)—populations separating around/surrounding an existing species. This could be due to genetic divergence within a population leading to separation in the surrounding region/area.

Allopatric populations may be separated geographically. They lead to speciation when genetic variations accumulate and reproductive barriers are formed. Parapatric speciation can occur on the boundary between two populations where some gene flow occurs, but at a rate too slow to overcome the divergence of the gene pools of two neighbouring

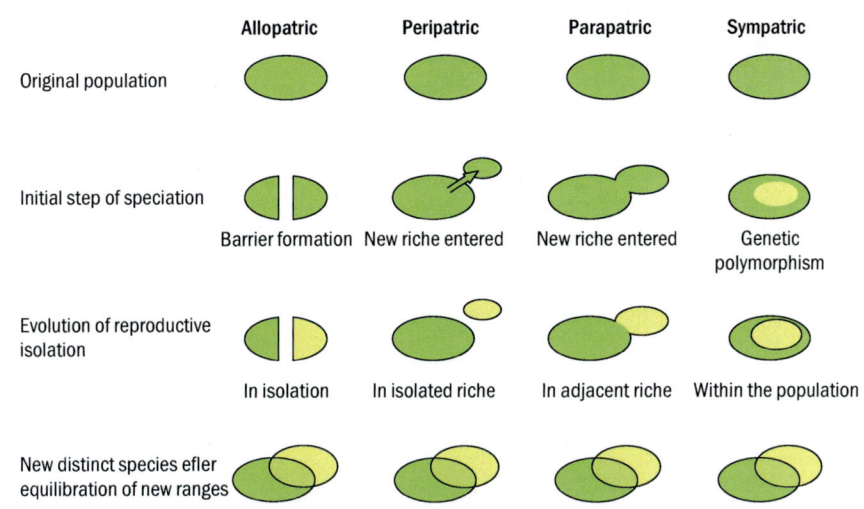

Figure 11 Comparison of allopatric, peripatric, parapatric, and sympatric speciation

Source < http://en.wikipedia.org/wiki/File:Speciation modes.svg >

populations. And finally, sympatric speciation occurs within a single population or among overlapping populations. This is dependent on mutations, hybridization, and polyploidy.

SPECIES DIVERSITY

The species richness in any habitat is used as an ecological unit to study the biodiversity of an area. It is an index essentially used as a measure for assessing the number of species in a sampling unit. It is very informative scientifically as it gives information about diversity of species. This leads to the understanding of the composition of floral or faunal community.

In nature, usually no community consists of species of equal abundance. There are four levels of inventory diversity (Whittaker 1977), and measures of species diversity can be divided into three categories (Magurran 1988): (1) species richness indices, (2) species abundance models, and (3) species proportional abundance-based indices.

Species Richness of India

India has two out of 34 biodiversity hotspots in the world (Figure 12) and is one of the 12 mega-biodiversity countries in the world.

The country is divided into 10 biogeographic regions starting from the cold mountains of the Trans-Himalayan region of Ladakh to the long western and eastern coastal belts having rich forests, mangroves, and sandy beaches.

Table 2 depicts various taxa showing the number of species in India, which is nearly 30% of the world's endemic flora and 62% of the known amphibian species in the world.

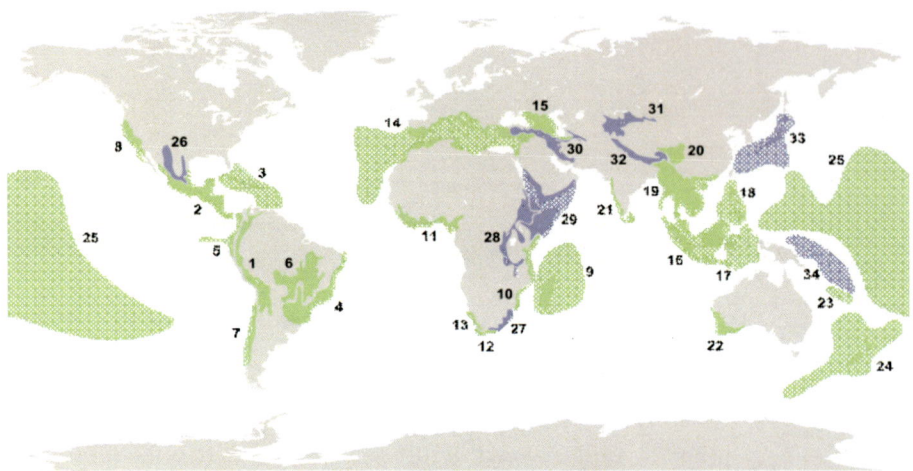

Figure 12 Biodiversity hotspots of the world

Source Myers, Mittermeier, Mittermeier, *et al.* (2000)

Table 2 Number and percentage of species of bacteria, fungi, plants, and animals in India

Taxon	Number of species	Percentage
Bacteria	85	0.08
Fungi	23 000	21.2
Algae	2 500	2.3
Bryophyta	2 564	2.4
Pteridophyta	1 022	0.9
Gymnosperms	64	0.06
Angiosperms	15 000	13.9
Insecta	53 430	49.3
Mollusca	5 050	4.7
Pisces	2 546	2.4
Amphibia	204	0.2
Reptilia	446	0.4
Aves	1 228	1.1
Mammalia	372	0.3
Total	108 276	100.0

Source Botanical Survey of India and Zoological Survey of India (1994)

EVOLUTION OF SPECIES, IDENTIFICATION, AND NOMENCLATURE TECHNIQUES

Understanding species is directly related to the advancements of scientific tools and techniques. These helped in more elaborate and deeper understanding of species.

CLIMATE CHANGE AND ITS IMPACT ON SPECIES

Climate change will bring positive as well as negative impacts on species. The effects will be positive for trees, which will grow faster due to increased CO_2, whereas negative for various animal and plant species (Figure 13), which would fail to cope with the rising temperatures and greenhouse gas concentrations, affecting their growth and survival. Changing climate will also influence the migration of species, which will change the habits and distribution of animals and will lead to extinction of non-coping plant species. It has been observed that high altitude species like *Quercus* spp. are getting replaced by much warmer *Pinus* spp. in the Uttarkashi area of Uttarakhand in the Indian Himalayas (Paul, Singh, and Das 2003).

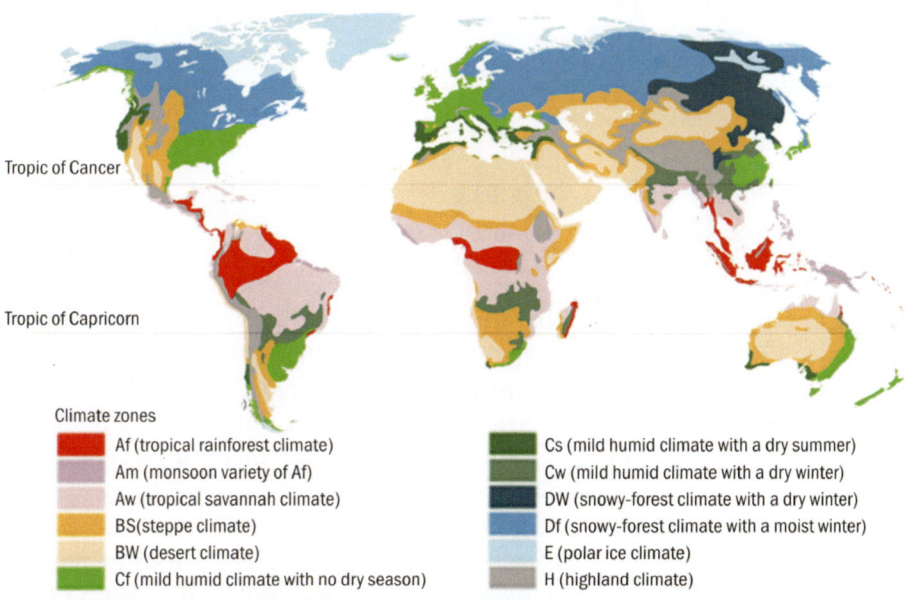

Figure 13   Various climatic zones of the world influencing vegetation

Source   <www.cid.harvard.edu>

A number of studies by various authors show that 15%–37% of a sample of 1103 land plants and animals would become extinct as a result of climate change by 2050 (Chris, Alison, Rhys, *et al.* 2004; Alan and Robert 2004).

SPECIES UNDER THREAT GLOBALLY

As per the International Union for Conservation of Nature (IUCN), 70% of plant species are under global threat along with a number of animal species (Figure 14). Invasive species, usually introduced from outside, are harmful to ecology and can destroy the local biodiversity. Invasive species cause huge loss to local endemic species and are the main cause of destruction of plant habitats and diseases and competition in animals.

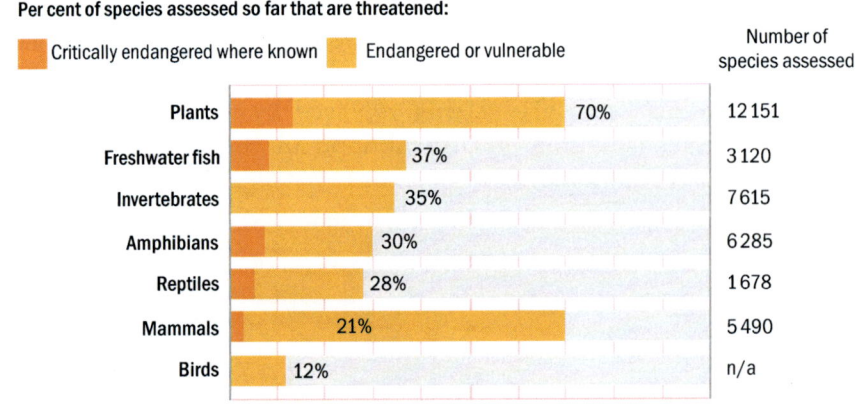

Figure 14 Species under threat globally
Source International Union for Conservation of Nature

In the USA alone, it is estimated that an economic loss of $79 billion was incurred because of 79 harmful alien species during 1906–91 (Stein and Flack 1996). In India, as per the Indian Council of Forestry Research and Education (ICFRE), there are nearly 75 identified forest invasive species (FIS), out of which 61 are of plants, 12 are of fungi, and 14 are of insects, which are a threat to the natural forest cover of the country.[2] In India, some of the invasive species like *Parthenium hysterophorus*, *Eupatorium adenophorum*, *Eupatorium odoratum*, *Mikania micrantha*, *Ageratum conyzoides*, and *Galinsoga parviflora* have already caused a huge loss to the local biodiversity (Raghubanshi, Rai, Gaur, *et al.* 2005). In the Indian flora, about 40% of the species are alien, of which

---
[2] Details available at <www.financialexpress.com/news/Invasive-species-are-threat-to-forest-cover-ICFRE/282361/0>

25% are invasive. As per the IUCN Red List 2011, the following plant (Table 3) and animal (Table 4) species are under various degrees of threat (IUCN 2011).

Table 3 Plants under threat

Name of plant species	IUCN category	Geographical distribution
Snowdrop (*Galanthus nivalis*)	Near to threatened	Across Europe
Pico de El Sauzal (*Lotus maculatus*)	Critically endangered	Endemic to the island of Tenerife, the Canary Islands, Spain
Centranthe À Trois Nervures (*Centranthus trinervis*)	Endangered	Endemic to Corsica
Trevo de Quatro Folhas (*Marsilea batardae*)	Endangered	Endemic to the Iberian Peninsula
Beta nana	Vulnerable	Endemic to the mountains of Southern and Central Greece

Source IUCN (2011)

Table 4 Animals under threat

Name of animal species Mammals	IUCN category	Geographical distribution
Arabian Oryx (*Oryx leucoryx*)	Vulnerable	Arabian Peninsula
Wallace's Tarsier (*Tarsius wallacei*)	Data insufficient	Indonesia
Siau Island Tarsier (*Tarsius tumpara*)	Critically endangered	Endemic to Siau Island, Indonesia
Northern Giant Mouse Lemur (*Mirza zaza*)	Vulnerable	Endemic to Madagascar
The Maned Three-toed Sloth (*Bradypus torquatus*)	Vulnerable	Restricted to the wet tropical forest on the Atlantic coast of Brazil
Reptiles		
Motagua Spiny-tailed Iguana (*Ctenosaura palearis*)	Endangered	Endemic to eastern Guatemala
Navassa Rhinoceros Iguana (*Cyclura onchiopsis*)	Extinct	Endemic to Navassa Island, West Indies
Bog Turtle (*Glyptemys muhlenbergii*)	Critically endangered	Endemic to the eastern USA
Wood Turtle (*Glyptemys insculpta*)	Endangered	Endemic to North America
Eastern Box Turtle (*Terrapene carolina*)	Vulnerable	Native to Canada, the USA, and Mexico
Anolis pogus	Vulnerable	Found only on the Caribbean Island of Saint Martin
Proboscis Anole (*Anolis proboscis*)	Endangered	Endemic to the western slopes of the Andes in Pichincha, Ecuador

Contd...

Table 4 Contd...

Name of animal species Mammals	IUCN category	Geographical distribution
Flap-necked Chameleon (*Chamaeleo dilepis*)	Least concern	Widely distributed and relatively abundant across southern and eastern Africa
Spiny-flanked Chameleon (*Trioceros laterispinis*)	Endangered	Endemic to Tanzania
Northern Pale-hipped Skink (*Celatiscincus similis*)	Endangered	Endemic to the Province Nord, New Caledonia
Gracile Burrowing Skink (*Graciliscincus shonae*)	Vulnerable	Endemic to Province Sud in New Caledonia
Eurydactylodes occidentalis	Critically endangered	Restricted to sclerophyll forest and closed mesophyll forest on the central west coast of Grand Terre
Crested Gecko (*Rhacodactylus ciliatus*)	Vulnerable	Restricted distribution; only occurs in Grand Terre and Ile des Pins, New Caledonia
Amphibians		
Dendrotriton chujorum	Critically endangered	Restricted to three sites in the northern region of the Sierra de los Cuchumatanes, Guatemala
Atelopus patazensis	Critically endangered	North-western Peru
Sri Lanka Petite Shrub Frog (*Pseudophilautus tanu*)	Endangered	Endemic to Sri Lanka
Green-eyed Bush Frog (*Raorchestes chlorosomma*)	Critically endangered	Only known from Munnar, Kerala, within the Western Ghats mountain range in India
Resplendent Shrub Frog (*Raorchestes resplendens*)	Critically endangered	Endemic to Kerala, India; restricted to a very small area on the Anamudi summit
Fishes		
Knipowitschia mrakovcici	Critically endangered	Only found in Lake Visovac, Croatia
Sea Trout (*Salmo trutta*)	Least concern	Widespread across much of Europe and has been introduced to North and South America, Africa, parts of Asia, New Zealand, and Australia
Pacific Hagfish (*Eptatretus stoutii*)	Data insufficient	Endemic to the north-eastern Pacific
Broad-gilled Hagfish (*Eptatretus cirrhatus*)	Least concern	Found in shallow to deep waters in Australia and New Zealand

Contd...

Table 4 *Contd...*

Name of animal species Mammals	IUCN category	Geographical distribution
Invertebrates		
Violet-spotted Reef Lobster (*Enoplometopus debelius*)	Data insufficient	Endemic to western Pacific
Mediterranean Slipper Lobster (*Scyllarides latus*)	Data insufficient	Found throughout most of the Mediterranean Sea and in the central eastern Atlantic Ocean
Banded Spiny Lobster (*Panulirus marginatus*)	Data insufficient	Endemic to Hawaii
Chapa (*Iberus gualtieranus*)	Endangered	Endemic to mainland Spain
Leptaxis minor	Endangered	Endemic to the Pico Alto complex on Santa Maria Island, Azores
Moreletina horripila	Least concern	Found on São Miguel, Terceira, São Jorge, Pico and Faial Islands, Azores
Suboestophora hispanica	Vulnerable	Endemic to the Spanish provinces of Alicante and Valencia
Helix pomatia	Least concern	Widespread in Central and Eastern Europe

Source IUCN (2011)

Global efforts are going on to conserve keystone species through a number of international species conservation agencies such as the International Union for Conservation of Nature (IUCN), United Nations Environment Programme (UNEP), United Nations Educational, Scientific, and Cultural Organization (UNESCO), Global Environment Fund (GEF), Nature Conservancy, and World Wildlife Fund (WWF). A number of species conservation acts are also in place to support the cause. But the onus lies on the *Homo sapiens,* the so-called intelligent human species, who are destroying the natural resources due to their greed and, in turn, eliminating species at a faster rate.

It is high time that man realize the importance of innumerable species of plants and animals provided to us by nature for the well-being of our planet earth and humans. Saving plant and animal species is saving ourselves, the human race!

BIBLIOGRAPHY

Alan P J and Robert P. 2004. **Global warming is altering the distribution and abundance of plant and animal species. Application of a basic** law of ecology predicts that many will vanish if temperatures continue to rise. *Nature* **427**: 107–109

Baum D A and Donoghue M J. 1995. **Choosing among alternative "phylogenetic" species concepts.** *Systematic Botany* **20**: 560–573

Chris D T, Alison C, Rhys E G, Michel B, Linda J B, Yvonne C C, Erasmus F N, Marinez F D S, Alan G, Lee H, Lesley H, Brian H, Albert S V J, Guy F M, Lera M, Miguel A O, Townsend P A, Oliver L P, Stephen E W. 2004. **Extinction risk from climate change.** *Nature* **427**: 145–148

Cronquist A. 1968. ***The Evolution and Classification of Flowering Plants***. London: Thomas Nelson (Printers) Ltd.

Davis J I. 1995. **Species concepts and phylogenetic analysis: introduction.** *Systematic Botany* **20**: 555–559

Davis P H and Heywood V H. 1963. ***Principles of Angiosperm Taxonomy***. New York: Robert E Krieger Publishing Co. Inc.

Doyle J J. 1995. **The irrelevance of allele tree topologies for species delimitation, and a non-topological alternative.** *Systematic Botany* **20**: 574–588

Grant V F. 1957. **The plant species in theory and practice.** In *The Species Problem*, edited by E Mayr, pp. 39–80. Washington, DC: American Association for Advancement of Science

Hull D L. 1977. **The ontological status of species as evolutionary units.** In *Foundational Problems in the Special Sciences*, edited by R Butts and J Hintikka, pp. 91–102. Dordrecht, Holland: D Reidel Publishing Co.

IUCN (International Union for Conservation of Nature). 2011. **IUCN Red List of Threatened Species. Version 2011.2.** Details available at < http://www.iucnredlist.org>

Luckow M. 1995. **Species concepts: assumptions, methods, and applications.** *Systematic Botany* **20**: 589–605

Magurran A E. 1988. ***Ecological Diversity and Its Measurement***. Princeton, New Jersey: Princeton University Press

McDade L. 1995. **Species concepts and problems in practice: insight from botanical monographs.** *Systematic Botany* **20**: 606–622

Myers N, Mittermeier R A, Mittermeier C G, da Fonseca G A B, Kent J. 2000. **Biodiversity hotspots for conservation priorities.** *Nature* **403**: 853–858

Olmstead R. 1995. **Species concepts and plesiomorphic species.** *Systematic Botany* **20**: 623–630

Paul V, Singh T P, and Das A. 2003. **Potential impact of climate change on forests: a case study in Uttaranchal.** In *Environmental Threats, Vulnerability, and Adaptation: case studies from India*. New Delhi: TERI

Raghubanshi A S, Rai L C, Gaur J P, Singh J S. 2005. **Invasive alien species and biodiversity in India.** *Current Science* **88**(4): 25

Randhawa M S. 1959. ***Zygnemaceae***. New Delhi: Indian Council of Agriculture Research

Sites J W Jr. and Crandall K A. 1997. **Testing species boundaries in biodiversity studies**. *Conservation Biology* **11**: 1289–1297

Stein B A and Flack S R (eds). 1996. *America's Least Wanted: alien species invasions of US ecosystems*. Arlington, Virginia: The Nature Conservancy

The College of Science, SIUC (Southern Illinois University Carbondale). 2007. **Species Concept, Chapter 13, Plant Biology/PLB 479/Lecture PLB479/SpeciesConcepts**. Details available at <www.plantbiology.siu.edu/PLB479/Lectures%20PLB479/SpeciesConcepts.html>

Wiens J J and Penkrot T A. 2002. **Delimiting species using DNA and morphological variation and discordant species limits in spiny lizards (*Sceloporus*)**. *Systematic Biology* **51**: 69–91

Whittaker R H. 1977. **Evolution of species diversity in land communities**. In *Evolutionary Biology*, Vol 10, edited by M K Hecht, W C Steere, and B Wallace, pp. 250–268. New York: Plenum Press

5

Modern Tools for Identification of Plants

Rajni Gupta and Ruchitra Gupta

INTRODUCTION

Taxonomy is the science of discovering, describing, naming, and identifying species and other taxa. Description and identification of species are fundamental to biology. Without taxonomy, biologists in various disciplines will not be able to identify different organisms. Hence, they will be unable to report their empirical findings or access available information on their target organisms without knowing their identities. Taxonomy lays the foundation for the construction of the tree of life, makes baseline data available for conservation and ecological studies, and provides humans the opportunity to take advantage of the underutilized resources offered by earth's biodiversity (Wilson 2004). Despite its importance as a foundation for other disciplines, taxonomy is one of the most neglected fields of research. Earlier, tools such as dichotomous keys, body-punched cards, and edge-punched cards were used for the identification of genera, species, and floras. Now computer programs have been developed for the identification of plant genera and species.

With the advancement in sequencing and computational technologies, DNA sequences have become the major source of new information to enhance our understanding of evolutionary and genetic relationships. The footprints of comparative sequence analysis are now apparent in almost all areas of biological sciences. Applications employed to assess the biological relationships with DNA sequences include molecular phylogenetics and population genetics.

WHAT IS DNA BARCODING?

DNA barcoding is based on the premise that a short standardized sequence can distinguish between individuals of a species, because interspecific genetic variation exceeds intraspecific variation (Hebert 2003). DNA barcoding systems are now being established for various groups of organisms, including plants, macroscopic algae, fungi, protists, and bacteria (Hajibabaei, Singer, Hebert, *et al.* 2007).

Identification of species through barcoding is usually achieved by the retrieval of a short DNA sequence (the barcode) from a standard part of the genome of the specimen under investigation. The barcode sequence from each unknown specimen is then compared with those available in the library of reference barcode sequences, which are derived from individuals of known identity. If the sequence of a species closely matches with the one in the barcode library, the specimen can be identified. DNA sequences are unique for each species. They can be viewed as genetic barcodes and have the potential to solve the problems inherent to the kind of taxonomy practised so far. With a possible nucleotide variation of four nitrogenous bases (A, T, C, G) at each site, there are 4^n probable codes for any given sequence ("n" nucleotides long), making it possible to identify every taxon. The survey of just 15 nucleotide positions can identify up to one billion species. The advantages of this method are that the identification can be performed quickly and at low cost. The other additional advantages include the possibility of identifying individuals at any stage of development and discriminating between morphologically identical species (Pires and Marinoni 2010).

DNA barcoding is a taxonomic system that uses a short stretch of a core DNA sequence to identify species. A region of approximately 648 bp of the mitochondrial gene cytochrome C oxidase 1 (COI) was initially proposed as the barcode source to identify and delimit all animal species (Consortium for the Barcode of Life [CBOL])[1]. The methodology involves the sequencing of that portion of DNA, followed by a comparison with other sequences previously deposited in a database. If the obtained sequence matches with any sequence in the database, the species can be identified (Hebert 2003). Although COI has been chosen for DNA barcoding in animals, it is not suitable for plants and fungi. Like other mitochondrial genes in these groups, it also evolves too slowly for species-level discriminations (Chase and Fay 2009). For this reason, alternative stretches of DNA have been proposed as target sequences for the barcoding of these organisms. Two regions of the plastid DNA have been recommended for terrestrial plants: *matK* and *rbcL* (CBOL Plant Working Group 2009). For fungi, the Internal Transcribed Spacer (ITS) region of the ribosomal DNA offers best results as target sequence. Researchers studying microorganisms were the first to practice a

---

[1] Details available at <www.barcoding.si.edu/DNABarcoding.htm>

genomic-mediated taxonomy (Patel, Leonard, Pan, *et al.* 2000). In fact, molecular tools cannot be considered as an innovation to taxonomy because they have already been used for a few decades to identify and distinguish species (Sperling, Anderson, and Hickey 1994). DNA probes on a single-stranded DNA molecule have been used to detect the presence of a complementary sequence. In this context, DNA barcoding can be seen as an attempt to revolutionize the study of biodiversity and has been proved to be a powerful tool in the identification of various species. The wide acceptance of the barcode of life reflects its scientific success since it was first proposed by Hebert, Stoeckle, Zemlak, *et al.* (2004) and Ward, Zenlak, Innes, *et al.* (2005). DNA barcoding research has been facilitated by the Barcode Life Data (BOLD) system—an online resource available to the scientific community.[2] This resource offers tools that allow researchers to perform neighbour-joining clustering, to store information on the different groups studied and to identify taxa using an updated sequence library, among other things.

ADVANTAGES OF BARCODES

DNA barcodes aid in gaining additional information about a species. Compiling a public library of sequences linked to named specimens, along with faster and cheaper sequencing, will make this new barcode key more useful. Barcode can be used to obtain different types of information, as follows.

- Barcoding can identify a species from bits and pieces. After the identification technique is confirmed, barcoding will be able to identify quickly the undesirable animal or plant material in processed food stuffs and detect commercial products derived from regulated species. It will help reconstruct food cycles by identifying fragments in stomach. It may also be able to identify roots sampled from soil layers.
- Barcoding can identify species at any stage of its lifecycle: from eggs and seed, through larvae and seedlings, to adults and flowers.
- A barcode of life provides a digital identifying feature, having a sequence of four nucleotides. A library of digital barcodes will provide an unambiguous reference, which will facilitate identification of different species across the globe and through centuries.
- A DNA barcode can distinguish among different species that look alike, uncovering dangerous organisms masquerading as harmless ones. It shall also enable a more accurate view of biodiversity.

---

[2] Details available at <www.boldsystems.org./views/login.phptt>

- Scientists can now use DNA barcoding for faster identification of known organisms and rapid recognition of new species.
- It shall make it possible to identify species, whether abundant or rare, native or invasive, engendering appreciation of biodiversity locally and globally.
- Barcoding links biological identification to advancing frontiers in DNA sequencing, miniaturization in electronics, and computerized information storage. Integrating those links will lead to portable desktop devices and ultimately to hand-held barcoders.
- Genetic similarities and differences among the two million species identified so far will provide a wealth of genetic details, helping to draw the tree of life on earth.
- The library of barcodes collects details of the millions of specimens available in museums, herbaria, zoos, gardens, and other biological repositories. Involvement of these institutions and their collections in DNA barcoding will strengthen their ongoing efforts to preserve earth's biodiversity.
- Compiling a library of barcodes linked to vouchered specimens and their binomial names will enhance public access to biological knowledge. Such a library shall help create an online encyclopaedia of life on earth, with a web page for every species of plants and animals.

DNA BARCODES TO IDENTIFY FLOWERING PLANTS

For the investigation of plant molecular systematics at the species level, the ITS region of the nuclear ribosomal cistron (18S–5.8S–26S) is the most commonly sequenced locus. This region has shown broad utility across photosynthetic eukaryotes and fungi and has been suggested as a possible plant barcode locus. Species-level discrimination has been validated in most phylogenetic studies that employ ITS, and a large body of sequence data already exist for this region. For example, more than 36 000 angiosperm sequences were available in the Gen Bank in December 2004. One advantage of the ITS region is that it can be amplified in two smaller fragments (ITS-I and ITS-2) adjoining the 5.8S locus, which has proved to be especially useful for degraded samples. The quiet conserved 5.8S region, in fact, contains enough phylogenetic signals for discrimination at the level of orders and phyla, although identification at this taxonomic level is not the concern of barcoding. Gen Bank blast searches with ITS data returned correct matches for the sequences in the Gen Bank. This success suggests that despite alignment concerns, current search algorithms will be fast and effective at using ITS for species-level identifications, given an adequate database for comparison. For all these reasons, ITS, even with its recognized limitations, is an effective locus for DNA barcoding in plants (Chase, Salamin, Wilkinson, et al. 2005; Table 1).

Table 1 Percentage of identity, as measured by the BLAST algorithm, at different taxonomic levels (inter/intra), and proportion of incorrect assignment of sequences at the specific and generic levels

DNA region	Taxonomic level	Identity	Wrong assignment	
			First occurrence	Last occurrence
ITS1 + 5.8S + ITS2	Species	95.75/97.20	6.59	45.73
	Genus	94.95/95.65	0.30	43.66
	Family	94.35/95.24		
	Order	92.98/93.12		
ITS1 + 5.8S	Species	97.24/97.43	0.84	43.65
	Genus	94.69/95.76	0	40.30
	Family	93.02/94.27		
	Order	92.07/92.54		
5.8S + ITS2	Species	95.48/97.21	3.57	40.77
	Genus	93.50/93.72	0.90	34.66
	Family	91.38/93.21		
	Order	90.75/91.79		
ITS1 + ITS2	Species	95.26/96.08	0.47	49.86
	Genus	93.98/94.42	1.13	46.28
	Family	93.20/92.03		
	Order	91.86/92.09		
ITS1	Species	96.43/96.72	1.11	6.79
	Genus	93.39/95.77	0.02	39.64
	Family	91.69/92.51		
	Order	91.11/91.34		
ITS2	Species	93.98/95.28	7.08	33.70
	Genus	93.78/93.10	0.10	51.68
	Family	91.79/89.46		
	Order	88.89/89.24		
5.8S	Species	97.18/99.80	32.76	62.94
	Genus	96.30/98.45	36.04	64.62
	Family	95.32/97.25		
	Order	93.25/95.28		
rbcL	Species	97.62/99.73	3.69	16.95
	Genus	95.25/97.13	0.23	
	Family	91.43/95.92		
	Order	88.81/92.65		

Source Chase, Salamin, Wilkinson, et al. (2005)

However, ITS has certain functional limitations for DNA barcoding of plants. Hence, a search for additional loci is warranted. For phylogenetic investigations, the plastid genome may offer plant barcoding what the mitochondrial genome has done for animals. It is a uniparentally inherited, non-recombining, and, in general, structurally stable genome. Universal primers are available for a number of loci and intergenic spacers that are evolving at various rates. The plastid locus most commonly sequenced by plant systematics for phylogenetic purpose is *rbcL*, followed by the *trnF* intergenic spacer, *matK*, *ndhF*, and *atpB*. *rbcL* has been suggested as a candidate for plant barcoding even though these have generally been used to determine evolutionary relationships at the generic level and above.

Besides *rbcL* and *atpB*, all other plastid loci mentioned above have been used at the species level with various degrees of success. Most of them require full-length sequences greater than 1 kb to yield enough sequence length to discriminate species. No region of the plastid genome has been found to have the high level of variation seen in most animal COI barcodes, although a few intergenic spacers have shown more promise than any plastid locus now in use. Kress, Wirdack, Zimmar, et al. (2005) chose two closely related flowering plants for comparison: *Atropa belladona* and *Nicotiana tabacum* (Table 2). The complete data are available for the plastid genomes of both species (29 additional complete plastid genomes spread across a wide range of plant groups are also available for comparison: algae, mosses and liverworts, ferns and

Table 2 Sampled loci in plastid genomes of *A. belladonna* and *N. tabacum* that were found to have base-pair sequence divergences greater than 2%

Locus	Sequence length, bp (Atropa/Nicotiana)	Indents for pairwise alignment	Per cent sequence divergence between Atropa and Nicotiana
trnK-rps16(B)	707/685	6	4.1
trnH-psbA(A)	412/453	8	3.9
rp136-rps8(I)	451/426	4	3.0
atpB-rbcL(H)	815/818	1	2.8
ycf6-psbM(D)	1091/1135	9	2.8
trnV-atpE(G)	494/485	2	2.6
trnC-ycf6(C)	681/670	4	2.4
psbM-trnD(E)	1099/1081	10	2.4
trnL-F(F)	363/357	2	2.2
rbcL	1434/1434 (Including stop codon)	0	0.83
ITS	622/628	7	13.6

Source <www.pnas.org/content/102/23/8369.full>

selective gymnosperms and angiosperms belonging to families *Fabaceae* and *Poaceae*). Kress, Wirdack, Zimmar, et al. (2005) aligned the *Nicotiana* and *Atropa* genomes and determined raw divergence levels across all genes, introns, and intergenic spacers. Plastid regions with raw sequence differences greater than or equal to 2% were categorized as the most variable segments and, therefore, the most promising of the plastid genome for DNA barcoding when normalized for length. The nuclear ITS region and plastid *rbcL* gene were used as baseline comparisons for these chloroplast test regions (Kress, Wirdack, Zimmar, et al. 2005).

In this comparison, nine intergenic spacers—*trnK-rps16, trnH-psbA, rp136-rps8, atpB-rbcL, Ycf6-psbm, trnV-atpE, trnC-YcF6, psbm-trnD,* and *trnL-F*—have been used. ITS also has a higher divergence value than any of the plastid regions, and *rbcL* is by far the lowest in divergence. Three spacers (*atpB-rbcL, ycf6-psbm,* and *psbM-trnL*) were slightly to moderately longer than the 800 bp cut-off. Wang, Yongui, Yiheng, et al. (2010) worked on the family Lemnaceae (commonly called duckweeds), which comprises 38 species in five genera. They are all aquatic plants growing on or below the surface of water all over the world. They are ideal for physiological, biochemical, and genomic studies because of their direct contact with medium, rapid growth, and relatively small genome sizes. The growth rate of certain species such as *Lemna minor* is affected by a wide range of environmental contaminants such as metals, nitrates, and phosphates. These species are used by the Environmental Protection Agency for testing water quality (proportional to the amount of contaminants present) (Meier, Shiyang, Vaidya, et al. 2006; Wang, Yongui, Yiheng, et al. 2010). It is necessary to establish a sequence database against which unknown species may be compared and tentative species identification can be validated. This database shall provide a high-resolution phylogenetic resource for the important plant monocot family. Polymerase chain reaction (PCR) products, DNA sequencing, and evolutionary studies have to be calculated for making comparative studies of DNA barcodes (Wang, Yongui, Yiheng, et al. 2010).

The CBOL Plant Working Group suggested 7^{14} leading candidates, 4 coding genes (*rpoB, rpoC, rbcL,* and *matK*), and 3 non-coding spacers (*atpF-atpH, psbk-psbl,* and *trnH-psbA*). Genomic DNA extracted from 97 ecotypes was subjected to PCR amplification with primer pairs based on the chloroplast sequence of *L. minor.*

PCR primers were also used for sequencing (≥95%) of all the barcode candidates except *matK* (71%). The maximal and minimal alignment lengths of PCR product for *rpoB, rpoC1, rbcL,* and *matK* were identical, whereas those of *atpF-atpH, psbK-psbI,* and *trnH-psbA* were quite variable, being in the range of 579–622, 185–576, and 286–504 bp, respectively.

It was not unexpected that the coding markers (*rpoB, rpoC, rbcL,* and *matK*) were conserved in PCR product length, while the non-coding spacers (*atpF-atpH, psbK-psbI,* and *trmH-psbA*) displayed more variability due to extensive insertions/deletions (Brain and Solomen 2007).

The accuracy of barcoding for the identification of species depends to a large extent on the barcoding gap between intra and interspecific sequence variations. Effective barcoding became weaker when inter and intraspecific distances overlapped. The rank order for the correct identification is *atpF-atpH* (92.85%) > *psbk-psbl* (84.7%) > *trnH-psbA* (82.5%) > *matK* (77.77%) > *rpoB* (77.5%) > *rpoC1* (70.58%) > *rbcL* (70.58%). Generally, the three non-coding spacers produced higher rates of successful identification than the four coding markers, *atpF-atpH* yielding the best result with 92.85% successful identifications (Meier, Shiyang, Vaidya, et al. 2006).

PRESENT SITUATION AND TENDENCIES

After numerous studies being conducted using DNA barcoding since 2003, this technique has become acceptable now. The method has been proved to be fast, reliable, and cheap, in both the discovery and the identification of diverse species (Radulovici, Archambault, and Dufresne 2010). The fact that DNA barcoding and taxonomy complement each other is also widely accepted by the modern scientific community. DNA barcoding has been considered an efficient aid to traditional taxonomy. Although morphology has been used for solving taxonomic problems, there is an increasing need to use molecular tools (Packer, Gibbs, Sheffield, et al. 2009). There is enough evidence to show that barcoding is not a substitute for traditional taxonomy but, in contrast, both fields are complementary. However, a factor has probably hindered the progress of an integrated approach. According to Ebach and Carvalho (2010), DNA barcoding does not adhere to best practice in taxonomy because it is not taxonomic in nature. Taxonomy, for instance, is purely morphological (requiring proper description of the organism) and not molecular (requiring understanding molecules). In this sense, the use of molecular data (such as barcoding) in the discovery of new species (which have to be described in morphological terms and not in molecular terms) is unable to completely replace traditional taxonomy (Ebach and Carvalho 2010).

FUTURE OF BARCODING

Molecular barcoding used for taxonomic purposes is already a reality (Blaxter 2003b). Description of new species is being published with a DNA sequence attached to the primary nomenclature (Sommer, Carta, Kim, et al. 1996), and this should be actively encouraged. The taxonomists preparing new taxa for publication should be welcomed by DNA-savvy biodiversity laboratories, which should be able to provide expertise at minimal cost. The methods for rapid sequence acquisition at minimal cost already exist at genome-sequencing centres and are easily adapted to taxonomic sampling. There needs to be a core plurality in the programme, with more than one sequence per specimen being a minimal aim. Museum collection curators need to assess the benefits and drawbacks of allowing the precious specimens to be sampled for DNA. To improve DNA preservation, there is a need to apply certain measures to prevent cross-contamination (Nadler 1999) and to investigate the changing storage conditions. With a sequence-based system, addition of an annexe to EMBL/Gene Bank that stores barcode data is simple to conceive, possibly difficult but not impossible to implement, and pressing in its urgency. Free standing efforts may also be viable (Hebert, Cywins, Ball, et al. 2003). A parallel effort is required to make the more traditional taxonomic literature accessible to all (Lee 2002; Oren and Stackebrandt 2002). A coordinated effort is needed in the molecular taxonomic community to fully investigate the variation of barcode targets within a taxon. It shall also be necessary to define the "taxon discrimination ability" of targets and develop new tools for the analysis of burgeoning data sets. The major contribution of molecular taxonomy shall be the ability of a few dedicated centres to produce hundreds of thousands of molecular tags per year so that they can be used to diagnose species. If the biotic component of this planet is to be identified and the endless forms are to be listed (if not understood), this is the only coherent way forward (Blaxter 2003a).

LIMITATIONS

One of the major concerns is the inclusion of molecular information into taxonomic aspects. Currently, DNA barcodes are being applied to two separate tasks: the first being the use of DNA data to distinguish between species and the second being the use of DNA data to discover new species. These two activities differ in the types and amount of data required. But there are some limitations to the utilization of barcodes.

A major issue that needs to be resolved is how to read the organismal barcode once it is generated. Most recently published approaches to DNA barcoding have utilized distance measures to make the inference as to species designation (Hebert 2003; Hebert, Stoeckle, Zemlak, et al. 2004). Distances are used in two major approaches. The first is a simple BLAST

(Altschul, Gish, Miller, *et al.* 1990) approach where a raw similarity score is used to determine the nearest neighbour to the query sequence. The second approach utilizes distances in tree building (Hebert, Ratnasingham, and deWaard 2003). Character-based approaches are more appropriate to DNA barcoding for both theoretical and practical reasons.

A major drawback of using distances in DNA barcoding is that all classical studies and taxonomic schemes can achieve the same result that barcodes are meant to accomplish. Both are character based, making the union of classical and DNA barcoding difficult, especially if the use of distances is continued in barcoding studies.

This shortcoming is related to the need for diagnostic characters that classical studies use to validate the existence of a species. Another shortcoming is that similarity scores often do not give the nearest neighbour as the closest relative (Koski and Golding 2001). Character-based methods have the logical advantage that when diagnostic character data are lacking, they will fail to diagnose, thus allowing for a degree of hypothesis testing. However, this is not possible when using distances. The third shortcoming involves the lack of an objective set of criteria to delineate taxa when using distances. A universal similarity cut-off to determine the status of the species will simply not exist because of the broad overlap of inter and intraspecific distances/variations (Goldstein, Desalle, Amato, *et al.* 2000). Desalle, Egan, Siddall, *et al.* (2005) suggested an alternative approach (including character-based phylogenetic analysis) as more appropriate for establishing or printing barcodes. The character-based approach is compatible with classical approaches, allowing the combination of classical, morphological, and behavioural information. Character-based approaches can avoid the nearest-neighbour problems of distances because they can reconstruct hierarchical relationships where common ancestry is inferred when two entities share derived characters. Neither BLAST nor neighbour-joining tree-building approaches allow for character-by-character diagnoses on branches of trees. Any such diagnosis would need to be based on maximum likelihood. The diagnosis of two separate entities in nature can be accomplished by the existence of a single character shared by a group of organisms to the exclusion of others, be it a DNA character or a morphological character.

Another controversial aspect of the DNA barcoding initiative is related to the number of individuals of each putative species to be included in the analysis. Classical taxonomic endeavours screen numerous individuals from multiple localities across the range of a given species to distinguish between intra and interspecies variation. This will help identify the characters uniquely shared among all members of a species. One or only a few individuals may not be representative of the species as a whole, especially for taxa with widespread distributions (Walsh 2000). For both distance-and character-based methods, an adequate number of individuals is required. However, neither a universal sample size is appropriate for all species, nor a universal geographic distance exists for determining the appropriate sampling strategy.

Strategic Value of DNA and Tissue Archiving

DNA barcoding requires assembling of tissue samples and subsequent isolation and archiving of genomic DNA. These archived samples can act as useful resources for phylogenetics, population genetics, and phylogeographic studies.

CONCLUSION

DNA barcoding contributes to taxonomic research, population genetics, and phylogenetics. In taxonomy, DNA barcoding can be used for routine identification of specimens, and it can also flag atypical specimens for comprehensive taxonomic investigation. In phylogenetic studies, DNA barcoding can be used for optimal selection of taxa and barcode sequences and can also be added to the sequence data set for phylogenetic analysis. On the basis of recent developments, we expect that the barcode databases will grow rapidly. Interestingly, some facilities are already processing more than 100 000 specimens per year. Consequently, the International Nucleotide Sequence Database (INSD), Gene Bank, EMBL, and DDBJ have adopted a unique keyword identifier (BARCODE) to recognize standard barcode sequences specified by the scientific community (CBOL). The introduction of DNA barcoding is a natural addition to the post-genomic era in which the whole genome sequencing has provided a vast amount of sequence information from a limited number of species. DNA barcodes can help expand our knowledge by exploring many more species rapidly and inexpensively.

BIBLIOGRAPHY

Altschul S F, Gish W, Miller W, Meyers E W, Lipman D J. 1990. **Basic local alignment search tool**. *Journal of Molecular Biology* **215**: 403–410

Blaxter M L. 2003a. **Molecular systematic: counting angels with DNA**. *Nature* **421**: 122–124

Blaxter M L. 2003b. **The promise of a DNA taxonomy**. *Philosophical Transaction of the Royal Society of London B* **359**: 669–679

Brain R A and Solomen K R. 2007. **A protocol for conducting 7 day daily renewal tests with *Lemna gibba*.** *Nature Protocols* **2**(4): 979–987

CBOL Plant Working Group. 2009. **A DNA barcode for land plants.** *Proceedings of the National Academy of Sciences* **106**(31): 12794–12797

Chase M W and Fay M F. 2009. **Barcoding of plants and fungi.** *Science* **325**(5941): 682–683

Chase M W, Salamin N, Wilkinson M, Dunwell J M, Kesanakurthi R P, Haider N, Savolainen V. 2005. **Land plants and DNA barcodes: short-term and long-term goals.** *Philosophical Transactions of the Royal Society of London B* **360**: 1889–1895

Desalle R, Egan M G, and Siddall M. 2005. **The unholy trinity taxonomy, species delimitation and DNA barcoding.** *Philosophical Transaction of the Royal Society* of London *B* **360**: 1905–1916

Ebach M C and Carvalho M R. 2010. **Anti-intellectualism in the DNA barcoding enterprise.** *Zoologie* **27**: 165–178

Goldstein P Z, Desalle R, Amato G, Vogler A P. 2000. **Conservation genetics at the species boundary.** *Conservation Biology* **14**: 120–131

Hajibabaei M, Singer G A C, Hebert P D N, Hickey D A. 2007. **DNA barcoding: how it complements taxonomy, molecular phylogenetics, and population genetics?** *Trends in Genetics* **23**: 167–172

Hebert P D N. 2003. **Biological identification through DNA barcodes.** *Proceedings of the Royal Society of London B* **270**: 313–321

Hebert P D N, Cywins K A, Ball S L, deWaard J R. 2003. **Biological identifications through DNA barcodes.** *Proceedings of the Royal Society of London B* **270**: 313–321

Hebert P D N, Penton E H, Burns J M, Janzen D H, Hallwachs W. 2004. **Ten species in one: DNA barcoding reveals cryptic species in the neotropica skipper butterfly *Astraples fulgerator*.** *Proceedings of the National Academy of Sciences* **101**: 14812–14817

Hebert P D N, Ratnasingham S, and deWaard J R. 2003. **Barcoding animal life: cytochrome oxidase submit divergence among closely related species.** *Proceedings of the Royal Society of London B* **270**: 596–599

Hebert P D N, Stoeckle M X, Zemlak T S, Francis C M. 2004. **Identification of birds through DNA barcodes.** *PLoS Biology* **2**: 1657–1663

Koski L B and Golding G B. 2001. **The closest BLAST hit is often not the nearest neighbour.** *Journal of Molecular Evolution* **52**: 540–542

Kress J W, Wirdack I J, Zimmar E A, Weigt L A, Janzen D H. 2005. **Use of DNA barcodes to identify flowering plants.** *Proceedings of the National Academy of Sciences* **102**: 8369–8374

Lee M S. 2002. **Online database could end taxonomic anarchy.** *Nature* **417**: 787–788

Meier R, Shiyang K, Vaidya G, Ng P K L. 2006. **DNA barcoding and taxonomy in *Diptera*: a tale of high interspecific variability and low identification success.** *Systematic Biology* **55** (5): 715–728

Nadler S A. 1999. **Nucleotide sequences from vintage helminths: fine wine or vinegar?** *Parasitology Today* **15**: 122

Oren A and Stackebrandt E. 2002. **Prokaryote taxonomy online: challenges ahead.** *Nature* **419**: 15

Packer L, Gibbs J, Sheffield C, Hanner R. 2009. **DNA barcoding and the mediocrity of morphology.** *Molecular Ecology Resources* **9**: 42–50

Patel J B, Leonard D G B, Pan X, Musser J M, Bergman R E, Nachamkin I. 2000. **Sequence based identification of *Mycobacterium* species using the micro seq. 500 16S rDNA bacterial identification system.** *Journal of Clinical Microbiology* **38**: 246–251

Pires A C and Marinoni L. 2010. **DNA barcoding and traditional taxonomy unified through integrative taxonomy: a view that challenges the debate questioning both methodologies.** *Biota Neotropica* **10**(2): 339–346

Radulovici A A E, Archambault P, and Dufresne F. 2010. **DNA barcodes for marine biodiversity: moving fast forward?** *Diversity* **2**: 450–472

Sommer R L, Carta L K, Kim S Y, Stiernberg P W. 1996. **Morphological, genetic, and molecular description of *Pristionchus pacificus* sp. N. (Nematode: Neodiplogastridae).** *Fundamental and Applied Nematology* **19**: 511–521

Sperling F A H, Anderson G S, and Hickey D A. 1994. **A DNA based approach to the identification of insect species used for postmortem interval estimation.** *Journal of Forensic Science* **399**: 418–427

Walsh P. 2000. **Sample size for the diagnosis of conservation units.** *Conservation Biology* **14**: 1533–1535

Wang W, Yongui W U, Yiheng Y, Marina E, Randall K, Jaochin M. 2010. **DNA barcoding of the Lemnaceae a family of aquatic monocots.** *BMC Plant Biology* **10**: 205

Ward R D, Zenlak S T, Innes B H, Last P R, Hebert P D N. 2005. **DNA barcoding Australian fish: species.** *Philosophical Transaction of the Royal Society of London B* **360**: 1847–1857

Wilson E O. 2004. **Taxonomy as a fundamental discipline.** *Philosophical Transaction of the Royal Society of London B* **359**: 739

6

Plant Taxonomy in Plant Genetic Resource Management

Anjula Pandey, D C Bhandari, and K Pradheep

Identity of a germplasm, one of the prerequisite information linked to all other fields of science, is established through knowledge of plant taxonomy. It has direct role in the utilization of genetic resources, and thus it is a strong tool in plant genetic resource (PGR) management. Unidentified and wrongly or incompletely identified germplasm, when added to the holding, provides no additional information, rather burdens the curators in its management and creates an unrealistic picture on germplasm holdings. Therefore, the PGR worker should be familiar with plant taxonomy and identification procedures to identify a material before conservation. The present chapter discusses the relevance of plant taxonomy in PGR management.

INTRODUCTION

The utilization of PGR requires a multidisciplinary approach through the involvement of researchers, conservators, policymakers, stakeholders, local communities, and other end users linked to the system. This field of science depends greatly on the correct identification of material to match it with the information that may be known about the taxon (Guarino, Ramanatha Rao, and Reid 1995). Correct identification achieved through knowledge of taxonomy not only helps its utilization but also establishes its links with different groups, species, and species complexes and differences among different levels of hierarchy. By furnishing correct information about a plant, an identifier helps all other fields of science in a basic way (Prithipalsingh 2006).

Note Information presented in this chapter is in reference to Indian Gene Centre and is focused on the activities undertaken at the NBPGR. The legal issues relating to PGRs are the views of authors in light of current developments and not of the NBPGR.

Unlike the traditional methodology used in identification of plants (Porter 1959), the PGR germplasm collections gathered for ex situ conservation are represented as seed, vegetative buds, tubers, rhizomes, and so on. The germplasm mainly represents crop plants; in addition, wild species, including crop relatives, forages, multipurpose trees and shrubs, and forestry species, are also represented. Growing out and re-identification of several large germplasm holdings have established that a significant proportion of the material was wrongly identified, identified up to the genus level or unidentified (Guarino, Ramanatha Rao, and Reid 1995). Materials incorporated in genebanks as unknown legume species or those identified only up to genus level (*Abelmoschus* sp., *Vigna* sp., *Oryza* sp., and so on) or group level (as cucurbits, orchids, legumes) are of limited conservation value and thus useless for the purpose of utilization.

The PGR management includes major activities such as germplasm exploration/collection, exchange/quarantine, characterization/evaluation and multiplication, conservation, and documentation. In India, with the growth of PGR science in the past three decades, there has been an increase in germplasm holdings, from native as well as introduced sources, in genebank at the National Bureau of Plant Genetic Resources (NBPGR). Germplasm exchange programmes meet the requirements of various indenters through material transfer agreement (MTA), this being a nodal institute authorized for international exchange of germplasm and for carrying out their quarantine processing deal with large material from native and introduced sources. For more than a decade, NBPGR has been continuously involved in capacity building in the field of PGR-related activities, including plant taxonomy. To develop taxonomic skill among students at the graduation and post-graduation levels, it runs teaching programmes affiliated to the Post Graduate School, IARI, New Delhi. The programmes emphasize on teaching taxonomy through lectures, field trainings, workshops, and brainstorming sessions at various levels.

The PGR activities deal with the problems of identity pertaining to correct botanical name (including correct nomenclature), synonyms, local/vernacular name, family, and so on. A good knowledge base is required to overcome these complications. The present account mainly deals with the Indian context and provides an array of information on the importance of taxonomy in PGR activities such as plant exploration, germplasm collection, exchange, characterization, quarantine, conservation, and documentation of PGR (Gadgil, Singh, Nagendra, et al. 1996).

PLANT EXPLORATION AND GERMPLASM COLLECTION

In plant exploration and germplasm collection activity, material is collected from native and exotic sources. Inaccurately identified germplasm may remain uncollected because it does not provide any useful information (Arora 1991). If the collected germplasm has wrong or incomplete identity, it may lead to spurious or incomplete results when studied and used.

Collecting Germplasm

The NBPGR undertakes germplasm collection and exploration (region/crop specific), in both collaborative and independent modes for gathering targeted materials (trait specific) from diversity-rich areas. The NBPGR carries out the task independently or in collaboration with state agricultural universities and crop-based institutes (Arora, Mehra, and Hardas 1975; Arora 1991). Also, collaborative research programmes are in place with non-agricultural universities for resolving problems in agricultural crops and vice versa for mutual benefit.

While collecting germplasm, identification of material requires sharp observation skills. Identification may be wrong or incomplete because of limitations of knowledge base, lack of specialization, and large amount of materials available for study. Hybridization, combined with polyploidy, is a common cause of confusion in taxonomic identity or wrong naming of taxa. To conclude on plant's identity and specific traits, many disciplines such as plant morphology, anatomy, embryology, palynology, physiology, ecology, genetics, and cytology play an important role (Stace 1984; Bajora, Hemp, Heel, *et al.* 2008). A taxonomist, by virtue of his experience and studies on plant characters, places a taxon in the correct family, followed by genus and species. Thus, correct observations followed by description, identification, and correct naming of a plant help to link it to its present and future value as a PGR. Diagnostic characters help in placing it to the nearest matching taxon based on similarity of major characters (particularly in closely allied genera, and weedy and wild relatives of crop plants). In case the plant characteristics do not match with those of existing species, probability could be of a new form/botanical variety, subspecies, species, and genus.

Collecting Herbarium Vouchers

Germplasm collectors gather herbarium vouchers to determine the identity of accession collected from an area. All germplasm of wild species/crop genepool and rare/endangered species, rare types need to be maintained as vouchers (Pandey, Bhandari, Bhatt, *et al.* 2003).

Vouchers may represent range of variation from mass samples gathered from type location, host species of diseased or pest samples, floristic records, less known/underutilized species collected from a site to any other experimental material kept for referral use.[1] Vouchers can be repeatedly referred for identification purpose, during regeneration cycles of germplasm characterization/evaluation and seed multiplication at different regeneration sites, thus adding further notes to the effect of environment on the phenotypic observations. Complementary collections such as edible parts and products of economic value can supplement the identification process, and make it faster.

Collecting Additional Information

During the collection of germplasm, information on the range of diversity distribution of a species, polymorphism, cytology, and other allied fields can help simplify identification exercise and avoid confusions. It is difficult to note down all details of a species during collection, but diagnostic characters noted serve as the most important clue. After detailed study, the data generated by proper description of plant are used for authenticating and validating a species. Supplementary information on local names/vernacular names, ecological data, associated flora, economic use, ethnobotanical notes, and so on is helpful. But one should not use this in isolation to establish the identity of germplasm. In seed/vegetatively propagated material, grow out tests play the safest and confirmatory role in establishing the identity of germplasm.

While collecting and handling germplasm, special notes on passport information on special features of fruits, perishable nature of material are required for separating seeds or propagules. The germplasm should accompany the special notes on characters, especially for tree species, bamboos, tuber crops, flowers, fruits/bunching behaviour, and so on. Range of variability of germplasm, even if not collected, must be mentioned in the accompanying data to assess the extent of variability. Photographic collections are advisable in case of difficult to collect materials (bulky), and also for vulnerable or rare types.

EXCHANGE AND QUARANTINE OF GERMPLASM

The NBPGR undertakes exchange of germplasm with more than 100 countries. The increasing import of germplasm (including transgenic and trial material) has enhanced the chances of introduction of new pests/pathogens. Every year, a large number of samples of different germplasm,

---

[1] Refer NBPGR-NHPC vouchers.

crops, or breeding material are introduced from all over the world and processed for quarantine clearance. Based on taxonomic knowledge of host and pathogens, the infested/infected samples are identified for proper treatment and salvaging. Only the pest-free germplasm is released to users. New pathogens identified during this process are to be treated on top priority for introduction and permitted only for restricted entry into the country or else detained. Need-based supportive research is conducted for developing user-friendly identification detection keys. PGR scientists dealing with exchange and introduction of germplasm should have significant knowledge of invasive and quarantine weeds (at least up to species level). They should be able to distinguish the species that are not covered under MTA, especially their propagules (including seeds), to avoid any chance of wilful or non-wilful default. Familiarity with synonymy can help avoid repeated introductions of same species into the country, thus protecting national interests.

EVALUATION OF GERMPLASM

The utilization of germplasm depends mainly on the identification of promising accessions through evaluation. The characterization and preliminary evaluation involve recording of highly heritable morphological characters, identifying promising accession(s), and separating the types at evaluation/utilization stage. Scoring of more new characters for plant breeders' use (Dhillon, Varaprasad, Srinivasan, *et al.* 2001) and evaluation of the same are carried out at different agro-ecological zones of the country through various regional stations and networks with other organizations. Identification of unique, distinct, newer, and superior genotypes in crops is considered to be most desirable.

Besides the classical methods of characterization (based exclusively on agro-morphology and physiological traits), more sensitive methods using molecular techniques are now used to distinguish between germplasm and varietal materials. These have proved to be more useful and powerful tools in establishing uniqueness and identity of a taxon (molecular taxonomy). Use of molecular marker technology for diversity analysis forms an integral part of characterization process (Dhillon, Varaprasad, Srinivasan, *et al.* 2001). This reduces the threat of biopiracy and infringement of plant varieties related to intellectual property right (IPR) issues through DNA fingerprinting of germplasm. The National Research Centre (DNAFP) (now called NGRC), located in the NBPGR, is responsible for the DNA fingerprinting of the released varieties and elite germplasm of crops. The research and development activities focus on marker development, gene tagging, molecular diagnostics, molecular phylogeny, and marker-assisted selections.

Evaluation of germplasm infraspecific variation (that is, subsp./var./form) needs to be handled and studied properly, especially in data analysis and interpretation of results. For distinct and unique type of materials, development of descriptors and descriptor states for database recording can be visualized with good knowledge on plant taxonomy. Knowledge on variation in morphological traits is a must for the development of crop descriptors. Visible traits linked with economic traits (for example, quality traits/pigmentation) may help sort out F_1 plants from their parents.

Agricultural practices like identification and removal of weedy/off-type plants, undesirable plant types (due to mechanical admixture in planting material), and introgressions (selection of rootstocks, pollinators for horticultural crops, and so on) require knowledge of plant characters and taxonomy. Also undesirable genetic contamination due to crossability with progenitors/related wild relatives, especially when the transgenic crops are being evaluated under experimental conditions, can be prevented with adequate knowledge.

CONSERVATION OF GERMPLASM

The NBPGR is the custodian of ex situ conserved native and exotic diversity available in the country through its regional stations and networks. Orthodox seeds are conserved in seed genebank (at –18°C): vegetative propagules as cultures in the in vitro genebank; and embryos, embryonic axes, dormant buds, and pollen in the cryogenebank (at –170°C to 180°C). Stored germplasm is routinely monitored for viability, quantity, and health as per genebank standards. Field genebanks generally maintain vegetatively/clonally propagated material at different national active germplasm sites situated at various institutions across the country.

On receiving a material, it is compulsory to ensure its identity prior to the proceeding for conservation. During the steps from collection to conservation of germplasm, there are chances of mechanical mix-up due to mishandling or some other reasons. In such cases, it is mandatory to validate it with the help of available expertise: facilities such as the National Herbarium of Cultivated Plants at the NBPGR, New Delhi, and supportive aids such as seed atlas/literature. Thus, there is a need to develop expertise to identify taxa through seed/propagules.

DOCUMENTATION AND INFORMATION BASED ON PGR

Database Management and Inventorization

The NBPGR has a computerized documentation system operational through an agricultural research information system. The data maintained in the central databases include botanical names (valid name), synonyms, vernacular names/common names, areas of collection, and status (wild/cultivated). Collection of other relevant information such as geographical location/availability of germplasm (exploration-based information), literature (monographs, floras, checklists), and availability of materials of PGR relevance has also been initiated (Zeven and de Wet 1982).

Information based on taxonomic knowledge, combined with floristic and exploratory studies, brought out a scientific monograph *Wild Relatives of Crop Plants in India* (Arora and Nayar 1984; Arora and Pandey 1996). It covered information on wild relatives of crop groups based on their agri-horticultural importance (number of species given in parenthesis), including cereals and millets (51), legumes (31), oilseeds (12), fibres (24), vegetables (54), fruits (109), spices and condiments (27), and miscellaneous (26). To build up collection and study crop plant taxonomy, a document *Check-list of Crop Plants* was published, based on exploratory studies, literature, and other parameters. It included over 480 crop species used widely for various purposes, including prioritization on the crop taxa of India (Nayar, Pandey, Venkateswaran, *et al.* 2003). Descriptors/descriptor states, which explain variability of morphological characters with unique traits recorded for germplasm registration, are visualized with good knowledge on plant taxonomy.

National Herbarium of Cultivated Plants

The National Herbarium of Cultivated Plants (NHCP) at the NBPGR, New Delhi, is a specialized herbarium that maintains a working collection of crop genepool comprising wild, weedy taxa, primitive landraces, obsolete cultivars, modern cultivars, and so on. The collection includes both indigenous and exotic plants, with a larger presentation of cultivated plants that provide the basis for PGR studies. In addition, the NHCP holds supplementary collections of seeds, economic products/carpological samples, drawings/illustrations, and photographic records. These are referred to during other PGR programmes, to validate germplasm frequently handled as seed material or seed propagules (as rhizomes, bulbs, bulbils, tubers, and others) and economic products. It serves as a reference source for identification purposes and for taxonomic, phytogeographic, and phylogenetic study of PGR. Its collections are complementary to that of the herbaria of the Botanical Survey of India (BSI), Kolkata; Forest Research Institute (FRI), Dehradun, Uttarakhand; and National Botanical Research Institute (NBRI), Lucknow, Uttar Pradesh.

Information Base

Changes in nomenclature are often associated with taxonomy. Synonymy of taxa is generally considered to create confusions in identification and, therefore, needs to be handled carefully while naming a taxon. Emphasis needs to be laid on aspects such as plant population dynamics, and its composition, means of reproduction and dispersal mechanism, inter-relationships, distribution pattern, and evolutionary path. Wider knowledge on variability pattern (in seed, seed coat, flowers, and others) is necessary. Similarly, factors responsible for phenotypic variation in a species (soils, irrigation, and temperature), polymorphism, distribution range, plasticity, crossability/swarm hybrids/intermediate forms, and other characteristics need to be understood. Inadequate information of these often leads to identification of wrong species. It is, therefore, important to possess significant knowledge about species identification and taxonomy. It is difficult to acquire in-depth knowledge on characters of all taxa/families. However, one can attempt to narrow down the major/economic plant families of much relevance to agri-horticultural crop groups. Out of over two dozen families that have greater relevance to the PGR management, it has been shown that families such as Poaceae, Cucurbitaceae, Malvaceae, and Fabaceae are significantly represented in the PGR holdings (Dhillon, Varaprasad, Srinivasan, *et al.* 2001; Karthikeyan 2000). Knowledge of unique/diagnostic characters of these families can help in understanding and providing clues to the nearest possible level of hierarchy during collection/conservation. To reduce time and simplify the efforts, a PGR worker can prepare a checklist including the characteristics of family, genus, and species to be explored. Identification aids, especially those including seed/propagules, need to be developed region-wise or crop group-wise.

Identification Methodology

The observations made on the characters of roots, tubers, rhizomes, bulbs, stem, leaves, and plant parts in some genera/species are useful key characters for identification (Porter 1959; Lawrence 1951). Some of the notable features are habit (woodiness or herbaceous), characters of stipule, leaf (morphology, leaf hairiness), root (nodules, aerial root system), fruit (capsule, pepo, berry), presence of exudates (milky juices or gums, resinous substances), and characters of specific parts in tree (bark, lenticels characters)/bamboos (spathe). The anatomical characters such as presence of clusters of tissues in stem (Cucurbitaceae), specialized structures (oleoresin ducts in rhizomes in Zingiberaceae), bast fibres of unique nature (Urticaceae), seed/seedling morphological characters (cotyledon number), presence of endosperm, embryo type, and aril on seed are useful key markers for taxonomic identification (Paria 1997).

Observations of petals (presence or absence; if present whether united or free) provide the primary subdivisions of dicots, such as apetalae (without petals), polypetalae (separated petals), and sympetalae (united petals); perianth characters; and ovary position (superior or inferior). For example, the family Cucurbitaceae is characterized by the presence of inferior ovary. Sometimes, fruit characters are quick parameters for identifying the plant families. For example, Cucurbitaceae (melon family), Poaceae (grass family), Malvaceae (marshmallow family), and Fabaceae (pea family) represent unique fruit type and are readily distinguishable. The type of placentation (marginal, free central, and basal) is generally associated with whole group or family and thus may help in establishing the relatedness among the genera.

Liaison with Other Organizations

The NBPGR functions independently and collaboratively with other PGR programmes/traditional institutes (BSI, NBRI, state universities, and others) in issues relating to taxonomy. Better collaboration with national organizations (BSI, FRI, regional/local herbaria, universities), researchers, and experts in various crop groups can support the PGR programmes for the country and can facilitate the common goal of taxonomy promotion in PGR.

MAJOR CONCERNS

While dealing with PGR management, the following points are of much concern with respect to taxonomy.
- While collecting the germplasm, it is not always possible for collectors to make ideal herbarium specimens for reference purpose.
- Generally, information on diagnostic characters of seed, rhizomes, bulbs, and tubers is lacking, changed, or modified in the material, when it reaches the identifier. This creates a problem in identifying materials, especially in vegetative propagated or ephemeral systems.

- Owing to lack of ecological, distributional or agronomic data, confirmation is difficult, particularly in case of wild species, related species, and wild and weedy types.
- Lack of data on flowering time, bunching of fruits, types of fruits, and other criteria in herbarium specimens makes identification a prolonged process.
- Germplasm (seed) received by the NBPGR is usually separated from the bulk collections (as fruit) while assembling, sorting, and indexing the material in the base camps. Absence of fruit characters makes it difficult to identify the material, especially in closely related/allied types.
- Mishandling or mismanagement of germplasm may increase the chances of mixing up of material at various levels (collection, evaluation/characterization and supply, conservation).
- Lack of expertise needs to be countered with emphasis on taxonomic research and training/teaching.

CONCLUSION

Plant taxonomy plays an important role in PGR science and serves those indirectly involved in research, management, and policymaking, and stakeholders in agricultural and non-agricultural fields. Institutions engaged in PGR management must focus on taxonomic research pertaining to species of Indian origin or where diverse material is added to the genebanks. The following areas need more emphasis.

- Building up PGR collection in prioritization of species/areas
- Identifying desirable traits through characterization/evaluation
- Developing atlas for identification of germplasm
- Developing identification keys primarily based on vegetative characters/propagules, especially in problem taxa
- Developing computerized/digitized identification tools/databases of major economic taxa in the Indian perspective
- Establishing linkages between organizations involved in ex situ and in situ conservation and developing documentary information/databases
- Spreading awareness about the value of taxonomy at various education levels

Keeping in view the importance of taxonomy in PGR programmes, taxonomic research should be given high priority to revive and rejuvenate interest of students and scientists to find new dimension in this field of science. Because the quality of plant germplasm includes its correct identity, the material collected and conserved in the genebank or supplied to various indenters must be authentic and correct. In-depth knowledge of taxonomy can help develop appropriate institutional and policy framework for handling IPR issues and safe-guarding national interests to meet the challenges of new legal regimes of the Convention on Biological Diversity (CBD), World Trade Organisation (WTO), Trade-Related Aspects of Intellectual Property Rights (TRIPS), Union for the Protection of New Varieties of Plants (UPOV), and International Treaty on Plant Genetic Resources for Food and Agriculture (IT PGRFA).

ACKNOWLEDGEMENTS

Authors express sincere thanks to the Director, NBPGR, for providing basic facilities. Acknowledgements are also due to our colleagues in the Division of Plant Exploration and Germplasm Collection for their help in completion of this work.

BIBLIOGRAPHY

Arora R K. 1991. **Plant diversity in the Indian gene centre**. In *Plant Genetic Resources: conservation and management,* edited R S Paroda and R K Arora, pp. 25–54. New Delhi: International Board for Plant Genetic Resources, Regional Office

Arora R K and Nayar E R. 1984. ***Wild relatives of crop plants in India***. New Delhi: National Bureau of Plant Genetic Resources [NBPGR Monograph No. 7]

Arora R K and Pandey A. 1996. ***Wild edible plants of India***. New Delhi: National Bureau of Plant Genetic Resources

Arora R K, Mehra K L, and Hardas M W. 1975. **The Indian Gene Centre: prospects for exploration and collection of forage grasses**. *Forage Research* **1**: 11–22

Bajora C S, Hemp A, Heel G, Nordel I. 2008. **A taxonomic and ecological analysis of two forest *Chlorophytum* taxa (Anthericaceae) on Mount Kilimanjaro, Tanzania**. *Plant System and Crop Evolution* **274**: 243–253

Dhillon B S, Varaprasad K S, Srinivasan K, Singh M, Archak, S, Srivastava U, Sharma G S. 2001. ***National Bureau of Plant Genetic Resources: a compendium of achievements***. New Delhi: National Bureau of Plant Genetic Resources

Gadgil M, Singh, S N, Nagendra, H, Subash C M D. 1996. ***In situ conservation of wild relatives of crop plants: guiding principles and a case study***. Bengaluru: FAO and Indian Institute of Science

Guarino L, Ramanatha Rao V, and Reid R. 1995. ***Collecting plant genetic diversity-technical guidelines***. Singapore and Cambridge, UK: CAB International, Colset Pvt. Ltd. and University Press. 748 pp.

Karthikeyan S. 2000. **A statistical analysis of flowering plants of India**. In *Flora of India*, Introductory Volume (Part 2), edited by N P Singh, D K Singh, P K Hajra, and B D Sharma. Kolkata: Botanical Survey of India

Lawrence G H M. 1951. ***Taxonomy of vascular plants***. New York: Macmillan Co. 823 pp.

Nayar E R, Pandey A, Venkateswaran K, Gupta R, Dhillon B S. 2003. ***Crop plants India: a check-list of scientific names***. New Delhi: National Bureau of Plant Genetic Resources. [Agro-biodiversity (PGR)-26. National agricultural technology project on sustainable management of plant biodiversity]

Pandey A, Bhandari D C, Bhatt K C, Pareek S K, Tomar A K, Dhillon B S. 2003. ***Wild relatives of crop plants in India: collection and conservation***. New Delhi: National Bureau of Plant Genetic Resources. 73 pp. [Agro-biodiversity (PGR)-26. National agricultural technology project on sustainable management of plant biodiversity]

Paria N. 1997. **Seedling morphology: its prospects and application in taxonomic study in relation to conservation of biodiversity**. In *Conservation and Economic Evaluation of Biodiversity*, edited by P Pushpangadan, K Ravi, and V Santosh, vol. 1, pp. 227–238. New Delhi: Oxford and IPH Publishing House

Porter C L. 1959. ***Taxonomy of flowering plants***. San Francisco: WH Freeman and Co. 452 pp.

Prithipalsingh. 2006. **Plant identification: use of taxonomic keys**. In *Training on Wild Relatives*. New Delhi: NATP, National Bureau of Plant Genetic Resources

Stace C A. 1984. ***Plant taxonomy and biosystematics (contemporary biology)***. UK: Edward Arnold (Publishers) Ltd. 279 pp.

Zeven A C and de Wet J M J. 1982. ***Dictionary of cultivated plants and their regions of diversity***. Wageningen, the Netherlands: Centre of Agricultural Publishing and Documentation

7

Indigenous Knowledge of Plants and Biopiracy in India

Piyush K Sharma

INTRODUCTION

The history of science holds a fascination for professionals and general public alike, and no work of science is regarded as complete without at least some historical consideration (Radford 1986). Folklore taxonomy has a significant role in the early development of scientific plant taxonomy and continues to be practised even today. Therefore, the scientific aspects of plant taxonomy need to be understood with reference to "folklore taxonomy" or "traditional indigenous knowledge". This enables us to obtain a better perspective of biodiversity, which has been referred to as "the planet's most valuable resource" (Prithipalsingh 1999). India has been a centre for many "ethnobotanical studies" for its rich biodiversity. It is, therefore, necessary to prevent "biopiracy" of its natural resources.

In India, local traditions, knowledge systems, institutions, and environmental conditions are important factors in the conservation of biodiversity. A large part of Indian economy is based on biodiversity. Therefore, biodiversity and indigenous knowledge (IK) are central to the economic security of our country. IK refers to the unique, traditional, and local knowledge that an indigenous community accumulates over time by living in a particular geographic area. The knowledge can be related to value of plants, plant genetic resources, plant products, and other natural resources existing within and developed around their habitat. These folklore practices have helped Indians identify plants and other natural resources from generations. The development of folklore taxonomy or indigenous knowledge systems (IKS), covering all aspects of life, including management of natural resources, has been a matter of survival to the people who generated these systems. IKS is based on experience, often tested over centuries of use, adapted to local culture, environment, and dynamics. Other names for IK (or closely related concepts) are "local knowledge", "indigenous technical knowledge", and "traditional ecological knowledge".

Biopiracy refers to the process by which the rights of indigenous cultures to natural resources and knowledge are erased and replaced by monopoly rights for those who have exploited IK and biodiversity. Biopiracy occurs when multinational companies (MNCs) make billions of dollars by claiming intellectual property rights (IPRs) to traditional knowledge and plant genetic resources. IPR regimes recognize and provide protection only to formal innovators and not to indigenous informal innovators. Therefore, formal innovators, who perform mere translations and minor modifications and then seek patents, claiming the knowledge as well as life forms as their private property, are pirating traditional knowledge evolved and utilized by informal innovators. Biopiracy through IPRs has arisen because of the devaluation and invisibility of IKS and the lack of existing protection of these systems. Therefore, protection of IKS as systems of innovation and prevention of biopiracy require widening of legal regimes beyond the existing IPR regimes such as patents. This discussion highlights the importance of folklore taxonomy or IKS—wealth of the nation and problem of biopiracy that India is facing in relation to many species of medicinal plants, plant genetic resources, and plant products. It also provides an overview of recent legal studies that clearly portray India's active role in generating knowledge based on a sophisticated understanding of the environment and devising mechanisms to conserve and sustain its biodiversity and natural resources.

MEANING AND VALUE OF INDIGENOUS KNOWLEDGE

In defining the concept of IK of plants, one must keep in mind practical as well as research needs. IK refers not only to the knowledge of indigenous people, but also to that of any other defined community. The concept of indigenous knowledge systems (IKS) delineates a cognitive structure that includes definitions, classifications, and concepts of physical, natural, social, and economic environments. To understand the indigenous practices, one must have knowledge and understanding of the concepts on which they are based. This is particularly relevant in cases where intervention (or improvement in indigenous practices in changing ecological and economic scenarios) is aimed at social sustainability. Rural people have an intimate knowledge of many aspects of their surroundings. Over centuries, people have learnt how to grow food and to survive, sometimes in difficult environments. They possess knowledge about what varieties of crops to plant, when to sow and weed, which plants are poisonous and which can be used for medicine, how to cure diseases, and how to maintain their environment in a state of equilibrium. IK covers a wide range of subjects such as agriculture, food preparation, education, institutional management, natural resource management, health care, and so on.

IK is a valuable resource for human development. It is stored in people's memories, used in their activities, and expressed in the form of local language and taxonomy, stories, songs, folklore, proverbs, dances, myths, cultural values, beliefs, rituals, community laws, agricultural practices, equipment, materials, plant species, and animal breeds. IK is shared and communicated orally by specific examples and through culture. Thus, indigenous forms of communication and organization are vital to local-level decision-making processes and for the preservation, development, and spread of knowledge of plants. It may be noted that IK is not confined only to tribal groups or original inhabitants of an area, or to rural people. Rather, any community can possess IK—rural or urban, settled or nomadic, and original inhabitants or migrants. Development efforts should, therefore, consider IK of plants and use it to its best advantage. Although more and more development professionals have started to realize the potential of IK, it remains a neglected field. A key reason for this is the lack of guidelines for recording and applying IK of plants. Without such guidelines, there is a danger that IK of plants will get eroded.

Importance of Indigenous Knowledge

An old African proverb says "when an old knowledgeable person dies, a whole library dies". The importance of IK is hidden in this proverb. IK has two powerful advantages over outside knowledge: it costs little or nothing at all and is readily available. IK is found to be socially desirable, economically affordable, and sustainable. It involves minimum risk to research users and is widely believed to conserve resources. Thus, IK provides the basis for problem-solving strategies for local communities.

IK is a valuable and sophisticated system of knowledge developed by *adivasi* and rural communities over a period of time. It has been developed based on human and animal health, home building, food and agriculture, textiles, handicrafts, natural resource management, and other factors. This vast repertoire of knowledge, which is still being developed, is transmitted from one generation to the next in oral form. Usually IK is treated as public information, freely available for use by anybody. IK is of immense importance for today's developing countries because it transforms "biodiversity" to "bioresources" by adding economic value to biodiversity. We may appreciate the beauty of some unknown flowers, but an *adivasi* will identify the plant if it has any important medicinal properties. They use roots, leaves, flowers, and other parts of certain plants for preparing medicine for arthritis, to flush out intestinal worms, and other illnesses. Indigenous and local communities have developed this form of utility-based knowledge with regard to local biodiversity, and they use it for meeting their needs. This IK now has a global appeal to an environmental and holistic health conscious population, particularly in the industrialized nations.

The rural and *adivasi* communities of India are the custodians of a wealth of knowledge developed over generations. This knowledge is under threat from neglect by our own policymakers. On the one hand, we are the victims of biopiracy; the west running away with our valuable knowledge.[1] On the other, we have failed to put in place a system to protect our IK and the local communities that have generated this important knowledge system. We need to be aware of its immense richness and find ways to incorporate the need for its conservation and sustainable use, into national policy.

Importance of Indigenous Knowledge in Agriculture

The importance of IK in agriculture and breeding new crops can be understood when one realizes that no rice, wheat, cotton, or mustard plants can be found lying around in forests. What are found in the forests are wild plants, out of which indigenous communities over generations have bred thousands of land races (that is, the local varieties). These land races are the foundation materials of modern plant breeding and global food security. These are the self-sown varieties that plant breeders use to breed other varieties and for which they seek special and exclusive privileges like Plant Breeders' Rights.[2] One can easily say that if the breeding of a crop variety involved 100 steps, IK contributed at least the first 70 or 80 steps and laboratory science contributed the next 20 or 30 steps (Anuradha 1999). Therefore, the credit, reward, and recognition for a new variety should be similarly shared.

These communities have also identified and managed a series of genes conferring valuable traits for commercial and domestic needs through a highly sophisticated system of crossing and selection. Genes for traits as diverse as disease resistance, pest resistance, high salt tolerance, resistance to water logging, and drought tolerance have been maintained in the repertoire of communities. IK provides the know-how

---

[1] Details available at <www.organicconsumers.org/patlink.html>
[2] Details available at <www.genecampaign.org>

for developing crop varieties suited to diverse climatic regions ranging from the cold desert in Ladakh to the scorching sands of Rajasthan and again from the flood-prone belts of Bihar to the coast line of Andhra Pradesh (Anuradha 1999).

Faced with the threat of global warming and climate change across agricultural zones, scientists are on the lookout for more heat-tolerant crop varieties.[3] To gather information, they go to deserts and hot regions and ask local farming communities about the varieties that grow in that region and that can withstand extreme heat. Armed with the IK, these scientists return to their laboratories and experimental farms and engage in a breeding and selection programme to produce the new variety for post global warming agriculture. Thus, the basic information needed for conducting research and breeding programmes to develop new and improved breed of crops is often derived from IK.

Indigenous Knowledge in Human Health

The role of IK in the realm of medicinal plants and herbal products is even more obvious than in the case of crop varieties. According to the All-India Coordinated Research Project on Ethno-botany, indigenous communities are acquainted with the use of over 9000 species of plants and specifically some 7500 species for healing purposes (Mashelkar 1999). This amount of knowledge is staggering, considering the fact that the allopathic system of medicine is based on the use of some 100 species of plants only. Even the Indian system of medicine comprising the *Ayurveda*, *Siddha*, and *Unani* traditions uses about 2500 species of plants in its various healing formulations (Acharya and Shrivastava 2008; Jain 2011). There is a need to properly document this knowledge existing among the indigenous communities. The younger generation lacks respect and appreciation for such knowledge. The absence of willing heirs to this knowledge has resulted in a precarious situation where the death of an IK holder can result in the loss of an entire tradition and knowledge system.

The global market for herbal products, with its appeal ranging from pharmaceuticals, nutraceuticals, and health foods to cosmetics, toiletries, and ethnic products, is estimated to touch $5 trillion by 2020 (Mruthyunjaya and Ranjitha 2004). This turnover is largely based on the know-how of locals and indigenous communities (Mashelkar 1999). According to the World Health Organization, traditional medicine serves the health needs of almost 80% of people in developing countries, where access to modern health care services and medicine is limited by economic and cultural reasons (Gadgil, Utkarsh, Chhatre, *et al.* 2000).

---

[3] Details available at <http://en.wikipedia.org/wiki/Climate_change_and_agriculture>

Indigenous Knowledge Leads to Finding new Compounds

Recent advances in biotechnology have increased the ability of scientists to investigate organisms at the molecular and genetic levels and to find ways to commercialize products developed from such investigations. This is recognized by the increasing number of companies involved in bioprospecting, that is, prospecting for biological materials such as plants with medicinal or other economically valuable properties like fibre or oil. With growing environmental consciousness, benign biological substitutes are being sought for certain categories of chemical products. Following the German ban on chemical Azo dyes in the textile sector, the search is on for suitable vegetable dyes for leather and textiles (Meyer 1998). Plant-based colouring agents are also being sought for the food-processing industry due to the rising incidence of allergies to chemical colours and additives. So a number of industry programmes are under way to research and record traditional uses of plants for commercial product development by companies. These companies, mostly from the pharmaceutical, agricultural, personal care, and cosmetic sectors, depend on IK as the primary source of information. A number of pharmaceutical companies rely extensively, and sometimes exclusively, on the knowledge of indigenous and local people when they screen forests for plants with medicinal value that can be turned into blockbuster drugs.[4]

Indigenous Knowledge in Conservation of Biodiversity

IK plays a key role in conservation and sustainable use of biodiversity. The traditional economic systems of indigenous people have a relatively low impact on biological diversity because they tend to utilize a great diversity of species, harvesting small numbers of each of them. They also try to increase the biological diversity of their territories to increase the variety of resources at their disposal and, in particular, reduce the risk associated with fluctuations in the populations of individual species.

Indigenous communities have several strategies for conservation such as establishing sacred groves, recognizing taboos relating to nature, and giving special status to medicinal plant species (Posey 2003). The degree of sanctity of the sacred groves (which may date back to several thousand years) varies: in some forests even the dry foliage and fallen fruits cannot be touched, and in others the deadwood may be picked up, but never the live trees or their branches. Many of these groves have

---

[4] Details available at <www.indianscience.org/reviews/27-%20current%20science%20biopiracy.pdf>

been turned into the "biosphere reserves" of today; they are the best examples of forests, which might have flourished in the region, and endangered plant species, many of which might have disappeared from the region outside the groves. For example, in the *Uttara Kannada* region of Karnataka (southern India), the only remaining natural strands of *Dipterocarpus* and a large patch of *Myristica indica* persist in a sacred grove of the Goddess Karikannama (Leelakrishnan 2006).

To sum up, the use of IK assures that the end user of specific development projects are involved in developing technologies appropriate to their needs. Learning from IK can improve understanding of local conditions and provide a productive context for activities designed to help the communities (WRI/IUCN/UNEP 1992). Yet IK is still an underutilized resource in development activities. It needs to be intensively and extensively studied and incorporated into formal research and extension practices to make rural development strategies more sustainable. Special efforts are needed to understand, document, and disseminate IK for preservation, transfer, or adoption elsewhere. In spite of growing recognition of the value of IK, even today, the holders of IK and primary conservers of biodiversity remain poor, while those who use their knowledge and material in breeding and biotechnological enterprises are rich. Our country needs a concurrent action to recognize and reward the conservers of biodiversity and holders of IK for their invaluable contributions (MoEF 2008). By recognizing the value of IK and by integrating it with frontier science and technology, we can conserve both the dying wisdom and the plant genetic resources under threat of extinction, thus ensuring a better present and a better future of the world (Tacconi 2001).

BIOPIRACY: PROBLEM FOR IPR LAW

The patent system was put in place by countries to reward inventors. This encouraged the inventors to keep inventing useful products and the society benefited from these products. To get a patent, a product had to be novel (that is, it should not already exist) and useful for the public. But over time, the original patent laws have undergone many changes at the behest of MNCs that exercise a great deal of influence in this area. In the USA, it is now possible to get patents not only for inventions but also for discoveries. Therefore, even things that exist and are not novel can be patented in the American system.[5] The patent system has been transformed in such a way that MNCs can now easily get patents on

---

[5] Details available at <www.plantpatent.com/faq.html>

imitations as well. This has implications for biopiracy because the US patent system allows the grant of patents on products derived from IK. Biopiracy is the illegal, unauthorized, and inequitable access to and use of biological resources and its derived products, as well as the associated traditional knowledge of indigenous people, especially through the use of IPRs to obtain exclusive rights (Shiva 1997).

There is an increasing demand from consumers in industrialized countries for herbal products. This has driven pharmaceutical companies to seek possible leads in indigenous systems of medicine and the information present with the traditional healers of indigenous and local communities. India and other developing countries rich in bioresources and IK are the favourite targets and victims of biopiracy. MNCs hire people who camp in villages and interact with local communities to identify plants and their local, indigenous use. India is a biologically diverse and genetically rich country, where a variety of medicinal plants, food crops, and other living things with strength, immunity, and curative qualities are available. These have huge commercial values.[6] Transnational corporations and other giant companies have realized this and are so racing with each other to manufacture pharmaceutical and agricultural products. The main ingredients of these products are genetic materials of natural origin ranging from soil microorganisms to animals and from food crops to genes developed by indigenous people. These companies are rushing to apply for patenting their new products containing collected genetic materials to prevent competitors from using them.[7] This scramble of corporations for getting patents on biological products, termed (by Martin Khor in his article "A Worldwide Fight Against Biopiracy and Patents on Life") "gene rush", poses a substantial

---

[6] North East hit by biopirates—rare seeds and herbs targeted: It was reported that biopirates are plundering the north-eastern part of India. At least two cases of herbal theft have been reported. Two Japanese scientists, collecting rare herbs and plants in north Sikkim reserved forest, were caught with a booty of rich plants and herbs. They had collected over three tonnes of rare herbs, mostly used in Japan as medicines. However, the government was unable to stop another scientist from the Czech Republic from smuggling out rare plant seeds last year from west Sikkim and Arunachal Pradesh. Now these seeds, claimed to be of Sino-Himalayan origin, are up for sale (Details available at <www.plantideas.com>). Officials said the government learnt about the matter when the herbs were demonstrated at different seminars in Europe. They were displayed as new disease-resistant plants of Sino-Himalayan origin. "It is often too late by the time government reacts", said V Shiva, Director of Research Foundation for Science, Technology and Ecology (Details available at <www.globalissues.org/article/191/food-patents-stealing-indigenous-knowledge>).

[7] Details available at <www.lawgenecentre.org>

threat to local communities and indigenous people.[8] They may, in future, have to pay high prices for materials that they had developed. This is "biopiracy", which means not only the smuggling of diverse forms of flora and fauna, but also the appropriation and monopolization of traditional population's knowledge and biological resources. Often in the process of biopiracy, the contribution of indigenous people is ignored and goes unrewarded because the "benefit-sharing" system under the patenting process fails to ensure them a share from the derived profits. Such unfair exploitation of biodiversity has only multiplied in the recent years with the facilitation of registering international trademarks and patents as well as international agreements on intellectual property.

Biopiracy: the Legal Perspective

Biopiracy, or the stealing of genetic material and knowledge from communities in gene-rich developing countries, is an exploding issue in Asia. Industrialized countries want exploitation and ownership rights over the biodiversity of the South. Biopiracy can be compared to the exploitation practised in colonial era, when countries like England and the Netherlands used to control crop resources in Asia to build up their trade empires around cotton, sugar, tea, rubber, pepper, and other resources. Liberalization of trade through fora like General Agreement on Tariffs and Trade (GATT) is driven by pressure from industrialized countries, which aim to dominate world markets. Winning monopoly control over Asia's biodiversity and IK through intellectual property laws is crucial to their strategy today. The Agreement on Trade-Related Aspects of Intellectual Property Rights (TRIPs) was signed at the end of the GATT Uruguay Round in 1994 and came into force in 1995.[9] It is administered by GATT's successor, the World Trade Organization (WTO). TRIPs was strongly resisted by the South, as it forces all WTO member states to extend IPRs to plant varieties, the basis of food security, and

---

[8] Details available at <www.genecampaign.org>

[9] The TRIPs Agreement sets out compulsory uniform standards for intellectual property protection throughout the world: patents, copyright, trademarks, and so on. It currently allows countries to exclude plants and animals from patent laws. However, all countries must provide titles of intellectual monopoly to "inventors" of microorganisms, microbiological processes and products, and plant varieties. Plant varieties must be either patentable or subject to "an effective *sui generis* system". [Agreement on Trade-Related Aspects of Intellectual Property Rights, Article 27.3(b)— *Members may also exclude from patentability*: plants and animals other than microorganisms, and essentially biological processes for the production of plants or animals other than non-biological and microbiological processes. However, Members shall provide for the protection of plant varieties either by patents or by an effective *sui generis* system or by any combination thereof.] Many governments interpret this *sui generis* option to mean plant variety protection, a special kind of patent developed in Europe for the corporate breeding industry (Gaia 1998).

health care. Until now, Asian countries have prohibited patents on life forms because corporate monopolies touching people's basic needs are dangerous. In addition, many Asian cultures are based on a holistic view of and respect for life, which Western technologies and property systems fundamentally disregard.

Bioprospecting, or collecting biological samples, can help medical and other scientific research. However, biopiracy, or illegal collection, can infringe on the sovereign rights of nations, decrease the economic health of indigenous communities and deplete or destroy valuable species. In earlier days, there was no law regarding obtaining samples of plants, microbes, and animals by scientists and other collectors. At the most the collector was required, in some instances, to obtain an informal permission from the local communities or landholders and, in cases of national lands, a permit. Therefore, scientists could collect specimens from anywhere in the world without any repercussions. Michael A Gollin, in his article "Law and Biopiracy", says that "take-and-run" describes the old approach to collecting, lately dubbed "biopiracy".[10] But now the developing countries, whose flora and fauna has been ruthlessly exploited by the industrialized nations, are raising protests (UNEP 1995). The practice of bioprospecting, or the search of plants that give improved crop yields or have pharmaceutical value, by industrialized nations is not necessarily a problem until the firms prospect without permission or expropriate the results of their investigations without payment or acknowledgement of the local people. According to Gollin, there are three sources of rules for biodiversity prospecting and natural product research: international laws, national laws, and a professional self-regulation.[11]

International Laws

Under the Convention on Biological Diversity (CBD),[12] sovereign national rights over biological resources are established and the member countries are committed to conserve them, develop them for sustainability, and

---

[10]Details available at <www.actionbioscience.or/g/biodiversity/gollin.html>

[11]Details available at <www.actionbioscience.or/g/biodiversity/gollin.html>

[12]31 International Legal Material 818 (1992). The UN Convention on Biodiversity was negotiated under the auspices of the United Nations Environment Programme. The preamble of the convention stated that the contracting parties are conscious of the intrinsic value of biological diversity and, thus, affirmed that the "conservation of biological diversity" is a "common concern of humankind" that "states have sovereign rights over their own biological resources" and "that states are responsible for conserving their biological diversity and for using their biological resources, in a sustainable manner". The preamble further affirmed that "biological diversity is being significantly reduced by certain human activities". The biodiversity treaty entered into force on 29 December 1993. India signed the Convention on 5 June 1992 and ratified it on 18 February 1994 (Rosencranz 2002).

share the benefits resulting from the use. A fair compromise regarding sharing of natural resources is arrived at through the "Access and Benefit Sharing Agreements". Under the convention, it is mandatory that the country providing the biological resources has prior information regarding what will be done with the resource, and what benefits will be shared. Bioprospectors, or collectors of natural products, must get permission to collect biological materials.

National Laws

Many countries have started practising their sovereign rights over biological resources as established under the CBD. In certain countries, including India, collection or export of biological materials without obtaining permission and satisfying some other conditions is considered poaching, as per the laws relating to biodiversity.

Professional Self-regulation

Meanwhile, many institutions and professional organizations have decided to implement natural products research policies for their members, which have quasi-legal or contractual status. Examples include botanical gardens, biotechnology companies, and professional groups.

Consequences of Breaking the Rules

The rules regarding informed consent and benefit sharing, if broken, have very serious consequences (UNEP 1998).

- The patents on such natural product inventions come under attack and are also cancelled if it is proved that all public knowledge about the species in question and its use are not fully disclosed.
- If a researcher removes biological material illegally from a source country and then profits from the material, the source country or affected person could recover all or some of the profits, based on a theory of misappropriation and related doctrines.
- Clean title to biological material should be obtained legitimately, and with prior informed consent from whoever had initial control over it. If there is no clean title, the value of the material is seriously reduced. The collector of an illegitimate sample will not be able to pass it on, in turn, to collaborators, partners, or third parties in the normal course of conduct for researchers.
- If the collector refuses to share the benefits, then the access to the samples may be denied.
- If a company is known to be evading the rules or is associated with biopiracy, its name will be tainted and further research possibilities

will dry up. Such a company may end up with weak patents, be exposed to equitable claims for profit sharing, lose sources of supply, and face the prospect of consumer and government boycotts, barriers to importation of biotechnology products, loss of market share, and financial penalties.

Cases of Attack on Patents Granted on Natural Plant Product Inventions

The Case of Basmati Rice

The biotechnology lobby, led by the US government, has been using trade negotiations to win strong protection for their markets and technologies worldwide. Despite talk about "free trade", intellectual monopolies are a form of protectionism. Companies complain that without legal ownership of their so-called innovations, they have no reason to invest in agricultural research in Asia. However, their arguments are upside down. IPR allows Northern companies to get ownership over plant seeds and knowledge developed by the South, to which they add very little and call it "new". Genetic engineering in rice is no more than adding a few genes to a plant, which already has ten thousands! In reality, the rights of the farmers and communities, who develop the knowledge and genetic diversity exploited by formally trained scientists, need protection. The rice economies and cultures of Asia are deeply threatened by IPR regimes as imposed by TRIPs.

A Texas company (Rice Tec) was granted patents over a hybrid rice that was crossed with traditional Indian Basmati rice and called "Texmati" in September 1997 by the US Patent Office (USPTO). The Indian government's Agricultural and Processed Food Products Export Development Authority (APEDA) filed for revocation of these patents on the grounds that the Basmati variety of fragrant rice has been grown for centuries in the central Indian Himalayan foothills.[13] This application for revocation of the patent succeeded in forcing Rice Tec to withdraw four out of twenty claims. This claim on Basmati rice by Rice Tec is said to be the most audacious instance of "biopiracy" by Western

---

[13]Hundreds of angry Indian farmers rallied in the streets of the capital to denounce a US patent on Basmati rice. Exasperated after several years of protest against American patents on the use of turmeric, neem, and other indigenous resources, Indian farmers are up in arms about a US monopoly claim on their own rice. "We have not done enough to protect our own treasures of this country", said Jaya Jetlie, general secretary of *Hind Mazdoor Kisan Panchayat*, an agricultural labour organization present at the rally. "If we lose our [rice] exports and lose whatever tradition and wealth we have, we will soon become a country where every pebble and every stone is owned by somebody else", she told reporters (quoted in Masako 1998).

transnational corporations. The manner in which Rice Tec established its patent demonstrates that it has ignored the contributions of local communities in the production of Basmati and that it does not intend to share the benefits.[14]

The Case of Neem

In India, *Azadirachta indica*, commonly known as the neem tree, is a symbol of Indian IK. This plant has been in use for medicine, toiletries, contraception, fuel, construction, fungicide, and so on from time immemorial. In 1971, US timber importer Robert Larson observed the tree's usefulness in India and began importing neem seed to his company's headquarters in Wisconsin. Over the next decade, he conducted safety and performance tests upon a pesticidal neem extract called Margosan-O and in 1985 received clearance for the product from the US Environmental Protection Agency. Three years later, he sold the patent for the product to the multinational chemical corporation, W R Grace and Co.

A challenge to the patent was made at the Munich office of the European Patent Office (EPO) by three groups: the EU Parliament's Green Party; Dr Vandana Shiva of the India-based Research Foundation for Science, Technology, and Ecology; and the International Federation of Organic Agriculture Movements. W R Grace's justification for patents was pivoted on the claim that these modernized extraction processes constitute a genuine innovation: "Although traditional knowledge inspired the research and development that led to these patented compositions and processes, they were considered sufficiently novel and different from the original product of nature and the traditional method of use to be patentable" (Nguyen 1998). The EPO, which administers patents under the European Patent Treaty, accepted the demanded invalidation of the patent on the ground that the fungicide qualities of neem and its use have been known in India for over 2000 years to make insect repellents, soaps, cosmetics, and contraceptives (Nguyen 1998).

The Case of Turmeric

In March 1995, two non-resident Indians associated with the University of Mississippi Medical Centre, Jackson, USA, were granted patents on turmeric. Turmeric has been used in India for thousands of years for healing wounds and rashes. Therefore, the New Delhi-based Council of Scientific and Industrial Research (CSIR) challenged the turmeric patents on the ground that they lacked novelty. The USPTO upheld the

---

[14]Details available at <www.navdanya.org/publications>

objection and cancelled the patents. It ruled that using the popular spice for medicinal purposes was not a new "invention", but a millennium-old Indian practice.[15]

Legal Challenges and Fight Against Biopiracy

There is a worldwide fight against biopiracy and patents on life. Groups as diverse as religious leaders, parliamentarians, and environment NGOs are intensifying their campaign against corporate patenting of living things. In March 1995, the Swiss Supreme Court, in a landmark decision, ruled that the *Manzana* variety of the Camomile plant may not be patented. It revoked the patent that the Swiss Patent Office had granted in 1988 to the German pharmaceutical company Degussa Asta Medica on its *Manzana* variety. A Swiss farmer Peter Lendi, president of the Bio-Herb Growers' Association, had brought the case to court. In February 1995, the EPO withdrew key parts of a patent granted to a Belgian company (Plant Genetic Systems) and a US company (Biogen Inc.) for genetically engineered herbicide-resistant plants. The patent was for plant cells made resistant to glutamine synthetase inhibitors by genetic engineering. It originally covered not only the gene that had been moved from bacteria to various plants but also all plant cells and plants that contain the gene. After a challenge by Greenpeace, the Patent Office's Appeal Board ruled that the patent may only cover genetically engineered genes and plant cells but cannot extend to a whole plant, its seeds, and future generations of plants grown from the cells.

Biopiracy has also been challenged by farmers and other indigenous people. In India, farmers' movements led by M D Nanjundaswamy of the Karnataka Farmers' Union are campaigning against the patenting of seeds and plants and the operation of foreign grain companies in the country.

Indigenous people's groups have held regional meetings in South America, Asia, and the Pacific to voice their opposition against granting patents to companies on plants and their genes. Also, at the UN Women's Conference in Beijing, 118 indigenous groups from 27 countries signed a declaration demanding "a stop to the patenting of all life forms", which is "the ultimate commodification of life, which we hold sacred".[16]

The Indian Parliament is against granting patents on life forms. In March 1995, India's Upper House of Parliament forced the government to defer indefinitely a patent amendment bill to bring the Indian Patent

---

[15]Details available at <www.navdanya.org/articles/turmeric.htm>

[16]Details available at <www.sciencebase.com/whats_new_on_sciencebase.html>

Act in line with the WTO's treaty on IPRs. The bill would have allowed for the patenting of life forms.

Religious groups and NGOs also challenged against granting such patents. In May 1995, leaders of 80 religious faiths and denominations (including the Protestants, Catholics, Muslims, Hindus, Buddhists, and Jewish) held a joint press conference in Washington announcing their opposition to the patenting of genetically engineered plants, animals and human genes, cells, and organs.[17] Environment and development NGOs have also been increasingly active. Groups such as the Third World Network and Rural Advancement Foundation International have been carrying out educational activities and also lobbying in the CBD.

The WTO dispute ruling is an attempt to put pressure on India to adopt US-style patent laws. However, the turmeric patent case makes it evident that the US patent system has its own weaknesses, which allow biopiracy to be practised as a rule. The withdrawal of the turmeric patent is only a first step in reversing biopiracy. The USA needs to revoke the patents on *Amla, Jar Amla, Anar, Salai, Dudhi, Gulmendhi, Bagbherenda, Karela, Rangoon-ki-bel, Erand, Vilayeti Shisham, Chamkura* on the basis of Indian IK and "prior art".

The USA also needs to change its patent laws that sanction biopiracy by its non-recognition of foreign "prior art". Patents should satisfy three criteria: novelty, non-obviousness, and utility. Novelty implies that the innovation must be new and cannot be part of "prior art" or existing knowledge. Non-obviousness implies that someone familiar in the art should not be able to achieve the same step. Most patents based on IK appropriation violate the criteria of novelty combined with non-obviousness because they range from direct piracy to minor tinkering involving steps obvious to anyone trained in the techniques and disciplines involved.

In the USA, the law has many distortions that facilitate the patenting process for companies such as in the pharmaceutical industry. One such distortion is the interpretation of "prior art". It permits patents to be filed on discoveries in the USA, despite identical ones being already in use in other parts of the world. Article 102 of the US Patent Law, which defines "prior art", does not recognize technologies and methods in use in other countries as "prior art". If knowledge is new for the USA, it is novel, even if it is part of an ancient tradition of other cultures and countries. This was categorically stated in the Connecticut Patent Law, which treated invention as "bringing in the supply of goods from foreign

---

[17]Details available at <www.thirdworldtraveler.com/Globalization/War_Against_Nature_VFTS.html>

ports" that is not yet of use among us. "Prior art" and "prior use" in other countries were, therefore, systematically ignored in US laws on monopolies, thus granting patents based on claims to invention. In a similar way, Section 102 of the US Patent Act of 1952 treats the use in the USA and publications in foreign countries as a "prior art"[18], but not the use in a foreign country. Hence, denial or non-recognition of "prior art" that exists and or is used elsewhere allows patents to be granted on the basis of new inventions. This is the basis of biopiracy of Indian knowledge systems and indigenous uses of biological resources being patented. US-style patent laws can only pirate IK, but cannot recognize or protect it. Article 102 thus enables the USA to pirate knowledge freely from other countries, patent it, and then fiercely protect this stolen knowledge as "intellectual property". Knowledge flows freely into the USA but is prevented from flowing freely out of it.

Indian Laws on Biopiracy

In India, the need to protect local communities and indigenous people, who have been developing and conserving the biological diversity from being ruthlessly exploited by transnational corporations, has been long felt. Therefore, amendments were made in the Patents Act, 1970, in 1999, 2002, and 2005. In addition, the Biological Diversity Act, 2002, and the Protection of Plant Varieties and Farmers' Rights Act, 2001, have been brought into effect to uphold the spirit of the CBD. Nevertheless, unfortunately these acts promote corporate hijack of biodiversity and knowledge, as well as the patenting of life forms, and have opened the door for biopiracy and biopatenting. The important relevant legal provisions of these acts are described as follows.

The Patents (Amendment) Act, 1999

The 1999 Act provided for the grant of exclusive marketing rights (EMRs) with regard to pharmaceuticals and food articles to those who apply for product patent in these areas pending the disposal of their patent

---

[18]Section 102 of the US Patent Act, 1952, which defines "prior art", reads as follows: 35 USC 102: Conditions of patentability: novelty and loss of right to patent. A person shall be entitled to a patent unless
  (a) The invention was known or used by others in this country or patented or described in a printed publication in this or a foreign country, before the invention thereof by the applicant for patent, or
  (b) The invention was patented or described in a publication in this or a foreign country or in public use or on sale in this country, more than one year prior to the date of the application for patent in the USA.

applications.[19] Under the amendment, these EMRs shall be granted if the applicant has obtained a product patent for that product in any other member country, which is a signatory to the new GATT Agreement. The only examination before the grant of such EMRs has been restricted to the matters mentioned under Sections 3 and 4 of the Indian Patents Act. Thus, a person having obtained a product patent in a member country would almost automatically be granted EMRs for the sale and distribution of that product in India by merely making a patent application here.[20] Although the amendment makes a provision for the grant of compulsory licence for marketing the product in India, there is no provision for the grant of compulsory licence for manufacturing the product here. The Act neither provides any safeguards against biopiracy of IKS, nor exempts plant-based medicines and drugs from being patented.

The Patents (Second Amendment) Act, 2002

The Patent (Second Amendment) Act, 2002, provides for changes in the scope of patentable inventions, grant of new rights and extension of the term of protection, provision for reversal of burden of proof in case of process patent infringement, and conditions for compulsory licenses. Two amendments have been made in the definition of what is not an invention, which has opened the floodgates of patenting genetically engineered seeds. First, in Section 3(i):

..."plants" have been omitted. According to Section 3(i), the following is not an invention: any process for the medical, surgical, creative, prophylactic [diagnostic therapeutic] or other treatment of human beings or any process for a similar treatment of animals or plants to render them free of disease or to increase their economic value or that of their products.

The omission of "plants" from this section implies that a method or process modification of a plant can now be counted as an invention and can hence be patented. The Second Amendment has also added a new Section 3(j), which allows production or propagation of genetically engineered plants to be counted as an invention and is, hence, patentable. Section 3(j) excludes as inventions:

...plants and animals...including seeds, varieties and species and essentially biological processes for production or propagation of plants and animals.

---

[19] Indian Patents Act, Section 24A. 1970

[20] The Patent (Amendment) Act, 2005, has repealed the provisions regarding EMRs, and accordingly Section 5 of the principal act dealing with process patent has also been omitted (refer The Patent (Amendment) Act, 2005).

However, the emergence of new biotechnologies is often used to define production of plants and animals through genetic engineering as not being essentially biological. Without a clear definition that all modifications of plants and animals is essentially biological, Section 3(j) opens the floodgates for patenting transgenic plants. By allowing patents on seeds and plants through Sections 3(i) and 3(j), the Second Amendment of the Patent Laws has jeopardized our seed and food security and hence our national security.

The Patents (Third Amendment) Act, 2005

The Indian Parliament passed the Patents (Amendment) Bill, 2005, which replaced the Patents (Amendment) Ordinance, 2004, issued by the Government of India in December 2004. The Patent (Amendment) Act, 2005, introduces product patent regime for food, chemicals, and pharmaceuticals. India was required to introduce product patent protection in these sectors from 1 January 2005, in accordance with the obligations under the TRIPs Agreement of the WTO. To fulfil this requirement, the Government of India had issued an ordinance in 2004. While introducing the Patents (Amendment) Bill, 2005, in the Parliament, the government introduced certain changes under the provisions of the ordinance. Both the Houses of the Parliament approved the bill with these amendments.

The relevant salient features of the Patents (Amendment) Act, 2005, are as follows.

- Extension of product patent protection to all fields of technology (that is, drugs, foods, and chemicals)
- Deletion of the provisions relating to EMRs (which would now become redundant) and introduction of a transitional provision for safeguarding EMRs already granted
- Introduction of a provision for enabling grant of compulsory license for export of medicines from plants to countries that have insufficient or no manufacturing capacity, to meet emergent public health situations, provided such importing country has either granted a compulsory license for import or allowed import of the patented pharmaceutical products from India by notification or otherwise (in accordance with the Doha Declaration on TRIPs and Public Health)
- Modification in the provisions relating to opposition procedures, with a view to streamlining the system by having both pre-and post-grant opposition in the patent office
- Addition of a new *proviso* in respect of mailbox applications so that patent rights in respect of the mailbox shall be available only from

the date of grant of patent, and not retrospectively from the date of publication
- After a patent is granted in respect of applications made under mailbox, continuation of manufacturing such products by the existing manufacturers after paying reasonable royalty to the patent holder
- Strengthening the provisions relating to national security to guard against patenting abroad of dual-use technologies relating to plants, plant gentic resources, and plant products

The third amendment to the Indian Patent Act raises fundamental issues that concern not only the health and well-being of our people but also their survival. The issues are wide ranging and include the availability of the essential drugs and medicines at affordable prices, monopolization of seed multiplication and exchange by MNCs, introduction of terminator seeds, technological dependency, and food security.

The amendment, which is of vital importance, is being rushed solely on the ground that the last date for introduction of product patents according to the TRIPs Agreement was 1 January 2005, which is not only unconvincing but also misleading. Moreover, appropriate legal solution can be found for the technical problem of giving effect to the amendment retrospectively.

The last decade has witnessed emergence of powerful public opinion against TRIPs not only in developing countries but also in Europe and America. The Doha Declaration on TRIPs and Public Health was an important pointer. Reputed economists who are otherwise supportive of the WTO system have gone on record underlining the fundamental inequity of TRIPs from the viewpoint of developing countries and the need to take steps to redress the imbalance (Sehgal 2006).

IPRs Under Indian Biological Diversity Act, 2002

The Biological Diversity Act (BDA) was enacted to provide for conservation, sustainable use of biodiversity and its components, and fair and equitable sharing of the benefits arising out of the use of biological resources and knowledge. Unfortunately, the BDA does not provide appropriate safety measures to deal with genetic pollution, nor does it reinforce a farmer's fundamental right to save seeds, which needs to be adequately protected by all IPR-related legislations. Section (6) of the Act states that no application for IPRs shall be made without the approval of the National Biodiversity Authority. However, Section 6(3) provides an exemption, which states the following.

The provisions of this section shall not apply to any person making an application for any rights under any law relating to protection of plant varieties enacted by the Parliament.

Exemption 6(3) in the Biological Diversity Act in effect says that companies can take varieties that farmers have evolved over ages with unique traits (for example, the aroma of Basmati rice and drought-resistant varieties), and patent the traits and qualities that are a result of farmers' breeding. The Act fails to do what it was designed to do—stop biopiracy. It has failed to recognize the legal standing of local communities and their inalienable rights to their biodiversity and collective innovation.

The Protection of Plant Varieties and Farmers' Rights Act, 2001

Concerns for agro-biodiversity conservation and management are one of the high priorities of biodiversity-rich nations like India. This legislation was necessitated by the commitment that India made in the TRIPs Agreement, and it opted for the *sui generis* system that aims at balancing farmers' and breeders' rights (Gupta 2003; TERI 1998). The Act provides for the establishment of an effective system to protect plant varieties and the right of farmers and breeders and to encourage the development of new plant varieties. It is necessary to recognize and protect the rights of farmers in respect of their contribution in conserving, improving, and making available plant genetic resources for the development of new plant varieties. It is also necessary to protect plant breeders' rights to stimulate investments for the R&D of new plant varieties, both in public and private sectors, for the accelerated agricultural development (Sahai 2003). Such protection will also facilitate the growth of seed industry to ensure farmers' access to high-quality seeds and planting materials.

The Act also goes beyond the requirement relating to the coverage of the number of plant genera and species. It recognizes the farmer not only as a cultivator but also as a conservator of all the agricultural gene pool and a breeder of several successful varieties. The Act makes provision that such farmers' varieties may be registered with the help of NGOs, so that they are protected against being scavenged by corporations. The rights of rural communities are also protected. The formulations of Section 39 allow farmers to sell seed in the way they have always done, however, with the restriction that these seeds cannot be branded with the breeders' registered name. This protects both the farmers' and the breeders' rights. The breeder is rewarded for his innovation by having control over the commercial market place and farmer is able to engage in his livelihood independently.

CONCLUSION

The biodiversity values may differ at the local, national, and international levels. Conservation of biodiversity is directly relevant to local residents for whom biological resources often represent their primary source of livelihood, medicine, and spiritual values (UNEP 2010). The Indian legal system has failed to address the issue of biopiracy effectively. India needs to ensure that its patent laws do not violate the public order or have a negative impact on plants, human health, and the environment. India has a legitimate method under international legal obligations to stop biopiracy and protect indigenous innovations. It needs to evolve legislation to protect its own plant innovations first, and only then foreign claims to plant innovations can be protected. The CBD allows India a mechanism to frame laws to prevent biopiracy. In particular, Article 8(j) recognizes that each contracting party shall "...subject to its national legislation, respect, preserve, and maintain knowledge, innovations, and practices of indigenous and local communities embodying traditional lifestyles relevant for the conservation and sustainable use of biological diversity and promote their wider application with the approval and involvement of the holders of such knowledge, innovations and practices and encourage the equitable sharing of benefits arising from the utilization of such knowledge, innovations, and practices".

India needs to take stock of her biodiversity-based economy for both ecological survival and economic and political survival. The most essential requirement for this is to understand taxonomy, including folklore taxonomy/IK. Biopiracy and patenting of IK is a double theft because, first, it allows theft of creativity and innovation and, second, the exclusive rights established by patents on stolen knowledge steal economic options of everyday survival based on our indigenous biodiversity and IK (Shiva 1997). Over a period of time, the patents can be used to create monopolies and make everyday products highly priced. The problem is deep and systemic. In addition, it calls for a systemic change, not a case-by-case challenge. If a patent system, which is supposed to reward inventiveness and creativity systematically, rewards piracy, if a patent system fails to honestly apply criteria of novelty and non-obviousness in the granting of patents related to IK, then the system is flawed and it needs to change. The patents on the anti-diabetic properties of *karela*, *jamun*, and *brinjal* once again highlight the problem of Biopiracy—the patenting of indigenous biodiversity-related knowledge. The US Patent No. 5900240 was granted recently to Cromak Research Inc. based in New Jersey. The assignees are two non-resident Indians: Onkar S Tomer and Kripanath Borah, and their colleague Peter Gloniski. The use of *karela*, *jamun*, and *brinjal* for control of diabetes is everyday knowledge and practice in India. Their use in the treatment of diabetes is documented in authoritative treatises like the "Wealth of India", the "Compendium of Indian Medicinal Plants", and the "Treatise on Indian Medicinal

Plants" (Soam and Rashmi 2006). This IK and use consists of "prior art". No patent should be given where "prior art" exists since patents are supposed to be granted only for new inventions based on novelty and non-obviousness. These criteria establish inventiveness, and patents are exclusive rights granted for inventions. The claim to the use of *karela* or *jamun* for anti-diabetic treatment as an invention is false since such use has been known and documented widely in India.

If biopiracy has to stop, the US patent laws must change and "Article 102 must be redrafted to recognize 'prior art' of other countries". This is especially important given that US patent laws have been globalized through the TRIPs Agreement of the WTO. Article 27.3(b) of the TRIPs Agreement most directly affects IK since it relates to living resources and biodiversity. TRIPs is based on the assumption that US-style IPR systems are "strong" and should be implemented worldwide. But in reality, the US system is inherently flawed in dealing with IK and is "weak" in the context of biopiracy; therefore, the review and amendment of TRIPs should begin with an examination of the deficiencies and weakness of Western-style IPR systems (Dhar, Chaturvedi, and Sinha 2000). There is a need to overhaul and amend a globalized IPR regime, which denies the knowledge and innovations of the Third World, which allows such innovations to be treated as inventions in the USA, and which legalizes monopolistic exclusive rights by granting of patents based on everyday common-place IK.

Since the CBD is also an international treaty, protecting IK via national biodiversity legislations does not violate our international obligations. In fact, removing the inconsistencies between TRIPs and CBD should be an important part of the international campaign for the review and amendment of TRIPs.

Some commentators have suggested that biopiracy happens because our knowledge is not documented, which is far from true. IK in India has been systematically documented, and this has made piracy easier. In addition, even the folk knowledge held orally by local communities deserves to be recognized as collective, cumulative innovation. The ignorance of such knowledge in the USA should not be allowed to treat piracy as invention. Piracy of IK will continue until patent laws directly address this issue, exclude patents on IK and trivial modifications of it, and create *sui generis* systems for the protection of collective, cumulative innovation. The protection of diverse knowledge systems requires a diversity of IPR systems, including systems that do not reduce knowledge and innovation to private property of monopolistic profits. Systems of common property in knowledge need to be evolved for preserving the integrity of IKS on which our everyday survival is based.

Neither TRIPs, nor the US patent law has scope for recognizing knowledge as a "commons", or recognizing the collective, cumulative innovation embodied in IKS. Therefore, to protect IK, TRIPs and US patent laws have to be changed. Only an overhaul of Western-style IPR systems with their intrinsic weaknesses will stop the epidemic of biopiracy. Moreover, if biopiracy is not stopped, the everyday survival of ordinary Indians will be threatened, because over time our IK and resources will be used to make patented commodities for global trade. Global corporate profits will grow at the cost of the food rights, health rights, and knowledge rights of one billion Indians, two-thirds of whom are too poor to meet their needs through the global market place. India should immediately start the movement for amendment of TRIPs and US patent laws. We need to show the quantum of potential loss to India due to biopiracy, in the form of both global and domestic markets, to protect our sovereignty and make our rightful claims with trading partners.

BIBLIOGRAPHY

Acharya D and Shrivastava A. 2008. *Indigenous herbal medicines: tribal formulations and traditional herbal practices*. Jaipur, India: Aavishkar Publishers Distributor

Anuradha R V. 1999. **Between the CBD and the TRIPs: IPRs and what it means for local and indigenous community**. Paper presented at the *Workshop on Biodiversity Conservation and Intellectual Property Regime*, RIS/Kalpavriksh/IUCN, New Delhi, 29–31 January 1999

Dhar B, Chaturvedi S, and Sinha P C. 2000. **Intellectual property rights and legal framework**. In *Convention on Biological Diversity: policy paper on crosscutting issue*, vol. 1. New Delhi: World Wildlife Fund

Gadgil M, Utkarsh G, Chhatre A, Ganguly P R, Seshasgiri R, Gokhale Y, Promod A. 2000. *People's Biodiversity Register: the Indian experience*, pp. 22–24. New Delhi: World Wildlife Fund

Gaia G. 1998. **Ten reasons to say no to UPOV**. In *Global Trade and Biodiversity in Conflict*. Switzerland: The World Conservation Union (IUCN)

Gupta S. 2003. **Intellectual property protection for plant innovation: a journey from UPOV to TRIPs**. *XXV Delhi Law Review* p. 131

Jain S K. 2011. *Dictionary of Indian Folk Medicine and Ethnobotany*, pp. 5–7. New Delhi: Deep Publications

Leelakrishnan P. 2006. *Environmental Law Case Book*. New Delhi: Lexis Nexis

Masako I. 1998. **India minister says to contest US basmati patent**. *Reuters*, New Delhi, 3 April 1998

Mashelkar R A. 1999. **Economics of knowledge**. Paper presented at *The 16th Dr C D Deshmukh Memorial Lecture*, India International Centre, New Delhi, 1999

Meyer H. 1998. **Precise precaution versus sloppy science: a case study**. Paper presented at *The 5th Meeting of the Ad Hoc Working Group on Biosafety*, Montreal, 17–29 August 1998 [Working Group on Biodiversity, Forum Environment and Development, Germany, Third World Network Briefing]

MoEF (Ministry of Environment and Forests). 2008. **Annual Report 2007/08**. New Delhi: MoEF

Mruthyunjaya and Ranjitha P. 2004. **Indian agricultural research system: structure, current policy issues, and future orientation**. *World Development* **34**(6): 854

Nguyen N H. 1998. **Organic agriculture in developing countries needs modern technologies**. *Biotechnology and Development Monitor* **31**(1): 383–384

Posey D. 2003. *Cultural and Spiritual Values of Biodiversity*. New Delhi: Intermediate Technology Publications

Prithipalsingh. 1999. **Biodiversity and taxonomy**. In *Biodiversity, Taxonomy, and Ecology*, edited by R K Tandon and Prithipalsingh. Jodhpur: Scientific Publishers

Radford A E. 1986. *Fundamentals of Plant Systematics*. New York: Harper and Row, Publishers Inc.

Rosencranz A. 2002. *Environmental Law and Policy in India*. New Delhi: Oxford University Press. 598 pp.

Sahai S. 2003. **India's Plant Variety Protection and Farmers' Rights Act, 2001**. *Current Science* **84**: 407–412

Sehgal R. 2006. **Access, benefit sharing and IPRs: an analysis of TRIPs Agreement and CBD for the ABS implementation and developing country concerns**. *X–XI National Capital Law Journal* (200-2006), p. 36

Shiva V. 1997. *Biopiracy: the plunder of nature and knowledge*. Cambridge, USA: South End Press

Soam S K and Rashmi H B. 2006. **Some reflections on patent search: a case study of medicinal plants of India**. *Journal of Intellectual Property Rights* **11**(3): 207–213

Tacconi L. 2001. **Biodiversity and ecological economics: participation, values and resource management**. *Earthscan* **33**(2): 17

TERI (Tata Energy Research Institute). 1998. **Convention of farmers and breeders: a forum for implementing farmers and breeders rights in developing countries**. New Delhi: TERI. [Draft treaty presented as an alternative to UPOV]

UNEP (United Nations Environment Programme). 1995. **Report of the Second Meeting of the Conference of the Parties to the Convention on Biological Diversity**. New York: UNEP [Conference of the Parties to the Convention on Biological Diversity: second meeting, Jakarta, November 1995]

UNEP (United Nations Environment Programme). 1998. **Report of the Fourth Meeting of the Conference of the Parties to the Convention on Biological Diversity.** New York: UNEP [Conference of the Parties to the Convention on Biological Diversity: fourth meeting, Bratislava, 4–15 May 1998]

UNEP (United Nations Environment Programme). 2010. **Report of the Tenth Meeting of the Conference of the Parties to the Convention on Biological Diversity.** New York: UNEP [Conference of the Parties to the Convention on Biological Diversity: tenth meeting, Nagoya, Japan, 18–29 October 2010]

WRI/IUCN/UNEP (World Resources Institute/The World Conservation Union/United Nations Environment Programme). 1992. *Global Biodiversity Strategy: guidelines to save, study and use earth's biotic wealth sustainably and equitably*, pp. 34. Washington, DC: WRI

8

Herbaria and Data Information Systems in Plant Taxonomy

Satish K Aggarwal

New species are discovered every day, usually in the lesser explored regions of the world such as tropical rain forests. New plant species, however, may be found in any place where wild plants grow. The botanist needs to show how the new plant differs from other species and give it a name, following the rules of the International Code of Botanical Nomenclature.

The name of each new species must also be represented by a specimen, called the type specimen, which describes a species. Specimens (including type specimens) are generally made by pressing the freshly collected plants, drying them, and then mounting them on sheets of paper following standard procedures. The herbarium specimens are suitably treated and carefully stored in a herbarium to ensure their longevity.

In normal practice, any unknown plant specimen is identified by comparing it with an already known specimen kept in a herbarium, by utilizing the available literature, and by comparing the description of the unknown plant with the published description.

WHAT IS A HERBARIUM?

A herbarium (plural herbaria) is a library of preserved plant specimens collected from far and wide, mounted on appropriate sheets, arranged according to a known system of classification, and preserved in pigeon holes of steel or wooden cupboards. It can also be expressed as a warehouse of information about plant biodiversity.

Herbaria are generally associated with botanical gardens and educational institutions. According to Fosberg and Sachet (1965), "a modern herbarium is a great filing system for information about plants both primary in the form of actual specimens and secondary in the form of published information, pictures, and recorded notes".

The specimens in a herbarium may be a whole plant or plant parts, which is usually be in a dried form, mounted on a sheet, but depending on the material it may also be kept in alcohol or other preservatives. Tournefort (1656–1708) first used the term "herbarium" for a collection of dried plants. This usage was taken up by Linnaeus, leading to the supersession of an earlier term "hortus succus" (Singh and Subramaniam 2008).

The botanist who invented the herbarium is Luca Ghini (1490–1556) from Bologna, Italy. He initiated the art of herbarium making by pressing and sewing specimens on sheets of paper. This art was disseminated throughout Europe by his students, who mounted the specimens and bound them into volumes. These herbarium specimens were portable and used as references by physicians who prepared medicines from plants.

The technique of herbarium making was a well-known practice at the time of Linnaeus, who practised mounting the dried specimens on single sheets and storing them horizontally, similar to the practice in use even today.

Herbaria range from small personal collections to large collections preserved in institutions of national and international stature with millions of specimens. Herbarium resources of the world may include as many as 272 800 926 specimens from 2639 herbaria located in 147 countries (Holmgren, Holmgren, and Barnett 1990); the numbers of specimens and herbaria have gone up over the years. *Index Herbariorum*, edited by Holmgren, Holmgren, and Barnett (1990), is an extremely useful index about the world's important herbaria. This index gives information about the location and curators of the world's herbaria, the number of specimens, and the abbreviation used to designate the respective herbarium. Tables 1 and 2 list the major herbaria of the world and India, respectively.

In addition, some other places such as the Indian Agricultural Research Institute, New Delhi, and many universities in Delhi, Jaipur, Bengaluru, Chandigarh, and other cities maintain their own herbaria.

Herbarium Specimen

A herbarium specimen consists of a pressed and dried plant permanently glued and/or strapped to a sheet of paper, called herbarium sheet, with a herbarium label. The herbarium sheet is of high quality and heavy weight, with the standard size of 28×42 cm^2. The size of the sheet limits the size of the plants for collection: in case of a small herb, the whole plant with root system can be mounted on one sheet, whereas bigger plant specimens may be folded in the shape of a V, Z or W. Herbarium

Table 1 Major herbaria of the world (in the order of number of specimens)

Herbarium	Abbreviation	Number of specimens
Museum National d'Histoire Naturelle, Paris, France	P	950 000
Royal Botanic Garden, Kew, England, UK	K	700 000
New York Botanical Garden, New York, USA	NY	700 000
Conservatoire et Jardin Botaniques, Geneva, Switzerland	G	600 000
Komarov Botanic Institute, Saint Petersburg (formerly Leningrad), Russia	LE	5 770 000
Missouri Botanical Garden, St. Louis, Missouri, USA	MO	5 400 000
British Museum of Natural History, London, UK	BM	5 200 000
Combined Herbaria, Harvard University, Cambridge, Massachusetts, USA (includes Arnold Arboretum, Farlow Herbarium, Gray's Herbarium, Herbarium of Oakes, Ames, and Oaks Ames Orchid Herbarium)	GH	5 005 000
Naturhistorisches Museum, Wein, Austria	W	5 000 000
US National Herbarium (Smithsonian), Washington, USA	US	4 368 000

Sources Holmgren, Holmgren, and Barnett (1990); Simpson (2006)

Table 2 Major herbaria of India (in the order of number of specimens)

Herbarium	Abbreviation	Number of specimens
Central National Herbarium, Kolkata	CAL	2 500 000
Herbarium of the Forest Research Institute, Dehradun	DD	330 000
Madras Herbarium, Coimbatore	MH	150 000
Herbarium of BSI, Pune	BSI	125 000
Herbarium of National Botanical Research Institute, Lucknow	LWG	115 000
Blatter Herbarium, St Xavier's College, Mumbai	BLAT	115 000
Herbarium of BSI, Eastern Circle, Shillong	ASSAM	115 000
Herbarium of BSI, Dehradun	BSD	65 000
Herbarium of BSI, Central Circle, Allahabad	BSA	50 000

specimens are still the most efficient and economical means of preserving a record of plant diversity.

A herbarium label is affixed to each specimen, usually at the lower right-hand corner. A label is 10–12 cm wide and 5–7 cm tall. It contains virtually all the information recorded in the field diary (*field notebook*) at the time of collection, which are as follows.
- Name of the institution/herbarium
- Scientific name
- Common name or vernacular name

- Family
- Locality
- Altitude
- Date of collection
- Collection (field) number
- Collector's name
- Field notes (habit, habitat, colour of the flower, dominance)

At times, another label (called annotation label or determination label) is appended on the left side of the herbarium label, which is usually 2 cm tall and 11 cm wide. The annotation label is used by an expert who wants to either record a name change or correct identification. In addition to the correction, the label carries the name of the expert and the date of change.

Roles of a Herbarium

A herbarium is a multifaceted facility, which serves various purposes. From being a safe place for storing pressed plant specimens, especially types, herbaria have become major centres of taxonomic study. Large herbaria houses millions of specimens from different parts of the world. Besides taxonomic research, they form an important link for research in other fields of study as well. In recent years, herbaria have been used to keep track of migration and extinction of species, providing the knowledge about which species are becoming rare or endangered due to human activities. The various roles played by herbaria are outlined below.

- Repository of dried plant specimens, which are protected against destruction by fungi and insects.
- Storehouse of type specimens—major herbaria keep type specimens in rooms with restricted access.
- Means of identification of specimens—newly collected specimens can be identified by comparison with the duly identified specimens.
- Providing help in compilation of flora, manuals, and monographs.
- Mapping current and past ecological and geographic distribution of plants to help in land care and bioprospecting.
- Preservation of a historical record of change in vegetation over time—when some plants become extinct in one area or may become extinct altogether, specimens preserved in a herbarium represent the only record of the plant's original distribution.
- Providing study materials on invasion biology and weed ecology.

- Repository of knowledge—properly collected specimens bear labels with abundant data on habitat, local names, uses, abundance/frequency of species, associated plants, colour of flowers, and other characters of the plant. Such data for the same species from different collections and different habitats, when carefully studied and analysed, provide valuable information on the range of variations, phytogeography, and ethnobotany, serving the valuable role of comprehensive data banks.
- Repository of *voucher specimens*—these are specimens on which palynological, chromosomal, chemosystematic, ultrastructural, micromorphological, or any other study have been conducted.
- Providing help in teaching and research—(1) training students in herbarium practices; (2) loaning herbarium specimens frequently to experts engaged in monographic work and exchanging duplicates among herbaria; and (3) acting as useful sources of plant DNA for use in molecular systematics and phylogenetics.

DATA INFORMATION SYSTEMS

In addition to housing plant collections, many herbaria today have initiated computerized data information systems to record and access the collected information of the plant specimens, as well as to access information from other collections worldwide. These data information systems (or database systems) refer to the organization, inputting, and accessing of information. The database systems can be supplemented with the recording of digitized images of plant specimens (further discussed in the section on virtual herbarium). The computerized database systems offer a number of advantages.

- They help in retrieving and summarizing information about plant collections.
- They help in the day-to-day organization of herbarium operations. For example, accession numbers (which may be scanned with bar codes) may automatically keep track of both outgoing and incoming loan returns.
- They help in directly connecting to the database of other herbaria electronically.
- They are invaluable in biodiversity studies as data information systems allow for the tabulation of presence, range, and distribution of taxa, which are important for studying endangered species.

VIRTUAL HERBARIUM

A virtual herbarium is a herbarium in a digitized form, that is, it contains a collection of digital images of preserved plants or plant parts.

It is, therefore, also known as the digital herbarium. All information, including photographs of herbarium specimens, is included in the virtual herbarium in electronic form. Specimens are not available in the physical form in a virtual herbarium. As the virtual herbarium contains information about the botanical name of the plant, date and place of collection, name of the collector, neighbouring plants, and habitat, the user is able to view the images, read about the information, and identify the plants through a virtual herbarium.

The concept of virtual herbarium came into existence during the 1970s when some herbaria in Australia and England started storage of data available on the herbarium labels in electronic format. Since 1995, there has been interlinking of herbarium databases in the developed countries. For instance, more than 70% of the specimens housed in various Australian herbaria have been databased and interlinked, providing comprehensive resources for (1) accurate depiction of geographic distribution and occurrence of plants, algae, and fungi, and (2) mapping of all plant, algal, and fungal species; this information is valuable for understanding the processes that threaten the vegetation and weed invasion.

Advantages of Virtual Herbarium

- Offers an easy and quick access method of sharing information about herbarium specimens among the users across the globe.
- Reduces cost and saves time considerably.
- Facilitates storing and retrieval of massive data on herbarium specimens.
- Helps in avoiding loss of or damage to herbarium specimens.
- Helps in developing conservation strategies for endangered species.

Important Virtual Herbaria

- Linnean Herbarium at Swedish Museum of National History[1]
- Australia's Virtual Herbarium[2]
- Virtual Herbarium of New York Botanical Garden[3]
- University of California Davis Herbarium[4]

---

[1] Details available at <www.linnaeus.nrm.se/botany/fbo/welcome.html.en>
[2] Details available at <www.anbg.gov.au/avh>
[3] Details available at <www.nybg.org/bsci/hcol>
[4] Details available at <www.herbarium.ucdavis.edu>

- Clayton herbarium from UK[5]
- Chinese Virtual Herbarium[6]

CONSERVATION AND LAW

Many plant species that were once common are now rare, primarily due to habitat destruction, exploitation by people, and competition with exotic plants (plants introduced from other parts of the world). Collectors, therefore, need to be conservation minded. Rare or threatened plants should not be collected; photographing the plant is a good alternative in such circumstances. In many countries, laws protect native as well as endangered plant species, and collection is allowed only with the proper permits.

International trade in plant material is covered in part by existing conservation legislation. The Convention on International Trade in Endangered Species of Wild Fauna and Flora (CITES) regulates shipment of many plant groups; CITES-listed plants should not be collected without proper authorization.

HERBARIUM ETHICS

Herbarium specimens are invaluable and often irreplaceable. These need to be handled with utmost care. Herbarium ethics refer to the rules that need to be followed so that the specimens are not damaged and remain in good condition for a long period of time.

Each herbarium should provide information to the users about (1) the arrangement of the collections kept; (2) any colour code adopted for the folders; (3) the families with their numbers; (4) details about any special collection such as ecological or economically useful plants; (5) loan procedures; (6) availability of microscopes; and (7) working days and hours.

The following rules and suggestions should be practised while handling the specimens.

- Specimens are fragile; the herbarium sheet should be kept flat.
- Books, other heavy objects, and elbows should not be placed on the sheets.
- Single folder or sheet should be lifted at a time.

---

[5] Details available at <www.nhm.ac.uk/botany/clayton/index.html>

[6] Details available at <www.cvh.org.cn>

- While carrying the sheets from the cabinet to the working table, a supporting sheet (such as a stiff board) should be kept below.
- Any broken part should be kept in a small packet in the same folder.
- Materials for dissection should be taken only sparingly and with permission.
- Nothing should be written on the herbarium sheets.

BIBLIOGRAPHY

Beaman J H, Rollins R C, and Smith A H. 1965. **The herbarium in the modern university: a symposium**. *Taxon* **14**: 113–133

Brenan J P M. 1968. **The relevance of national herbaria to modern taxonomic research**. In *Modern Methods in Plant Taxonomy*, edited by V H Heywood. London: Academic Press Inc.

Fosberg F R and Sachet M H. 1965. **Manual for tropical herbaria**. *Regnum Vegetabile* **39**: 1–132

Funk V A. 2002. **The importance of herbaria**. *Plant Science Bulletin* **49**: 94–95

Hicks A J and Hicks P H. 1978. **A selected bibliography of plant collection and herbarium curation**. *Taxon* **27**: 63–69

Holmgren P K, Holmgren N H, and Barnett L C (eds). 1990. ***Index Herbariorum: the herbaria of the world***. New York: New York Botanical Garden

Jain S K and Rao R R. 1977. ***A Handbook of Field and Herbarium Methods***. New Delhi: Today and Tomorrow Printers and Publishers

Lawrence G H. 1951. ***Taxonomy of Vascular Plants***. New Jersey: Prentice Hall

Simpson M G. 2006. ***Plant Systematics***. New York, USA: Elsevier Academic Press

Singh G. 2004. ***Plant Systematics: theory and practice***, 2nd edn. New Delhi: Oxford and IBH

Singh H B and Subramaniam B. 2008. ***Field Manual on Herbarium Techniques***. New Delhi: National Institute of Science Communication and Information Resources (NISCAIR)

9

Phylogenetic Systematics

Manoj M Lekhak, Anil Kumar, and Shrirang R Yadav

PHYLOGENY, SYSTEMATICS, AND TAXONOMY

Keeping in mind the statement "systematic botany is an unending synthesis", by Constance (1964), it would be interesting to know about the developments in understanding the fascinating discipline of phylogenetic systematics.

At the outset, it is very important that the meanings of certain terms such as phylogeny, systematics, and taxonomy are clearly understood. This will help in getting into the nuances of phylogenetic systematics.

The term phylogeny was coined by Ernst Haeckel in 1866. It refers to the history of a species and its relationships to other species (in Greek *phyl* refers to tribe and *gen* refers to origin or descent).

There is no general agreement as to the definition of systematics and taxonomy. Capricious use of these and related terms is a very old plague of the discipline. The early documented use of the term *systematics* (as systematic botany) can be traced back, at least, as far back as Linnaeus (Stuessy 1990).

The Swiss botanist de Candolle (1813), in the herbarium at Geneva, coined the term taxonomy (as taxonomie) to refer to the theory of plant classification. From this time until the publication of Darwin's *Origin of Species* (1859), the terms taxonomy and systematics were used synonymously, although the latter was used much more frequently. Authors like Mason (1950) have treated taxonomy as a broad field of biological investigation involving four main concerns: systematics (comparative study of the organisms), taxonomic system, nomenclature, and documentation. Simpson (1961), Heywood (1967), Mayr (1969), and Ross (1974) treated "systematics" as a more inclusive field covering the scientific study of the diversity and differentiation of organisms and the relationships existing between them. They consider taxonomy a part of systematics. Simpson (1961) defined taxonomy as "the theoretical study of classification, including its bases, principles, procedures, and rules". However, as Hawksworth and Bisby (1988) observed, with the increasingly wide arrays of fields currently investigated by taxonomists to substantiate classificatory arrangements, "the separation in practice between taxonomy and systematics as defined above has become obsolete".

BIOLOGICAL CLASSIFICATIONS

A classification is generally regarded as the primary product of systematic effort, although two students rarely agree to their concepts of classification. To classify means arranging objects into groups according to their characteristics. Linnaeus' classification of animals and plants was a hierarchy, as are all classifications in zoological and botanical textbooks.

In principle, a hierarchical classification could be developed either in a descending or in an ascending way. Andreas Caesalpino developed downward classification by applying the Aristotelian *principium divisionis* to the universal set of plant species known to him (Caesalpino 1583). A first choice between two opposite traits (for example, trees and herbs) allows partitioning of the universal set into smaller subsets. The procedure is then repeated until a satisfactory level of discrimination has been reached. This is the method used in identification keys, where couplets of contrasting traits are presented for choice, allowing recognition of increasingly smaller sets of species. Downward classification is very useful as an operational tool to facilitate identification and is rightly used. However, the artificiality of this method is evident. Size and circumscription of the groups, as well as the arrangement of species within the key, depend on the order we follow in introducing the different characters in the discriminating couplets.

Things are different in upward classification. Ray (1690) might have been the first author to suggest (in his *Synopsis methodica stirpium Britannicarum*, 1690) that a classification should be built by putting together individual species into increasingly larger units (in modern terms: genera, then families, orders, classes, and so on). Although the original method was often obscured by a widespread pragmatic concept of classification (presumably following Caesalpino's footsteps), Ray's concept of an upward classification gained re-formulation of classificatory procedures in evolutionary terms.

In an upward classification, it is necessary to identify the elementary units to be classified and a hierarchical structure that allows recognition of several, increasingly more inclusive ranks. Moreover, we need criteria for grouping the elementary units into classes, grouping the lower, less inclusive classes into higher, more inclusive ones, and finally, for giving them specified ranks.

Mayr (1982) and Knight (1981) give preliminary accounts of the history of systematics.

In the system developed by Linnaeus, the elementary units of classification are species, their existence is an expression of God's will, and their relationships mirror a divine design. The naturalist's job is to trace this design from the evidence he can find in nature. Ranks can be decided later by the principle formalized by Georges Cuvier (1799–1805): genera are distinguished according to the differences in organization, whereas orders and classes are separated by differences in the organs most important to the animal (Cuvier 1805). The highest level in classification separates animals with wholly different, non-comparable body plans. In this way, Cuvier could recognize, within the animal kingdom, four irreducible embranchements (type, phyla): Radiata, Articulata, Mollusca, and Vertebrata. Using Cuvier principle, the taxonomic ranks are determined by idealistic morphology. The embranchements are, therefore, defined in terms of differences in the nervous system as the (ideally) most important trait of anatomical organization.

The same criterion for ranking taxa according to differences in organization was advocated by Agassiz (1857) and other 19th century authors.

The advent of the evolutionary theory led to the reinterpretation of species and the relationships between them by reference to natural phenomena such as interbreeding and common descent. Common descent (with modifications) explains the hierarchical pattern of similarities exhibited by different species, as recognized in purely descriptive terms by the pre-evolutionary systematists. Roughly speaking, similarities may be understood as proofs of relatedness in the phylogenetic tree. Living things can be classified into genera, families, and so on according to some pattern of affinities from which we can infer their genealogical relationships.

Soon after the publication of Darwin's (1859) *Origin of Species,* Ernst Haeckel (Haeckel 1866, 1874) released his well-known, aesthetically attractive, drawings of phylogenetic trees. Although a theoretical refoundation of biological classification seemed to occur, the enthusiastic beginnings laid down by Haeckel failed to revolutionize our science. It was not until several decades later that the Darwinian and Haeckelian heritage was welded together and extensively received by systematists.

EVOLUTIONARY SYSTEMATICS

Evolutionary systematics is one of the most problematic areas revolved around the concept(s) of species. Merging of different research traditions in systematics as well as in palaeontology and genetics prompted the development of "The New Systematics", the major achievements of which are mirrored in a book of the same name (Huxley 1940).

For a long time, these developments, what Mayr (1982) calls microtaxonomy, or the science of species, ran quite independently from the course of macrotaxonomy, or the science of classification. Theoretical comments ranged from a denial that supraspecific taxa were natural entities to vague statements that a phylogenetic classification is more natural. How do we reconstruct phylogeny? How do we represent it in a formal classification? For many years, the most comprehensive theoretical essays on comparative biology as well as the otherwise useful textbooks on biological systematics did not advance the theoretical background of systematics.

During the 1940s and 1950s, systematics polarized around a few leading biologists (especially Ernst Mayr and George Gaylord Simpson). Later, with the development of different approaches to biological systematics, it became fashionable to identify a school of evolutionary systematics corresponding to the views of Mayr, Simpson, and others.

NUMERICAL TAXONOMY

Robert R Sokal and P H A Sneath were the first to develop a new theoretical and practical approach to biological systematics, as illustrated in their book *Principles of Numerical Taxonomy* (Sokal and Sneath 1963). Sneath and Sokal (1973) gave a more detailed account. Sokal and Sneath's numerical taxonomy developed from an operational attitude, by explicitly refusing to incorporate dubiously retrievable phylogenetic information into taxonomy, "at face value", regarding as many characters as possible. To underline their methods compared with others, Sokal and Sneath's philosophy has been defined as phenetic, that is, directly dependent on the overall similarity of the characters (the phenotypes).

In phenetics, therefore, there is no place for homology, history-dependent concepts such as ancestry, or evolutionary change. Also, theory-laden biological terms, such as species, are not used in phenetics. The branching diagrams found in works on phenetics are not phylogenies but simple clustering of operational taxonomic units (OTUs). The criteria for clustering are equally provided by those other than biological.

A phenetic analysis begins with the construction of a data matrix with suitably coded data for n characters in m OTUs. From this data matrix, a distance matrix is calculated, which gives a measure of overall similarity (calculated over all n characters) between each individual pair of OTUs. This distance matrix is used to group the OTUs using a suitable algorithm, which produces the branching diagram of similarity or phenetic relatedness (the dendrogram). Formal groups (that is, more or less inclusive clusters) are arbitrarily obtained by cutting the dendrogram at one or more conventional levels of similarity.

This work requires the use of several algorithms, and computer programs have been developed to carry out the calculations. Phenetics, as a system of biological systematics excluding phylogeny and homology, has progressively declined.

HENNIG'S PHYLOGENETIC SYSTEMATICS

Graphical illustration of the diversification of life with the help of "tree" became popular with the theory of evolution. Darwin (1859) illustrated the extinction of species and successive speciations with a tree graph, but he did not develop phylogenetic trees. In 1864, Fritz Müller applied out-group comparison and ontogenetic criterion and pointed out evolutionary novelties (Müller 1864). The Russian palaeontologist Woldemar Kowalewski also used evolutionary novelties to differentiate between groups of species. He constructed ground patterns for groups such as the ancestor of ungulates. However, these approaches lacked a clearly formulated methodology, and the application of methodical principles was inconsequent. Parker (1883), from New Zealand, had already mapped characters on a phylogenetic tree of rock lobsters. In Italy, Daniele Rosa (1918) discussed the essential principles of cladistics (including the monophyly of the taxa, avoidance of paraphyletic groups, and the end of a species after speciation among others). However, it is not known whether W Hennig was aware of Rosa's work. Other authors (such as E Meyrick, A Dendy, and J W Tutt) used some aspects of phylogenetic systematics in their work. Konrad Lorenz introduced an argumentation scheme prior to Hennig (Lorenz 1941). Many scientists contributed to establishing the relationship between phylogeny and systematics and developing necessary methods. However, Hennig, in his analyses of insect phylogeny, very clearly and convincingly established the significance and advantages of his phylogenetic argumentation. These studies led to the acceptance of the method and the terminology and their propagation by an important part of the scientific community. In his influential book

Phylogenetic Systematics, Hennig did not mention about the application of the principle of parsimony to the reconstruction of phylogenetic trees, which many contemporaneous systematists consider to be the heart of cladistics, in the form of an instruction for data analyses. This principle is rather implied between the lines and was first specified explicitly by cladists (Kluge and Farris 1969; Farris 1970) and by later authors who described Hennig's approach in a more concise way (Ax 1984, 1988).

In his first book on the theory of phylogenetic systematics, Hennig (1950) explained how the boundaries of species have to be defined along the time axis. He further explained that supraspecific taxa only have a relation to reality when they are monophyletic and that monophyla comprise the last common stem species and their descendants. Hennig introduced the term "paraphyletic" only in 1966. In 1950, although he had not yet introduced the tools necessary for the identification of monophyla, which we know now, he had chosen the right way that led to later improvements. He perceived the significance of Haeckel's biogenetic law for the determination of character polarity and the criterion of character complexity for the evaluation of the probability of homology. Hennig (1949) first introduced the terms "apomorphic" and "plesiomorphic" for taxa and not for characters, a usage that is methodically not convenient. He elucidated the present significance of these terms together with the introduction of the prefixes aut-, syn-, and sym- (Hennig 1953). The much-read English version of his book *Phylogenetic Systematics* (a translation by Davies and Zangerl), published in 1966, contains several improved definitions and explanations (Hennig 1966).

CONTRASTING SYSTEMATIC SCHOOLS

Between the late 1960s and 1970s, the attitudes of most students towards the principles and methods of systematic biology polarized around the leading figures of three different "schools": of Hennig's phylogenetic systematics or cladistics, Sokal and Sneath's numerical taxonomy, and Mayr and Simpson's evolutionary systematics. The debates between these schools have often been acrimonious, as expressed in some books and journals, particularly *Systematic Zoology* (Hull 1988). The new ideas also led to the foundation of new societies or research groups. In 1967, the first Numerical Taxonomy Conference was held in Lawrence, Kansas. Such conferences, born out of the Society of Systematic Zoology, initially incorporated all numerical approaches. Eventually, however, the burgeoning cladistic school grouped around a new Willi Hennig Society, founded in 1981. This group also brought out the journal *Cladistics* (first issue published in 1985).

However, it is quite difficult to identify a set of qualifying principles common to all students assigned to a given school by their opponents. Evolutionary systematists, in particular, are often recognized in strictly negative terms, as students refusing to accept either the phenetic approach of Sokal and Sneath or the strictly phylogenetic classification advocated by Hennig. In due course of time, Hennig's suggestions for the reconstruction of phylogeny have progressively become incorporated within the mind and language of the adherents of the evolutionary systematic school. However, strong contrasts remain regarding the acceptance of the so-called paraphyletic taxa, which are rejected by cladists but not by the followers of evolutionary systematics.

Mayr and Ashlock (1991) have made serious efforts to clarify the differences between evolutionary and cladistic systematics, to conclude, as expected, with a desperate defence of the first approach.

The aim of cladistics is to construct cladograms (branching diagrams depicting the genealogical relationships between species or other suitable terminal taxa), whereas that of evolutionary biology is to construct phylograms (branching trees incorporating a phenetic measure of morphological "advance" of the individual branches in addition to cladistic information). In evolutionary systematics, comparisons of organisms take all homologous features into account for grouping and ranking taxa, whereas cladistics selectively searches for orderly sets of synapomorphies (that is, shared derived [advanced] characters, as opposed to simple siomorphies, or shared ancestral characters).

Accordingly, cladistics only accepts monophyletic taxa, which are "complete" phylogenetic units comprising all the taxa issued from a given hypothesized ancestor, irrespective of their degree of divergence. Evolutionary systematics, on the other hand, takes into account phenetic information in addition to phylogenetic information. Accordingly, this school does not reject the so-called paraphyletic groups.

THE NEED TO STUDY PHYLOGENY

Knowing the pattern of descent, in the form of a cladogram, can be viewed as an important end in itself. The branching pattern derived from a phylogenetic analysis may be used to infer the collective evolutionary changes that have occurred in ancestral/descendant populations through time. Thus, knowledge of phylogenetic relationships may be invaluable in understanding structural evolution and gaining insight into the possible functional, adaptive significance of hypothesized evolutionary changes. The cladogram can also be used to classify life in a way that directly reflects evolutionary history. Cladistic analysis may also serve as a tool for inferring biogeographic and ecological history, assessing evolutionary processes, and making decisions in the conservation of threatened or endangered species.

AIMS OF PHYLOGENETIC SYSTEMATICS

Phylogeny, the evolutionary history or pattern of descent of a group of organisms, is one of the primary goals of systematics. Phylogenetic systematics, or cladistics, is a branch of systematics concerned with inferring phylogeny. Ever since Darwin laid down the fundamental principles of evolutionary theory, one of the major goals of the biological sciences has been the determination of life's history of descent. This phylogeny of organisms, visualized as a branching pattern, can be determined by an analysis of characters from living or fossil organisms, utilizing phylogenetic principles and methodology. Phylogeny is commonly represented in the form of a cladogram or *phylogenetic tree*—a branching diagram that conceptually represents the best estimate of phylogeny (Figure 1). The lines of a cladogram are known as lineages or clades, which represent the sequence of ancestral-descendant populations through time, ultimately denoting descent. Thus, as previously reviewed,

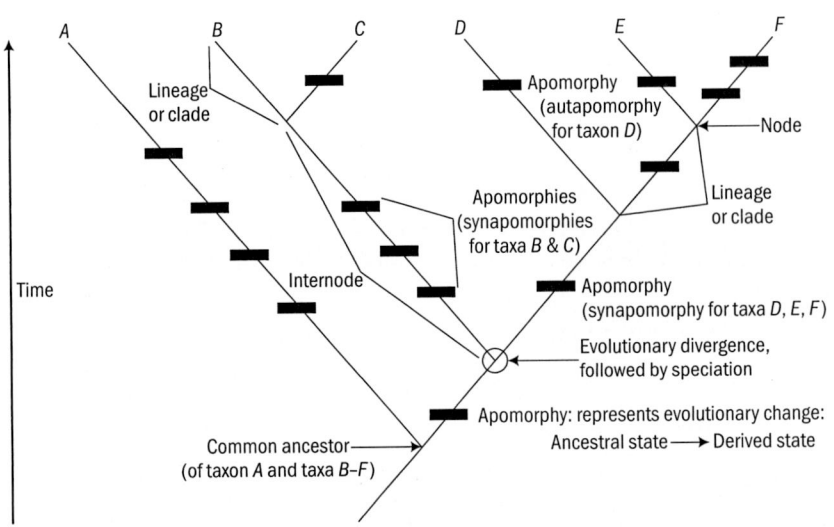

Figure 1 Example of a cladogram or phylogenetic tree for taxa A–F

Note Apomorphies indicated by thick hash marks
Source Redrawn from Simpson (2010)

cladograms have an implied, but relative, timescale. Any branching of the cladogram represents lineage *divergence* or *diversification*—the formation of two separate lineages from one *common ancestor*. (The two lineages could diverge into what would be designated separate species; the process of forming two species from one is termed *speciation*.) The *point* of divergence of one clade into two (where the most common ancestor of the two divergent clades is located) is termed a *node*; the region between two nodes is called an *internode*. Evolution may occur within lineages over time and is recognized as a change from a pre-existing *ancestral* (also called *plesiomorphic* or *primitive*) condition to a new *derived* (also called apomorphic or *advanced*) condition. The derived condition, or *apomorphy*, represents an evolutionary novelty. An apomorphy that unites two or more lineages is known as a *synapomorphy* (*syn*, together) and that occurs within a single lineage is called an *autapomorphy* (*aut*, self).

TAXON SELECTION

The study of phylogeny begins with the selection of *taxa* (taxonomic groups) to be analysed, which may include living and/or fossil organisms. Taxon selection includes both the group as a whole, called the study group or ingroup, and the individual unit taxa (OTUs). The selection of a particular taxon among many is decided by the necessity of previous classifications or phylogenetic hypotheses. The ingroup is often a traditionally defined taxon for which there are competing or uncertain classification schemes, the objective being to test the bases of those different classification systems or to provide a new classification system derived from the phylogenetic analysis. The OTUs are previously classified members of the study group and may be species or taxa consisting of groups of species (for example, traditional genera, families). Sometimes named subspecies or even populations, if distinctive and presumed to be on their own evolutionary track, can be used as OTUs in a cladistic analysis. In addition, one or more out-group OTUs are selected. An out-group is a taxon closely related to but not a member of the ingroup. Out-groups are used to root a tree. Some caution should be taken in choosing taxa for study. First, the OTUs must be well circumscribed and delimited from one another. Second, the study group itself should be large enough to include all closely related OTUs in the analysis. Both OTUs and the group as a whole must be assessed for *monophyly* before the analysis begins. In summary, the initial selection of taxa in a cladistic analysis, both study group and OTUs, should be questioned beforehand to avoid the bias of blindly following past classification systems.

CHARACTER ANALYSIS OF PLANTS

Description

Description is fundamental to any systematic study. It is the characterization of the attributes or features of taxa using any numbers or types of evidence. A systematist may make original descriptions of a group of taxa or rely partly or entirely on previously published research data. In any case, it cannot be overemphasized that the ultimate validity of a phylogenetic study depends on the descriptive accuracy and completeness of the primary investigator. Thorough research and a comprehensive familiarity with the literature on the taxa and characters of concern are prerequisites to a phylogenetic study.

Character Selection and Definition

After taxa are selected and the basic research and literature survey are completed, the next step in a phylogenetic study is the actual selection and definition of *characters* and *character states* from the descriptive data. Character is an attribute or feature; character states are two or more forms of a character. Generally, those features that (1) are genetically determined and heritable (termed intrinsic), (2) are relatively invariable within an OTU, and (3) denote clear discontinuities from other similar characters and character states should be utilized. However, an element of subjectivity is added to the study by the selection of a finite number of characters from the virtually infinite number that could be used. Thus, it is important to realize that any analysis is inherently biased by the selected characters and how the characters and character states are defined.

Generally, two types of characters are used in phylogenetic analyses. The first is "morphological", essentially equivalent to non-molecular features such as organ morphology, anatomy, embryology, palynology, and some aspects of reproductive biology. The second type is "molecular", derived from genetic data such as DNA sequences. Usually, morphological features are the manifestation of numerous inter-coordinated genes; evolution occurs by a change in one or more of those genes. Providing a precise definition of a feature, in terms of characters and character states, may be problematic.

Molecular characters may be less "subjective" than morphological ones, but they are not foolproof. Polymorphisms or uncertainties in base determination may occur for DNA sequence data.

Character State Discreteness

Since phylogenetic systematics involves the recognition of an evolutionary transformation from one state to another, an important requirement of character analysis is that character states be discrete or discontinuous from one another. Molecular characters and their states are usually discrete. For some non-molecular, qualitative characters such as corolla colour, the discontinuity of states is clear; for example, the corolla is red in some taxa and blue in others. But for other features, character states may not be clearly distinguishable from one another. This lack of discontinuity often limits the number of available characters and is often the result of variation of a feature either within a taxon or between taxa. Because cladistic analysis requires character states to be clearly discrete from one another, they must be evaluated for discontinuity. A standard way to evaluate character state discontinuity is to perform a statistical analysis, by comparing the means, ranges, and standard deviations of each character for all taxa in the analysis (including outgroup taxa). Additional statistical tests, such as ANOVAS, t-tests, or multivariate statistics, may be used as other criteria for evaluating character state discontinuity.

Character Correlation

Character correlation is an interaction between separate characters that are actually components of a common structure—the manifestation of a single evolutionary novelty. Two or more characters are correlated if a change in one always accompanies a corresponding change in the other. When characters defined in a cladistic analysis are correlated, they should be either combined into one character or scaled such that each component character gets a reduced weight in a phylogenetic analysis.

Homology Assessment

A critical concept to cladistics is that of homology, which can be defined as similarity resulting from common ancestry. Characters or character states of two or more taxa are homologous if the same features were present in the common ancestor of the taxa. Taxa with homologous features are presumed to share, by common ancestry, the same or similar DNA sequences or gene assemblages that may, for example, determine the development of a common structure such as a flower. Homology may also be defined with reference to similar structures within the same individual; two or more structures are homologous if the DNA sequences that determine their similarity share a common evolutionary history. For example, carpels of flowering plants are considered to be homologous with leaves because of a basic similarity between the two in form, anatomy, and development. Their similarity is thought to be the result of a sharing of common genes or gene complexes of common origin that direct their development.

Similarity between taxa can arise not only by common ancestry but also by independent evolutionary origin. Similarity not resulting from homology is termed homoplasy (also sometimes called analogy). Homoplasy may arise in two ways: convergence and reversal. Convergence is the independent evolution of a similar feature in two or more lineages. Thus, liverwort gametophytic leaves and lycopod sporophytic leaves evolved independently as photosynthetic appendages; their similarity is homoplasious by convergent evolution. Reversal is the loss of a derived feature with the re-establishment of an ancestral feature. For example, the reduced flowers of many angiosperm taxa, such as *Lemna*, lack a perianth. Comparative and phylogenetic studies have shown that flowers of these taxa lack the perianth by secondary loss, that is, via a reversal, reverting to a condition prior to the evolution of a reproductive shoot having a perianth-like structure. The determination of homology is one of the most challenging aspects of a phylogenetic study and may involve a variety of criteria.

Generally, homology is hypothesized on the basis of some evidence of similarity, either direct similarity (for example, of structure, position, or development) or similarity via a gradation series (for example, intermediate forms between character states). Homology should be assessed for each character of all taxa in a study, particularly of those taxa having similarly termed character states. For example, both the cacti and stem-succulent euphorbs have similarity between their spines. But their structural and developmental dissimilarity indicates that they are homoplasious and had independent evolutionary origins (with similar selective pressures, that is, protection from herbivores). This hypothesis necessitates a redefinition of the characters and character states, such that the two taxa are not coded the same. For molecular data (DNA sequence), homology must be assessed by alignment of the sequences to evaluate the homology of individual base positions. In addition, gene duplication can confound comparison of homologous regions of DNA.

Character State Transformation Series

Transformation series or morphocline represents the hypothesized sequence of evolutionary change, from one character state to another, in terms of direction and probability. For a character with only two character states, known as a binary character, only one transformation series exists. Characters having three or more character states, known as multistate characters, can be arranged in transformation series that are either ordered or unordered. An unordered transformation series allows for each character state to evolve into every other character state with equal probability, that is, in a single evolutionary step. An ordered transformation series places the character states in a predetermined sequence that may be linear or branched. Ordering a transformation series limits the direction of character state changes. For example, the evolution of monosulcate pollens from pantoporate condition (or vice versa) takes two evolutionary steps and necessitates passing through the intermediate condition, tricolpate pollens; the comparable unordered series takes a single step between monosulcate pollens to pantoporate condition (and between all other character states). The rationale for an ordered series is the assumption or hypothesis that evolutionary change proceeds gradually, such that going from one extreme to another most likely entails passing through some recognizable intermediate condition. In cladistic analyses, it is preferred to code all characters as unordered unless there is compelling evidence for an ordered transformation, such as the presence of a vestigial feature in a derived structure. For example, a unifoliolate leaf might logically be treated as being directly derived not from a simple leaf but from a compound leaf, evidence being the retention of a vestigial, ancestral petiolule.

Character Weighting

Certain characters are sometimes weighted in phylogenetic analyses. This weighting reflects the assumption that certain characters should be more difficult to modify than others. One might hypothesize that leaf anatomy is less likely to change compared to leaf hairiness, and therefore a change in leaf anatomical character could be counted as equivalent to two changes in pubescence for the purposes of counting steps in the tree.

Such weighting decisions can easily become subjective or arbitrary, which may bias the outcome of the study towards finding particular groupings. To avoid the possibility of bias, weighting decisions should be based on an objective criterion. One approach is to do a preliminary phylogenetic analysis with all characters assigned equal weights. The results of this analysis will identify which characters have the least homoplasy on the shortest tree(s), and these characters can then be given more weight in subsequent analyses—a process known as *successive weighting*. Another approach is to base weights on the knowledge of the underlying genetic basis of characters. For example, in DNA sequence analyses, transversions (purine → pyrimidine or pyrimidine → purine changes) are weighted over transitions (purine → purine or pyrimidine → pyrimidine changes) because transitions are known to occur more frequently and are easier to reverse. Restriction site gains may be weighted over restriction site losses because there are fewer ways to gain a restriction site than to lose one.

Assigning greater weight to a character has the effect of listing it more than once in the character–taxon matrix, to possibly override competing changes in unweighted characters. In practice, character weighting is rarely done, in part because of the arbitrariness of determining the amount of weight a character should have. A frequent exception, however, is molecular data, for which empirical studies may justify the rationale for and degree of weighting.

Polarity

Polarity is the designation of relative ancestry to the character states of a morphocline. A change in character state represents a heritable evolutionary modification from a pre-existing structure or feature (termed *plesiomorphic*, ancestral, or primitive) to a new structure or feature (*apomorphic*, derived, or advanced). For example, for the character ovary position, with character states superior and inferior, if a superior ovary is hypothesized as ancestral, the resultant polarized morphocline would be superior → inferior. The designation of polarity is often one of the more difficult and uncertain aspects of a phylogenetic analysis, but also one of the most crucial. In cladistic literature (Eldredge and Cracraft 1980; Wiley 1981), generally three main criteria are discussed for determining character polarity: out-group comparison (Figure 2), palaeontological or stratigraphical correlation, and ontogenetic correlation.

Character Step Matrix

Character state transformation determines the number of steps that may occur when going from one character state to another. Computerized phylogeny reconstruction algorithms available today permit a more precise tabulation of the number of steps occurring between each pair of character states through a *character step matrix*. The matrix consists of a listing of character states in the top row and left column; intersecting numbers within the matrix indicate the number of steps required, going from states in the left column to those in the top row. For example, the character step matrix of Figure 3(a) illustrates an ordered character state transformation series, such that a single step is required when going

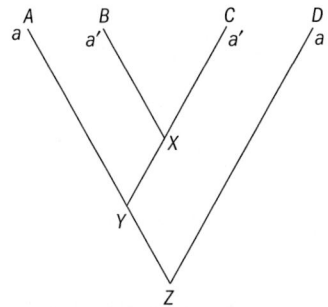

Figure 2 Out-group comparison

Note To determine the polarity of the character **a** within the taxon comprising A, B, and C, a related group ("out-group") D is also taken into consideration. Within A plus B plus C, **a** is present in two different character states (**a** and **a'**). State **a**, which is also present in out-group D, is regarded as plesiomorphic (primitive), whereas **a'** is regarded as apomorphic (derived). The transition from **a** to **a'** is postulated to have occurred between Y (the common ancestor of A, B, and C) and X (the common ancestor of B and C).

$$
\begin{array}{c|cccc}
 & 0 & 1 & 2 & 3 \\ \hline
0 & 0 & 1 & 2 & 3 \\
1 & 1 & 0 & 1 & 2 \\
2 & 2 & 1 & 0 & 1 \\
3 & 3 & 2 & 1 & 0 \\
\end{array}
\quad
\begin{array}{c|cccc}
 & 0 & 1 & 2 & 3 \\ \hline
0 & 0 & 1 & 1 & 1 \\
1 & 1 & 0 & 1 & 1 \\
2 & 1 & 1 & 0 & 1 \\
3 & 1 & 1 & 1 & 0 \\
\end{array}
\quad
\begin{array}{c|cccc}
 & 0 & 1 & 2 & 3 \\ \hline
0 & 0 & 1 & 2 & 3 \\
1 & \infty & 0 & 1 & 2 \\
2 & \infty & \infty & 0 & 1 \\
3 & \infty & \infty & \infty & 0 \\
\end{array}
\quad
\begin{array}{c|cccc}
 & 0 & 1 & 2 & 3 \\ \hline
0 & 0 & 1 & 5 & 5 \\
1 & 1 & 0 & 5 & 5 \\
2 & 5 & 5 & 0 & 1 \\
3 & 5 & 5 & 1 & 0 \\
\end{array}
$$
(a)　　　　　　　(b)　　　　　　　(c)　　　　　　　(d)

Figure 3 Character step matrices for (a) ordered character; (b) unordered character; (c) irreversible character; and (d) differentially weighted character

Source Redrawn from Simpson (2010)

from state 0 to state 1 (or state 1 to state 0), two steps are required when going from state 0 to state 2, and so on. The character step matrix of Figure 3(b) shows an unordered transformation series in which a single step is required when going from one state to any other (non-identical) state. Figure 3(c) illustrates an ordered but irreversible transformation series, disallowing a change from a higher state number to a lower state number (for example, from state 2 to state 1) as such transformation requires a large number of step changes (symbolized by "∞"). Character step matrices are most useful with specialized types of data. For example, the matrix of Figure 3(d) could represent DNA sequence data, where 0 and 1 are the states for the two purines (adenine and guanine), and 2 and 3 are the states for the two pyrimidines (cytosine and thymine). In this the matrix change from one purine to another purine or one pyrimidine to another pyrimidine is given one step, as it is more likely to occur, whereas a change from a purine to pyrimidine or vice versa is given five steps, it being quite unlikely. The latter change is given more weight in a cladistic analysis.

Character × Taxon Matrix

Prior to cladogram construction, characters and character states for each taxon are tabulated in a *character × taxon matrix*. The characters and character states must be assigned a numerical value to analyse the data using computer algorithms. Therefore, character states are assigned non-negative integer values, typically beginning with 0. The states are numerically coded in sequence to correspond with the hypothesized transformation series for that character. In the character × taxon matrix, polarity is established by including one or more out-group taxa as part of this matrix and by subsequently rooting the tree by placing the out-groups at the extreme base of the final most parsimonious cladogram.

CLADOGRAM CONSTRUCTION

Apomorphy

The primary tenet of phylogenetic systematics is that derived character states, or *apomorphies*, which are shared between two or more taxa (OTUs), constitute evidence that these taxa possess them because of common ancestry. These shared derived character states, or synapomorphies, represent the products of unique evolutionary events that may be used to link two or more taxa in a common evolutionary history. Thus, by sequentially linking taxa together based on their common possession of synapomorphies, the evolutionary history of the study group can be inferred. The character × taxon matrix supplies the data for constructing a phylogenetic tree or cladogram. For example, Figures 4(a) and 4(b) illustrates construction of the cladogram for the five species of the hypothetical genus *Xid* from the character × taxon matrix. First, the OTUs are grouped together as lineages arising from a single common ancestor above the point of attachment of the out-group (Figure 4[c]). This unresolved complex of lineages is known as a polytomy. Next, *derived* character states are identified and used to sequentially link sets of taxa (Figures 4[d] and 4[e]). In this example, synapomorphies include (1) the derived states of characters 1 and 3 that group together *X. nigra*, *X. purpurea*, and *X. rubens*; (2) the derived state of character 4 that groups together *X. alba* and *X. lutea*; (3) the derived state "four stamens" of character 5, which is found in all ingroup OTUs and constitutes a synapomorphy for the entire study group; and (4) the derived state "two stamens" of character 5 that groups *X. nigra* and *X. purpurea*. The derived state of character 2 is restricted to the taxon *X. lutea* and is, therefore, an *autapomorphy*. Autapomorphies occur within a single OTU and are not informative in cladogram construction. Finally, the derived state of character 6 evolved twice, in the lineages leading to both *X. alba* and *X. purpurea*; these independent evolutionary changes constitute homoplasies due to convergence. One important principle is illustrated in Figure 4(e) for character 5, in which the derived state "four stamens" is an apomorphy for *all* species of the study group, including *X. nigra* and *X. purpurea*. Although the latter two species lack the state "four stamens" for that character, they still share the evolutionary event in common with the other three species. The lineage terminating in *X. nigra* and *X. purpurea* has simply undergone additional evolutionary change in this character, transforming from four to two stamens (Figure 4[e]).

Recency of Common Ancestry

Cladistic analysis allows for a precise definition of biological relationship. *Relationship* in phylogenetic systematics is a measure of *recency of common ancestry*. Two taxa are more closely related to one another if they share a common ancestor that is more recent in time than the common ancestor they share with other taxa. In the earlier example of Figure 4(e), it is evident that *X. nigra* and *X. purpurea* are more closely related to one another than either is to *X. rubens* because the former two species together share a common ancestor (S) that is more recent in time than the common ancestor (R) that they share with *X. rubens*. Similarly, *X. rubens* is more closely related to *X. nigra* and *X. purpurea* than it is to either *X. lutea* or *X. alba* because the former three taxa share a common ancestor (R) that is more recent in time than the common ancestor shared by all five species (Q).

Monophyly, Paraphyly, and Polyphyly

An important concept in phylogenetic systematics is that of monophyly, or monophyletic groups. A *monophyletic group* consists of a common ancestor plus *all* descendants of that ancestor. The rationale for monophyly is based on the concept of recency of common ancestry. A group can generally be identified as monophyletic when members of that group share one or more unique evolutionary events. For example, four monophyletic groups can be delimited from the cladogram of Figure 4(e); these are circled in Figure 4(f). Note that all monophyletic groups include the common ancestor plus all lineages derived from the common ancestor, with lineages terminating in an OTU. The two descendent lineages from one common ancestor are known as *sister groups* or *sister taxa*. For example, in Figures 4(e) and 4(f), sister group pairs are (1) *X. lutea* and *X. alba*, (2) *X. nigra* and *X. purpurea*, (3) *X. nigra* + *X. purpurea* and *X. rubens*, and (4) *X. lutea* + *X. alba* and *X. nigra* + *X. purpurea* + *X. rubens*.

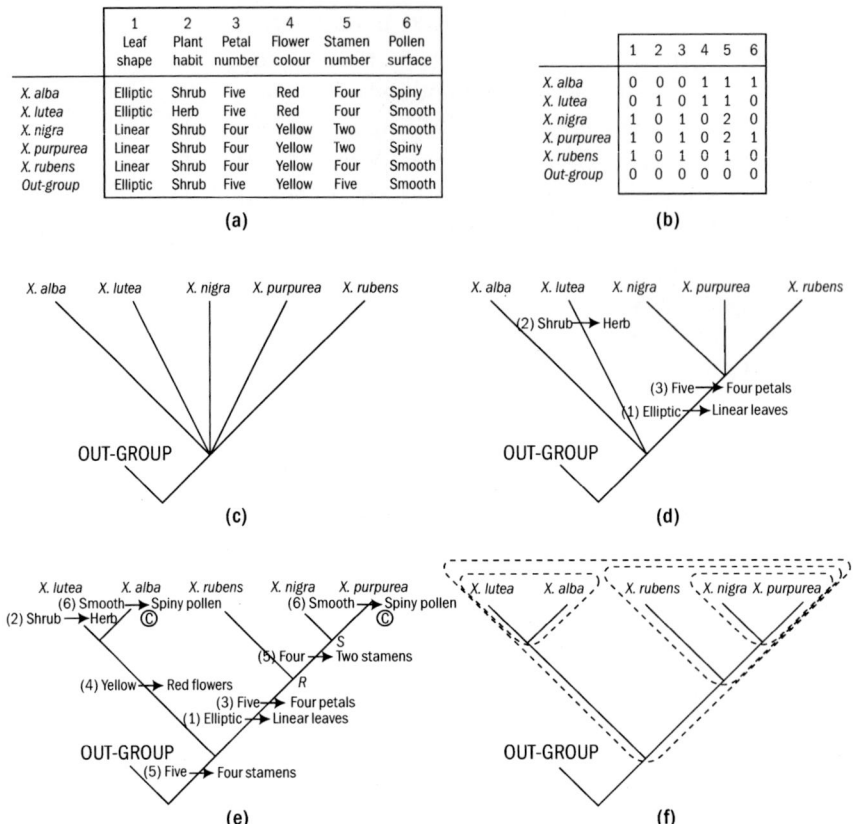

Figure 4 Character × taxon matrix for five species of the hypothetical genus *Xid* plus an out-group taxon (left column), showing six characters (top row) and their character states (inner columns): (a) character state names listed; (b) characters and character states converted to numerical values; (c) unresolved cladogram; (d) addition of characters 1 and 3; (e) most parsimonious cladogram, with addition of other characters. [common ancestors *Q*, *R*, *S*, and *T* shown for illustrative purposes]; and (f) cladogram at E, with all monophyletic groups circled

Source Redrawn from Simpson (2010)

The converse of monophyly is paraphyly. A *paraphyletic group* includes a common ancestor and some, but not all, known descendants of that ancestor. For example, in Figure 4(e), a group including ancestor *Q* and the lineages leading to *X. lutea*, *X. alba*, and *X. rubens* alone is paraphyletic because it has left out two taxa (*X. purpurea* and *X. nigra*), which are also descendants of common ancestor *Q*. Similarly, a *polyphyletic group* contains two or more common ancestors. For example, in Figure 4(e), a group containing *X. lutea* and *X. purpurea* alone could be interpreted as polyphyletic as these two taxa do not have a single common ancestor that is part of the group. Paraphyletic and polyphyletic groups are not natural evolutionary units and should be abandoned in formal classification systems. Their usage in comparative studies of character evolution, evolutionary processes, ecology, or biogeography is likely to bias the results. In addition, paraphyletic groups cannot be used to reconstruct the evolutionary history of that group. A good example of a paraphyletic group is the traditionally defined dicots. Most recent analyses show that some members of the dicots are more closely related to monocots than to other dicots. So the term dicot should not be used in formal taxonomic nomenclature.

Parsimony Analysis

The number of possible dichotomously branching cladograms increases dramatically with a corresponding increase in the number of taxa. For two taxa, there is only one cladogram; for three taxa, three dichotomously branched cladograms can be constructed; and for four taxa, 15 dichotomously branched cladograms are possible. The formula for the number of trees is $\Pi (2i - 1)$, with Π being the product of all the factors $(2i - 1)$ from $i = 1$ to $i = n - 1$, where n is the number of OTUs. The number of trees keeps on increasing with the number of OTUs. The number of trees is even greater when the additional possibilities of reticulation or polytomies are taken into account. Since there are generally many possible trees for any given data set, one of the major methods of reconstructing phylogenetic relationships is the *principle of parsimony* or *parsimony analysis*. This principle states that of the numerous possible cladograms for a given group of OTUs, the one (or more) exhibiting the fewest number of evolutionary steps is accepted as being the best estimate of phylogeny. The principle of parsimony is actually a specific example of a general tenet of science known as Ockham's razor: do not generate a hypothesis more complex than is demanded by the data. The rationale for parsimony analysis is that the simplest explanation minimizes the hypotheses for which there is no direct evidence. In other words, of all possible cladograms for a given group of taxa, the one (or more) implying the fewest number of character state changes is accepted. Minimization of the total number of character state changes minimizes the number of homoplasious reversals or convergences. The principle of parsimony is a valid working hypothesis because it minimizes uncorroborated hypotheses, thus assuming no additional evolutionary events for which there is no evidence. Parsimony analysis can be illustrated as follows. Various computer programs (algorithms) are used to determine the most parsimonious cladogram from a given character × taxon matrix. Some of the widely used programs are PHYLIP (Felsenstein 1989), NONA (Goloboff 1993), and PAUP* 4.0 (Swofford 2000).

Unrooted Trees

An unrooted tree is a branching diagram that minimizes the total number of character state changes between all taxa. Unrooted trees are constructed by grouping taxa from a matrix in which polarity is not indicated (in which no hypothetical ancestor is designated), perhaps because the polarity of one or more characters cannot be ascertained. In an unrooted tree, no evolutionary hypotheses are implicit because no assumptions of polarity are made. Note that monophyletic groups cannot be recognized in unrooted trees because relative ancestry (and therefore an out-group) is not indicated. The character state changes noted on the unrooted tree simply denote evolutionary changes when going from one group of taxa to another, without reference to direction of change. After an unrooted tree is constructed, it may be rooted and portrayed as a cladogram. Rooting is effectively done by including one or more out-groups in the analysis and placing these out-groups at the base (the root) of the tree.

Character Optimization

Optimization of characters refers to their representation (or plotting) in a cladogram in the most parsimonious way, such that the minimal number of character state changes occur. Character state evolution can be optimized in either of two equally parsimonious ways: *Acctran* (accelerated transformation) optimization that hypothesizes an earlier initial state change with a later *reversal* of the same character and *Deltran* (delayed transformation) optimization that hypothesizes two later, *convergent* state changes. Note that when alternative character optimization exists, there are nodes in the cladogram that are *equivocal*, that is, for which the character state cannot be definitively determined. Optimization is automatically performed by computer algorithms that trace characters and character states.

Polytomy

Occasionally, the relationships among taxa cannot be resolved. A *polytomy* (also called a polychotomy) is a branching diagram in which the lineages of three or more taxa arise from a single hypothetical ancestor. Polytomies arise either because data are lacking or because three or more of the taxa were actually derived from a single ancestral species.

In case of a polytomy arising via missing data, there are no derived character states identifying the monophyly of any two taxa among the group. For example, from the character × taxon matrix of Figure 5(a), the relationships among taxa W, X, and Y cannot be resolved; synapomorphies link none of the taxon pairs. Thus, W, X, and Y are grouped as a polytomy in the most parsimonious cladogram (Figure 5[b]). The other possible reason for the occurrence of a polytomy is that all the taxa under consideration diverged independently from a single ancestral species. Thus, no synapomorphic evolutionary event links any two of the taxa as a monophyletic group. The occurrence of a polytomy in phylogenetic analysis should serve as a signal for the reinvestigation of taxa and characters, perhaps indicating the need for continued research.

	1	2	3	4	5
W	1	1	1	0	0
X	1	1	0	1	0
Y	1	1	0	0	1
Z	1	0	0	0	0
OUT-GROUP	0	0	0	0	0

Figure 5(a) Hypothetical data set

Source Redrawn from Simpson (2010)

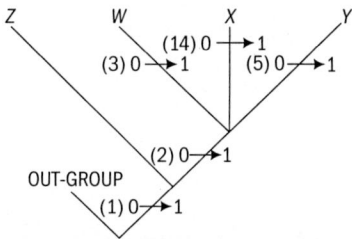

Figure 5(b) Resultant tree from data set of Figure 5(a); note the polytomy of lineages to W, X, and Y

Source Redrawn from Simpson (2010)

Reticulation

The methodology of phylogenetic systematics generally presumes the dichotomous or polytomous splitting of taxa, representing putative ancestral speciation events. However, another possibility in the evolution of plants is *reticulation*, the hybridization of two previously divergent taxa forming a new lineage. A reticulation event between two ancestral taxa (E and F) is exemplified in Figure 5(d), resulting in the hybrid ancestral taxon G, which is the immediate ancestor of extant taxon X. Most standard phylogenetic analyses do not consider reticulation and would yield an incorrect cladogram if such a process had occurred. For example, the character × taxon matrix of Figure 5(c) is perfectly compatible with the reticulate cladogram of Figure 5(d). However, the methods of phylogenetic systematics would construct the most parsimonious *dichotomously branching* cladogram of Figures 5(e) or 5(f), which show homoplasy and require one additional character state change than Figure 5(d).

Reticulation among a group of taxa should always be treated as a possibility. Data, such as chromosome analysis, may provide compelling evidence for past hybridization among the most recent common ancestors of extant taxa. A good example of this is the evolution of durum and bread wheat (*Triticum* spp.) via past hybridization and polyploidy (Figure 5[g]).

	1	2	3	4	5	6
W	1	0	0	1	0	0
X	1	1	1	0	1	0
Y	0	1	1	0	0	1
OUT-GROUP	0	0	0	0	0	0

Figure 5(c) Hypothetical data set

Source Redrawn from Simpson (2010)

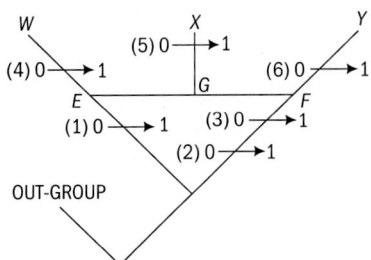

Figure 5(d) Cladogram exhibiting reticulation that is compatible with data set of Figure 5(c)

Source Redrawn from Simpson (2010)

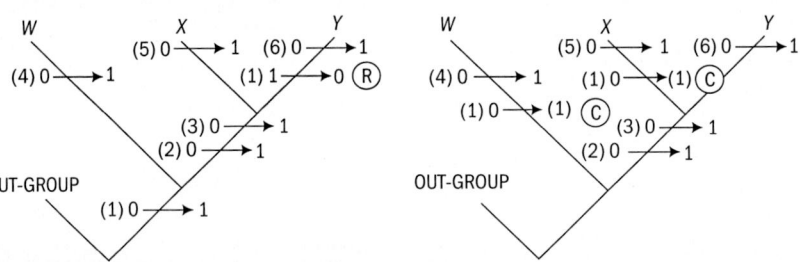

Figure 5(e; f) Dichotomously branching cladograms arising from data set of Figure 5(c), showing two alternative distributions of character state changes

Source Redrawn from Simpson (2010)

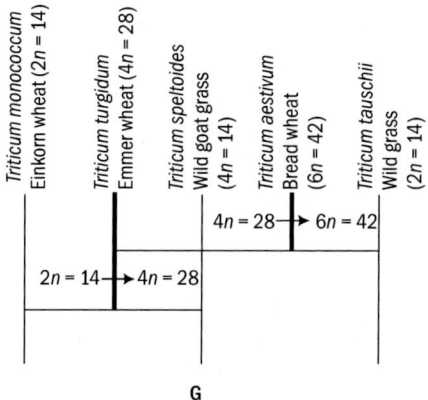

Figure 5(g) Evolution of wheat via ancestral hybridization and polyploidy
Source Redrawn from Simpson (2010)

Consensus Trees

Parsimony analyses often find multiple trees, all with the same length but with different linkages among the taxa. Sometimes, different methods of analysis will find trees showing different topologies and, therefore, different evolutionary histories for the same taxa. In addition, studies using different kinds of characters (for example, gene sequences and morphology) may find different trees. Rather than viewing and discussing each of these trees, it is usually convenient to visualize only one tree that is compatible with all equally most parsimonious tree. A consensus tree is a cladogram derived by combining the features in common between two or more cladograms. There are several types of consensus trees. One of the most commonly portrayed is the *strict consensus tree*, which collapses differences in branching pattern between two or more cladograms to a polytomy. Another type of consensus tree is the 50% majority *consensus tree* in which only those clades that occur in 50% or more of a given set of trees are retained. Consensus trees may be valuable for assessing the robust clades, that is, clades having strong support.

Long-branch Attraction

Long-branch attraction was identified originally by Felsenstein (1978) as a potential problem for phylogenetic analysis (Figure 6). If there are great differences in the rates of character evolution among lineages such that some lineages are evolving much more rapidly than others, and if the characters have only a limited number of character states, then long branches can be connected to each other in a tree irrespective of whether they are actually closely correlated or not. This problem is particularly acute with DNA sequence data for which each character has four possible states and mutation rates widely vary. This phenomenon occurs because numerous random changes, some of which appear in parallel in the rapidly evolving lineages, outnumber the changes that provide information about common ancestry.

This situation can affect all methods of tree construction. With the correct model of evolution, however, maximum likelihood methods are less afflicted by this problem. Long-branch attraction is basically a sampling problem and may be alleviated by including taxa that are related to those terminating the long branches.

Maximum Likelihood and Bayesian Analysis

Parsimony analyses remain very common in phylogenetic analysis, but for analyses using DNA sequences as characters, maximum likelihood and Bayesian analysis are becoming more common. Maximum likelihood, like parsimony methods, also evaluates alternative trees (hypotheses of relationship), but considers the probability, based on some selected model of evolution, that each tree explains the data. The tree that has the

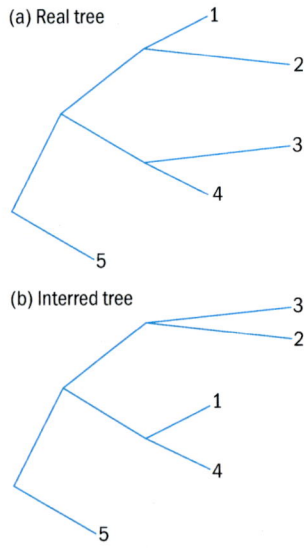

Figure 6 A simulated example of long-branch attraction: (a) the real tree of the relationship among five taxa, with two taxa (2 and 3) having long evolutionary branches; (b) an inferred tree of the taxa in which 2 and 3 are artificially grouped because of the phenomenon of long-branch attraction

Source Redrawn from Barton, Briggs, Eisen, et al. (2007)

highest probability of explaining the data is preferred to trees having a lower probability. Maximum likelihood is used in practice for molecular sequence data. With this approach, the probabilities are considered for every individual nucleotide substitution in a set of sequence alignments. It is commonly understood that transitions occur three times more frequently as compared to transversions. Thus, if C, T, and A occur in one column (representing one site), the sequences with C and T (pyrimidines) are more likely to be closely related than the sequence with A (Purine). Using objective criteria probability for each site and every possible tree that describes the relationship of sequences, the tree with the highest aggregate probability is selected as representation of a true phylogenetic tree.

Another more recent method is Bayesian analysis, which is based on posterior probability, utilizing a probability formula devised by T Bayes in 1763. Bayesian inference calculates the posterior probability of the phylogeny, branch lengths, and various parameters of the data. In practice, the posterior probability of phylogenies is approximated by sampling trees from the posterior probability distribution, using algorithms known as the Markov chain Monte Carlo (MCMC) or the Metropolis-coupled Markov chain Monte Carlo (MCMCMC). The results of a Bayesian analysis yield the probabilities for each of the branches of a given tree (derived from the 50% majority consensus tree of sampled trees). Maximum likelihood analyses take a long time to run, and bootstrap analyses require a high-performance computer. Bayesian methods estimate support for the tree at the same time as the tree is computed and so they are faster.

Assessing Homoplasy

Major difficulties with the construction of phylogenetic relationships derive from the widespread occurrence of what comparative morphologists have long described as parallelisms and convergences. There is also the possibility of reversal of character states. These events are the cause of similarities that are neither synapomorphies nor symplesiomorphies, but are homoplasies. If significant homoplasy occurs in a cladistic analysis, the data might be viewed as less than reliable for reconstructing phylogeny. Several indexes have been proposed to measure the amount of homoplasy in a given phylogenetic tree. The consistency index (CI) c was introduced by Kluge and Farris (1969), whereas the retention index r and the rescaled CI rc were both suggested by Farris (1989). Their simple formulations are as follows.

$c = m/s$ with $s = m + h$

$r = g-s/g-m$

$rc = r.c$

Here m is the minimum amount of change that the character may show on any tree (it is a function of the number of possible states and for numerical characters, it may be regarded as the range of the character); s is the amount of change in the character required in the least homoplastic way, by the considered tree; h is the fraction of change that must be attributed to homoplasy; and g is the greatest amount of change that the character may require on any tree (that is the greatest possible value of s).

A CI close to 1 indicates little to no homoplasy, whereas that close to 0 is indicative of considerable homoplasy.

Cladogram Robustness

A way to evaluate cladogram robustness is bootstrap. Bootstrapping is a method that re-analyses the data of the original character × taxon matrix by selecting (resampling) characters *at random*, such that a given character can be selected more than once. In resampling, some characters are given greater weight than others, but the total number of characters used is the same as that of the original matrix. This resampled data are then used to construct the most parsimonious cladogram(s). Many sequential bootstrapping analyses are generated (often 100 or more runs), and all most parsimonious cladograms are determined. From all these most parsimonious trees, a 50% majority consensus tree is constructed; the percentages placed beside each internode of the cladogram represent the percentage of the time (from the bootstrap runs) that a particular clade is maintained. A bootstrap value of 70% or more is generally considered a robustly supported node. The rationale for bootstrapping is that differential weighting by resampling of the original data will tend to produce the same clades if the data are good, that is, reflect the actual phylogeny and exhibit little homoplasy. One problem with the bootstrapping method is that it technically requires a random distribution of the data, with no character correlation. These criteria are almost never verified in a cladistic analysis. However, bootstrapping is still the most used method to evaluate tree robustness.

Another method of measuring cladogram robustness is the so-called jacknife (or jacknifing), which is similar to bootstrap but differs in that each randomly selected character may only be resampled once (not multiple times), and the resultant resampled data matrix is smaller than the original.

A second way to evaluate clade confidence is by measuring clade decay. A decay index (also called Bremer support) is a measure of how many extra steps are needed (beyond the number in the most parsimonious cladograms) before the original clade is no longer retained. Thus, if a given cladogram internode has a decay index of 4, the monophyletic group arising from it is maintained even in cladograms that are four steps longer than the most parsimonious. The greater the decay index value, the greater the confidence in a given clade.

CLADOGRAM ANALYSIS

A typical cladistic analysis may involve the use of DNA sequence data from one or more genes plus morphological (that is, non-molecular) data. Often, separate analyses are done for (1) each of the gene sets individually, (2) all molecular data combined, (3) morphological data alone, and (4) a combined analysis utilizing all available molecular and morphological data. It has been demonstrated that utilizing the totality of data often results in the most robust cladogram. The strict consensus tree of this combined analysis generally represents the best estimate of phylogenetic relationships of the studied group. From the most robust cladogram(s) derived from cladistic analyses, it is valuable to trace all character state changes. In addition, all monophyletic groupings should be evaluated in terms of their overall robustness (for example, bootstrap support) and the specific apomorphies that link them together. Homoplasies (convergences or reversals) should also be noted. A homoplasy may represent an error in the initial analysis of that character, which may warrant reconsideration of character state definition, intergradation, homology, or polarity. Thus, cladogram construction should be viewed not only as an end in itself, but also as a means of pointing out those areas where additional research is needed to resolve the phylogeny of a group of organisms satisfactorily. Cladograms represent an estimate of the pattern of evolutionary descent, both in terms of the recency of common ancestry and in the distribution of derived (apomorphic) character states, which represent unique evolutionary events.

Phylogenetic Classification

Cladograms also serve as a basis for classification. The pattern of evolutionary history portrayed in a cladogram may be used to classify taxa phylogenetically. A phylogenetic classification may be devised by naming and ordering monophyletic groups in a sequential, hierarchical classification, sometimes termed an *indented* method. The hierarchically arranged monophyletic groups may be assigned standard taxonomic ranks. For example, for the most parsimonious cladogram of Figure 7(a), a possible classification of hypothetical genus *Xid* is seen in Figure 7(b). Note that in this example, each named taxon corresponds to a monophyletic group (Figure 7[a]) and that these groups are sequentially nested such that the original cladogram may be directly reconstructed from this classification system. Two taxa of the same rank (for example, sections *Rubens* and *Nigropurpurea*) are automatically sister groups. Each higher taxon above (for example, subgenus *Luteoalba*) would also include automatically created lower taxa (for example, species *Xid alba* and *Xid lutea* in this case).

An alternative, and often more practical, means of deriving a classification scheme from a cladogram is by *annotation*. Annotation is the sequential listing of derivative lineages from the base to the apex of the cladogram, each derivative lineage receiving the same hierarchical rank. The sequence of listing of taxa may be used to reconstruct their evolutionary relationships. For example, an annotated classification of the taxa from Figure 7(a) is seen in Figure 7(c). In this case all named taxa are monophyletic, but taxa at the same rank are not necessarily sister groups. The particular rank at which any given monophyletic group is given is arbitrary and is often done to conserve a past, traditional classification. A recent trend in systematics is to eliminate ranks altogether or, alternatively, to permit unranked names between the major rank names. In either case, the taxon names, minus ranks, would still retain their hierarchical, evolutionary relationship (as in Figure 7[d]). This most common type of phylogenetic classification is sometimes termed *node based*, because it recognizes a node (common ancestor) of the cladogram and all descendants of that common ancestor as the basis for grouping (Figure 7[e]). In some cases, it may be valuable to recognize a *stem-based* group, which includes the stem (internode) region just above a common ancestor plus all descendants of that stem (Figure 7[e]). A stem-based group might be useful, for example, in that it might include a well-defined and a corroborated node-based monophyletic group, plus one or more extinct, fossil lineages that contain some, but not all, of the apomorphies possessed by the node-based group. A third general type of phylogenetic classification is apomorphy-based in which all members

Phylogenetic Systematics

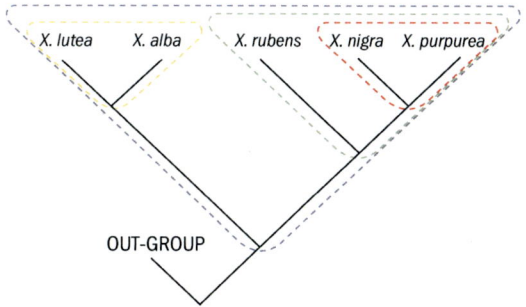

Figure 7(a) Cladogram from Figure 4(e)
Source Redrawn from Simpson (2010)

Genus *Xid* (including all 5 species)
    Subgenus *Luteoalba* (2 species)
        X. alba
        X. lutea
    Subgenus *Rubenigropurpurea* (3 species)
        Section *Rubens* (1 species)
            X. rubens
        Section *Nigropurpurea* (2 species)
            X. nigra
            X. purpurea

Figure 7(b) Indented classification scheme
Source Redrawn from Simpson (2010)

Genus *Xid* (including all 5 species)
    Subgenus *Luteoalba* (2 species)
        X. alba
        X. lutea
    Subgenus *Rubens* (1 species)
        X. rubens
    Subgenus *Nigropurpurea* (2 species)
        X. nigra
        X. purpurea

Figure 7(c) Annotated classification scheme
Source Redrawn from Simpson (2010)

Xid
    Luteoalba
        X. alba
        X. lutea
    Rubens
        X. rubens
    Nigropurpurea
        X. nigra
        X. purpurea

Figure 7(d) Indented but rankless classification scheme
Source Redrawn from Simpson (2010)

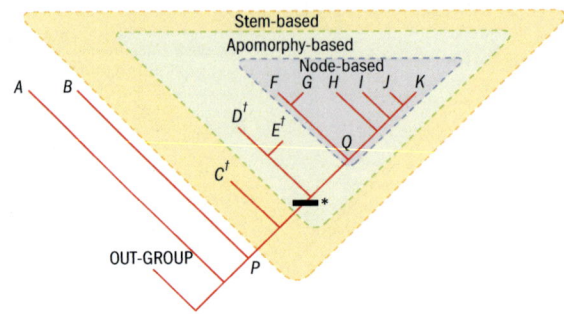

Figure 7(e) Cladogram illustrating node-based, apomorphy-based, and stem-based classification.

†Extinct taxon
*Major evolutionary change, used as the basis for an apomorphy-based group
Source Redrawn from Simpson (2010)

of a monophyletic group that share a given, unique evolutionary event (illustrated by the asterisk in Figure 7[e]) are grouped together.

Based on the tenets of phylogeny, there have been many classification systems, some noteworthy systems being those provided by Takhtajan (2009), Cronquist (1988), Dahlgren (1989), Thorne and Reveal (2007), and Angiosperm Phylogeny Group (APG). We shall briefly discuss the phylogenetic classification proposed by APG, the latest being APG III that appeared in the *Botanical Journal of Linnaean Society* (APG 2009). This classification is continuously upgraded in the angiosperm phylogeny website (APweb) by P F Stevens.[1]

The APG classification is not a complete formal classification of angiosperms and recognizes only families and orders, leaving many major nodes unnamed or giving these only informal names (magnoliids, monocots, lamiids, and others). This was done partly for practical reasons; the deeper nodes in many cases were weakly supported or unresolved and it was unwise to name these until they were better supported. Since the time of the first APG classification (APG 1998), many of the unclear relationships in the angiosperm tree have been robustly resolved, and it is now possible to provide a system of formally named higher taxa (Chase and Reveal 2009). A summarized phylogenetic tree showing relationships of the orders of angiosperms is shown in Figure 8.

---

[1] Details available at <www.mobot.org/MOBOT/reasearch/APweb>

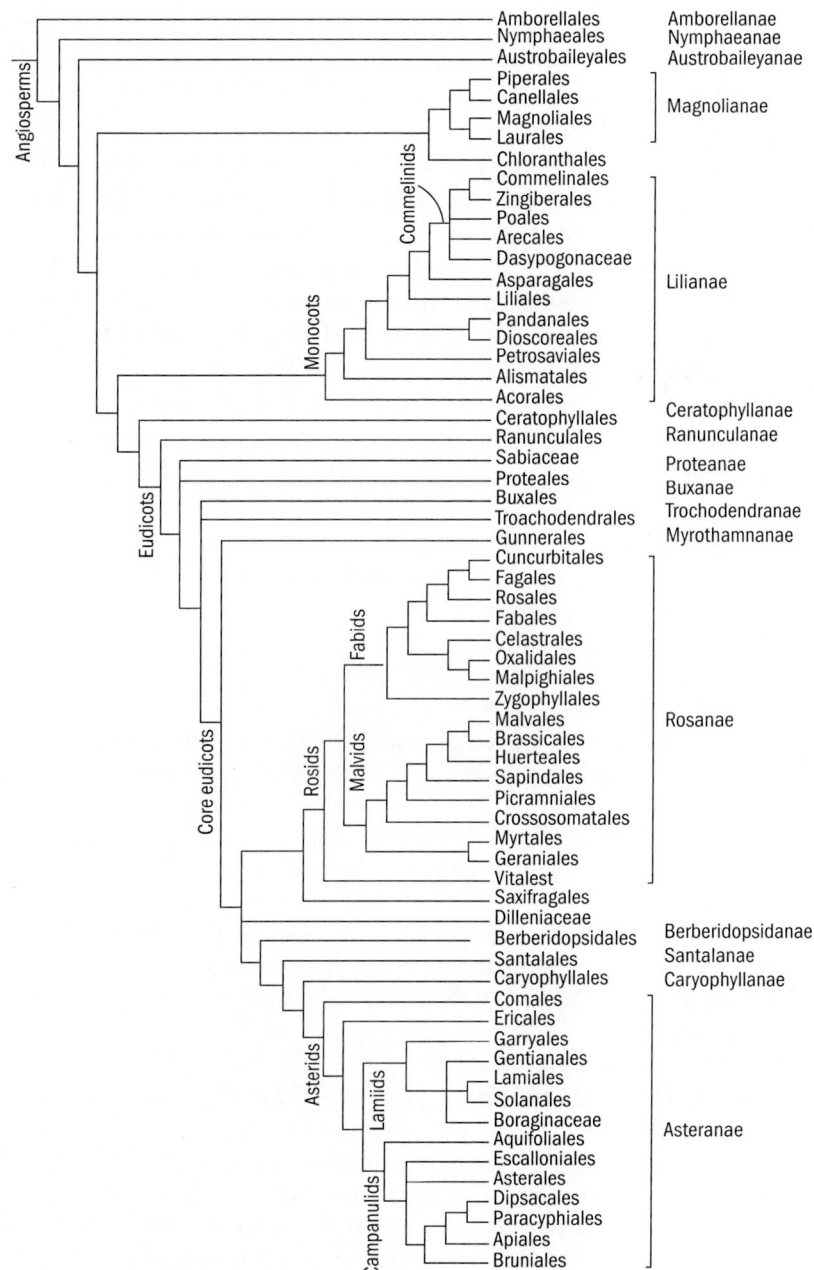

Figure 8 The APG III (2009) tree showing relationships of the orders of the angiosperms

Source Chase and Reveal (2009)

The publication of the third APG classification (APG 2009) for the orders and families of flowering plants has resulted in the need for a revised systematic listing of the accepted families. Linear sequences of families are required for herbarium curators who wish to arrange collections systematically rather than alphabetically, and there are currently a wide range of systems in use. Over time, in most herbaria, as the understanding of relationships changed, these classifications have become outdated and no longer represent the best estimate of relationships. This has led to the linear sequence of families based on APG classification, the latest being LAPG III based on APG III (Haston, Richardson, Stevens, *et al.* 2009). These authors have listed a total of 413 families, with the first and the last being Amborellaceae and Apiaceae, respectively. This system has now been adopted in textbooks (for example, Judd, Campbell, Kellogg, *et al.* 2007), dictionaries (for example, *The Plant Book*, 2008), and the general literature. In fact, several European herbaria (for example, K, E, BM, P, G, and the Dutch herbaria collectively) are in the process of reorganizing their collections along APG lines (APG 2009).

BIBLIOGRAPHY

Agassiz L. 1857. **Essay on classification**. In *Contributions to the Natural History of the United States*, vol. 1. Boston, Massachusetts: Little Brown and Co.

APG (Angiosperm Phylogeny Group). 1998. **An ordinal classification for the families of flowering plants**. *Annals of the Missouri Botanical Garden* **85**: 531–553

APG (Angiosperm Phylogeny Group). 2009. **An update of the Angiosperm Phylogeny Group classification for the orders and families of flowering plants: APG III**. *Botanical Journal of the Linnean Society* **161**: 105–121

Ax P. 1984. *Das Phylogenetische System-G*. Stuttgart: Fischer Verlag

Ax P. 1988. *Systematik in der biologie-UTB-G*. Stuttgart: Fischer Verlag

Barton N H, Briggs D E G, Eisen J A, Goldstein D B, Patel N H. 2007. *Evolution*. New York: Cold Spring Harbor Laboratory Press de Candolle A P. 1813. *Theorie Elementaire de la Botanique*. Paris: Déterville

Caesalpino A. 1583. *De Plantis*. Florantiae: G. Marescottum

Chase M W and Reveal J L. 2009. **A phylogenetic classification of the land plants to accompany APG III**. *Botanical Journal of the Linnean Society* **161**: 122–127

Constance L. 1964. **Systematic botany: an unending synthesis**. *Taxon* **13**: 257–273

Cronquist A. 1988. *Evolution and classification of flowering plants*, 2nd edn. New York: New York Botanical Garden

Cuvier G. 1805. *Lecons d'Anatomie Comparee*. Paris: Baudoin

Dahlgren G. 1989. **The last Dahlgrenogram: system of classification of dicotyledons**. In *The Davis and Hedge Festschrift*, edited by K Tan, pp. 249–260. Edinburgh: Edinburgh University Press

Darwin C. 1859. *On the Origin of Species by Means of Natural Selection*. London: Murray

Eldredge N and Cracraft J. 1980. **Phylogenetic Patterns and the Evolutionary Process: methods and theory in comparative biology**. New York: Columbia University Press

Farris J S. 1970. **Methods for computing Wagner trees**. *Systematic Zoology* **19**: 83–92

Farris J S. 1989. **The retention index and the rescaled consistency index**. *Cladistics* **5**: 417–419

Felsenstein J. 1978. **Cases in which parsimony and compatibility methods will be positively misleading**. *Systematic Zoology* **27**: 401–410

Felsenstein J. 1989. **PHYLIP 3.2 manual**. Berkeley: University of California Herbarium

Goloboff P A. 1993. **NONA, Version 1.5.1**. Argentina: Tucumán

Haeckel E. 1866. **Generelle Morphologie der Organismem**. Berlin: G Reimer

Haeckel E. 1874. **Anthropogenie oder Entwicklungsgeschichte des Menschen (Keimes-und Stammesgeschichte)**. Leipzig: Engelmann

Haston E, Richardson J E, Stevens P F, Chase M W, Harris D J. 2009. **The Linear Angiosperm Phylogeny Group (LAPG) III: a linear sequence of the families in APG III**. *Botanical Journal of the Linnean Society* **161**: 128–131

Hawksworth D L and Bisby F A. 1988. **Systematics: the keystone of biology**. In *Prospects in Systematics*, edited by D L Hawksworth, pp. 3–30. Oxford: Clarendon Press

Hennig W. 1949. **Zur Klarung einiger Begriffe der phylogenetischen Systematik**. *Forsch Fortschr* **25**: 136–138

Hennig W. 1950. **Grundzüge einer Theorie der Phylogenetischen Systematik**. Berlin: Zentralverlag

Hennig W. 1953. **Kritische Bemerkungen zum phylogenetischen System der Insekten**. *Beitrage zur Entomologie Sonderheft* **3**: 1–61

Hennig W. 1966. **Phylogenetic Systematics**. Illinois: University of Illinois Press

Heywood V H. 1967. **Plant Taxonomy**. London: Edward Arnold

Hull D L. 1988. **Science as a Process: an evolutionary account of the social and conceptual development of science**. Chicago: Chicago University Press

Huxley J (ed). 1940. *The New Systematics*. Oxford: Oxford University Press

Judd W S, Campbell C S, Kellogg E A, Stevens P F, Donoghue M J. 2007. *Plant systematics: a phylogenetic approach*, 3rd edn. Sunderland, MA: Sinauer

Kluge A G and Farris J S. 1969. **Quantitative phyletics and evolution of anurans**. *Systematic Zoology* **18**: 1–32

Knight D. 1981. *Ordering the World: a history of classifying man*. London: Burnnet Books

Lorenz K. 1941. **Vergleichende Bewegungsstudien an Anatinen**. *Journal fur Ornithologie Ergänzungsbd* **3**: 194–293

Mabberley D J. 2008. *Mabberley's Plant Book: a portable dictionary of plants, their classifications, and uses*, 3rd edn. Cambridge, UK: Cambridge University Press

Mason H L. 1950. **Taxonomy, systematic botany, and biosystematics**. *Madrono* **10**: 193–208

Mayr E. 1969. *Principles of Systematic Zoology*. New York: McGraw Hill

Mayr E. 1982. *The Growth of Biological Thought: diversity, evolution, and inheritance*. Cambridge, MA: Harvard University Press

Mayr E and Ashlock P D. 1991. *Principles of Systematic Zoology*, 2nd edn. New York: McGraw-Hill

Müller F. 1864. *Für Darwin*. Leipzig: Engelmann

Parker T J. 1883. **On the structure of head in *Palinurus*, with special reference to the classification of the genus**. *New Zealand Journal of Science* **1**: 584–585

Ray J. 1690. *Synopsis Methodica Stirpium britannicarum*. London: Sam Smith.

Rosa D. 1918. *Ologenesi: Nuova teoria dell'evoluzione e della distribuzione geografica dei viventi*. Firenze: Bemporad e Figlio

Ross H H. 1974. *Biological Systematics*. Massachusetts: Addison-Wesley

Simpson G G. 1961. *Principles of Animal Taxonomy*. New York: Columbia University Press

Simpson M G. 2010. *Plant Systematics*, 2nd edn. London: Academic Press

Sneath P H A and Sokal R R. 1973. *Numerical Taxonomy*. San Francisco: WH Freeman

Sokal R R and Sneath P H A. 1963. *Principles of Numerical Taxonomy*. San Francisco: W H Freeman

Stuessy T F. 1990. *Plant Taxonomy: the systematic evaluation of comparative data*. New York: Columbia University Press

Swofford D L. 2000. **PAUP*: phylogenetic analysis using parsimony and other methods**, Version 4.0 [Beta-test edition], distributed by Sinauer Associates, Sunderland, MA.

Takhtajan A. 2009. **Diversity and classification of flowering plants**, 2nd edn. Berlin: Springer

Thorne R F and Reveal J L. 2007. **An updated classification of the class Magnoliopsida ("Angiospermae")**. *Botanical Review* **73**: 67–181

Wiley E O. 1981. **Phylogenetics: the theory and practice of phylogenetic systematics**. New York: Wiley

10

Plant Anatomy in Relation to Taxonomy

Rajni Gupta and Kusum Shukla

INTRODUCTION

The foundations of comparative and systematic plant anatomy were well established by the mid 18th century. Phenotypic plasticity was fairly well understood, especially in leaf anatomy and also in growth rate-related variation in wood and bark structure. Systematic anatomists attempted to distinguish between characteristics that were purely diagnostic and phylogenetically informative. Janossonius (1950), as a representative of the old school, may be forgiven for confusing the issue by considering woody families closely related if they ended up together in his artificial identification keys to Javanese woods. The foundations of phylogenetic wood anatomy laid in 1918 by Bailey and Tupper were still unshaken in the early 1950s. The allegedly proven irreversibility (from long to short fusiform cambial initials) of xylem elements from tracheids with large bordered pits via vessels with multiple perforations to those with simple perforations (for sap transport) and from tracheids to fibres with increasingly reduced pit borders (for mechanical support) constituted a major transformation series in anatomy of the angiosperms. This constituted a major criterion to allow or deny phylogenetic derivations among and within putatively basal orders in the angiosperms. The irreversibility, of at least parts of the "major trends in wood evolution", was first challenged by the results of ecological wood anatomy (Carlquist 1966; Baas 1986). These studies showed that environmental factors, especially macroclimatic ones (temperature, season water availability, and so on), exhibited such general and strong correlations with wood structure that the total irreversibility of evolutionary trends, for instance, the shortening of cambial initials, could no longer be defended. The comparative study of plant structure, morphology, and anatomy has always been the backbone of plant systematics, which endeavours to elucidate plant diversity, phylogeny, and evolution. During the second half of the

20th century, systematics and structural studies greatly profited from new techniques and methods (Endress, Baas, and Gregory 2000). The aim of the present review is to provide a glimpse of major developments in structural plant systematics in the second half of the 20th century.

MAJOR ANATOMICAL ISSUES

A major issue attracting the attention of botanists is whether the angiosperms were primitively vessel-less or not. Early cladistic analyses had strongly argued against the cornerstone of Baileyan theory (Young 1981; Stevenson 1973). Carlquist (1996) stated (from a biological point of view) that "several origins of vessel-bearing wood type from a primitively vessel-less stock is far more likely than a few reversions of vessel-bearing woods to vessel-less ones", which is dictated by parsimony. This conservative anatomical view now finds support in the position of vessel-less *Ambroella* at the root of combined molecular gene-trees of the angiosperms (Soltis, Soltis, Chase, *et al.* 2000). For early convincing evidence, the fossil record should unearth vessel-less angiosperms existing earlier than those known so far. The common occurrence of vessel-bearing angiosperms with fairly advanced woods (in the middle and upper cretaceous [Wheeler and Baas 1991]) and the oldest vessel-less angiosperm fossil woods (from the upper cretaceous) remains a mystery (Poole and Francis 2000).

SEED ANATOMY AND TAXONOMIC RELATIONSHIPS OF *TETRACENTRON* AND *TROCHODENDRON*

The homoxylic genera *Tetracentron* Oliv. and *Trochodendron* Sieb. and Zucc., included in monotypic families, Tetracentraceae and Trochodendraceae, respectively, occupy a relatively stable and unquestioned position among primitive, lower Hamamelididae. They are usually considered a connecting link between the putatively archaic Magnoliidae and apetalous "amentiferous" Hamamelididae (Dahlgren 1989; Cronquist 1992). In the original description of *Trochodendron* (Siebold 1838), the genus was placed in the Winteraneae of Magnoliaceae Juss. Oliver (1889) assigned the newly described genus *Tetracentron* to Magnoliaceae (Tribe Trochodendreae). Seeman (1864) and Tieghem (1900) emphasized the anomalous position of *Trochodendron* in Magnoliaceae and suggested its segregation into a distinct family. Prantl (1888) was the first to assign a family name, Trochodendraceae, according to the rules of botanical nomenclature. Tiegham (1900) accepted the family status for *Tetracentron*. According to ICBN, Smith (1945) published a family name, Tetracentraceae, for the first time. Hallier (1901, 1912) removed the genera from Magnoliaceae and placed them into distinct families—Hamamelidaceae and Trochodendraceae. Nast and Bailey (1946) continued to treat both of them in Ranales, but this was rejected by Takhtajan (1966) and Cronquist (1968). The inclusion of *Trochodendron* and *Tetracentron* in Hamamelididae is supported by modern studies of floral morphology and anatomy (Endress 1986). Barabe, Bergeron, and Vincent (1982) demonstrated a highly isolated, basal position of *Trochodendron* and *Tetracentron* within their subclass, but continued to treat them as natural hamamelids. On the basis of rbcL molecular analysis, *Trochodendron* and *Tetracentron* are now considered part of a monophyletic line sister to "Paleoherbs" (Soltis 1997) and a relationship with other typical hamamelids is questioned. Melikian (1991), who studied seed anatomy of both genera, partly supported the putative phylogenetic relationship of *Trochodendron* and *Tetracentron* with Hamemelidiales. Nast and Bailey (1945) corrected Tieghem's (1900) descriptions of Trochodendraceous seed coats as bitegmic mesotestal and reported an exotegmic construction of seed coat for both genera. Melikian (1973) described the seed coats of both as strongly endotestal, supporting (in part) the putative phylogenetic relationships of both. However, Mohana Rao (1983) described it as exotegmic. As different interpretations of seed coat types for *Trochodendron* and *Tetracentron* were suggesting different systematic affinities, Doweld (1998) re-evaluated seed anatomy of *Tetracentron* and *Trochodendron* with detailed carpological descriptions. A few carpological descriptions for Trochodendrales had been made earlier, but *Tetracentron* has not been studied at all. Mature fruits and seeds were obtained from the V L Komarov Botanical Institute Herbarium (St Petersburg) and D Syreiseikov Herbarium (Moscow). The study of the carpology and seed anatomy of *Trochodendron* and *Tetracentron* supports the standpoint of Smith (1945, 1972) and his successors (Cronquist 1992; Takhtajan 1997) that both genera deserve a distinct familial rank (Teighem 1900) and should be placed in single-order Trochodendrales (Takhtajan 1997). Differences in spermoderm and pericarp anatomy supplement the previously recognized differences in vegetative and reproductive features between the two genera. The most significant differences in seed coat anatomy are the endotestal and exotegmic versus exotegmic construction and different cuticular sculpturing of seed. Significant differences in pericarp anatomy are (1) the lack of differentiation of the mesocarp into two zones, (2) a reduced number of layers in the exocarp and mesocarp, and (3) an enlarged sclerendocarp within *Tetracentron*. The new observation of seed coat type and pericarp structure of *Trochodendron* and *Tetracentron* allows a re-evaluation of the systematic position of both genera. Based on seed coat and pericarp features, Trochodendraceae and Tetracentraceae were assigned to the order Trochodendrales, super order Trochodendranae (Takhtajan 1997). It was also removed from Hamamelididae and placed in Dilleniidae, super order Dillenianae (Doweld 1998).

VEGETATIVE ANATOMY: TRENDS AND RESULTS

Arguably the most significant reference book on systematic plant anatomy of the 20th century—C R Metcalfe and L Chalk's *Anatomy of the Dicotyledons*—was published in 1950. The comprehensive survey of the vegetative anatomy of all dicot families (as they were recognized at that time) included all information gathered since the late 19th century, initially by L Radlkofer, the father of systematic anatomy and micromorphology. He had confidently proclaimed (in 1883) that plant systematics would be dominated by the anatomical method for the next hundred years (Baas 1982). Metcalfe and Chalk's (1950) magnum opus was a thoroughly revised version of Solereder *Systematische Anatomie der Dicotyledonen* (1899, 1908). Later, Metcalfe (1960, 1971) initiated the *Anatomy of the Monocotyledons* series, single handedly dealing with the grasses and sedges.

Later, Metcalfe worked on the *Anatomy of the Dicotyledons* with others; four volumes of this work have been published (Metcalfe and Chalk 1979, 1983; Metcalfe 1987; Cutler and Gregory 1998). Attempts to standardize terminology and define individual characteristics clearly have been a significant contribution to the use of anatomical diversity patterns in systematics. For dicot and gymnosperm wood anatomy, the International Association of Wood Anatomists contributed very important standards (IAWA Committee 1989). Matias, Soares, and Scateva (2007) studied the systematic consideration of petiole anatomy of species of *Echinodorus* Richard (Alismataceae) from north-eastern Brazil. Members of the family Alismataceae are emergent or floating herbaceous plants inhabiting both stagnant and flowing water environments. The family comprises 11 genera and approximately 75 native species in the temperate and tropical zones (Hayens and Holm Nielsen 1994). *Echinodorus* and *Sagittaria* are the only neotropical genera of this family, which demonstrate the greatest species diversity (Fasset 1955). *Echinodorus* has 26 species occurring predominantly in the South American tropics, a region considered the primary centre of diversity for the genus (Lot and Novelo 1984). Howard (1962) developed the complete classification system of petiole vein patterns for arbored (growing in a shady place formed by trees or shrubs) dicotyledonous species, taking into consideration the modifications occurring along their entire length. Based on this system, Tomlinson (1982) observed that the family Alismataceae had petiole vascular bundles arranged in arcs. According to Stant (1964), the vascular patterns in Alismataceae seem to be correlated with the size of the plant. The genera *Wisneria, Baldellia,* and *Luronium* have simple systems with one to three vascular bundles, while in other genera, the vascular bundles are dispersed into one or more arcs with a large number of bundles. Erwin and Stockey (1989) used the anatomical characteristics of the petiole to describe the fossil genus (*Helophyton*). They described five series of vascular bundles placed concentrically, and a large number of vascular bundles for each series. Populations of *Echinodorus* are frequently found in temporary lakes of north-eastern Brazil. Identification of species is often difficult in sympatric areas because of their generally similar morphology. Anatomical characteristics can often be useful in distinguishing the various species. Matias, Soares, and Scateva (2007) investigated the petiole anatomy of the most representative species of *Echinodorus* in north-eastern Brazil, with the objective of analysing the taxonomic characteristics most useful in defining their taxa. The aerenchyma in petioles of species of *Echinodorus* has laticiferous ducts along the entire length of the petiole as well as vascular bundles arranged in concentric series. The higher number of vascular bundles present in the most common species in this semi-arid region, such as *E. glandulosus, E. palaefolius,* and *E. pubescenes,* may help these amphibious aquatic plants to utilize the available water resources in these ephemeral environments efficiently during the short wet season. Lacunae present in the aerenchyma are found to be responsible for providing oxygen to the submerged organs (Visser, Voesenek, Vartapetian, *et al.* 2003) and also for helping in floatation by buoying the petiole and allowing the leaves to remain above the surface of the water during flood. Fabbri, Rua, and Bartoloni (2005) observed a significant increase in the size of the lacunae present in the aerenchyma in populations of *Paspalum* exposed to flooding, presumably favouring their competitive position in these aquatic environments. Williams and Barber (1961) observed that in Nymphaeaceae, the reticulate form of aerenchyma allows maximum petiole expansion with only a small increase in the total volume of tissue, providing adequate support for the leaves. This gives mechanical strength to leaves when water level is low and petioles are fully exposed. Latex-secreting structures are also present in other taxa of Alismatales, although Dahlgren, Clifford, and Yeo (1985) classified these structures into different anatomical categories. The petioles of all the species have latex-secreting ducts. This is a specialization that Dahlgren and Rasmussen (1983) considered "an autapomorphic character of Alismataceae". Latex-secreting structures are also present in other taxa of Alismatales, but these are classified in different anatomical categories. The lack of ontogenic studies does not allow us to establish homologous relationships between the latex-secreting structures within the different families of Alismatales (Les and Haynes 1995). The use of anatomical characteristics for taxonomic purposes allows the identification of fragmented material and herbarium specimens, in addition to supplying useful information for establishing inter-relationship between taxa at the species and supra-species levels (Metcalfe 1968). This can be observed in other groups of monocotyledons such as *Heliconia*, where characteristics such as the arrangement of fibre bundles in the abaxial region of the leaf, the size of the air canal in the mesophyll, and the form of the bracts provide important characteristics.

The arrangement of vascular bundles in series or in arcs is a distinguishing characteristic of interspecific anatomical patterns for most of the species examined, except *E. paniculatus* and *E. lanceolatus*. The latter two species also have scapes with very similar anatomical patterns, and their seeds have micro-morphological reticulate foveolate surface patterns. Additionally, they share many macro-morphological characteristics as they have basically the same habit. Only the presence of glands on the fruit of *E. lanceolatus* distinguishes it from *E. paniculatus*. Meyer (1935) was the first to use the arrangement of the vascular bundles in series during taxonomic studies of representatives of the family Alismataceae. Later, this characteristic was used to classify the fossil species of *Heleophyton helobiaeoides* (Erwin and Stockey 1989). It shows the similarity between *H. helobiaeoides*, *E. palaefolious*, and *E. glandulosus*: all three species have five series of vascular bundles and a higher number of vascular bundles in each arc.

LEAF ANATOMY IN GRASS SYSTEMATICS

Although the subfamilies and tribes of the Gramineae are generally distinguished by morphological characteristics of the spikelet and inflorescence, it has been known for 75 years or more that additional characteristics may contribute to a more natural and phylogenetic arrangement of the major taxa. Avdulov (1931) was the first to organize the results of such studies and use them in conjunction with a number of new characteristics, especially those of chromosome size and basic number. Prat (1932) used the microscopic characteristics (leaf epidermal of cells) as reflecting true relationships more exactly than spikelet characteristics. At present, characteristics of chromosomes, root hairs, stem apices, the first seedling leaf, embryo structure, physiology, reserve carbohydrates, nucleoli, geographical distribution, leaf epidermis, and anatomy have been found useful in characterizing the major taxa in the family (Brown, Heimsch, and Emery 1957). Duval-Jouve (1875) was the first to attempt the use of leaf anatomy for systematics. The characteristic used was the position of the bands of bulliform cells in relation to the nerves—for example, the presence of bulliform cells over the tertiary nerves in the Paniceae and Andropogoneae and the existence of bulliform cells in both upper and lower epidermis of Paniceae. The nature of the sheath that surrounds each vascular bundle (the inner microtome sheath that has the characteristics of an endodermis) has been reported to be either present in all grasses (Duval-Jouve 1875) or present in some grass groups but absent in others. External to the parenchyma sheath, the chlorenchyma tissues show various cellular arrangements. Subsequent anatomical studies of grass leaves have added little importance of the systematics of leaf anatomy. Pee-Laby (1898) noted the concentration of chlorophyll in parenchyma sheath cells of certain genera. Lohauss (1905) pointed out striking anatomical differences in leaf anatomy within such groups of the classical tribe, Festuceae.

Avdulov (1931) recognized two basic types of leaf anatomy in the grass family. The type I anatomy has a thick-walled mestome sheath, connected by sclerenchyma to the upper and lower epidermis; a poorly developed parenchyma sheath; and irregularly arranged chlorenchyma. This type of anatomy is found in Festuceae, Agrostideae, Hordeae, Aveneae, and Phalarideae. Prat (1932) called this the festucoid type. Type II anatomy is characterized by the large size of the parenchyma sheath cells that separate the xylem from the sclerenchyma next to the upper epidermis and by the radial arrangement of the chlorenchyma cells. This type of anatomy is the characteristic of Paniaceae, Andropogoneae, Maydeae, Chlorideae, and Zoyxieae, and it is called the Panicoid type of leaf anatomy by Prat. Brown (2011) reported the grass leaf anatomy relative to new concepts of grass taxonomy. The characteristics of the parenchyma sheath cells, whether small or large, empty, or with simple or specialized plastids, are considered significant. The presence or absence of a mesotome sheath is recognized as an important characteristic.

The arrangement of the mesophyll chlorenchyma cells—whether irregularly arranged or somewhat radially arranged in a single layer—seems to be a fundamental characteristic. Brown and Emery (1957) found that parenchyma sheath cells have either typical chloroplasts or highly specialized starch plastids, a characteristic that may be of systematic significance (Reeder 1957). A second significant anatomical feature is the arrangement of cells immediately outside the parenchyma sheath. Chlorenchyma is arranged in a very regular pattern in the

Chlorideae and related tribes, somewhat less regularly in genera such as *Andropogon* and *Panicum*, and in a very irregular manner in grasses such as *Poa* and *Elymus*. The inner bundle sheath has cell-wall thickening and physiological activity characteristic of an endodermis. This endodermis, which is present in some grasses but absent from others (Schwendener 1890), represents a third diagnostic characteristic.

The genera studied may be consolidated into six major groups on the basis of the following characteristics.

- Festucoid type—well-developed, thick-walled endodermis surrounded by a very distinct parenchyma sheath. The cells are small and thin walled, and chloroplasts are present.
- Bambusoid type—endodermis with specialized parenchyma sheath cells and modified chlorenchyma cells. The parenchyma sheath cells have thick walls, and are round or elliptical in cross section. Tribes are Danthonieae, Streptochaeteae, Unioleae, Bambuseae, and Oryzeae.
- Arundinoid type—lack of chloroplast in enlarged parenchyma sheath cells. Chlorenchyma cells are densely packed, and endodermis is poorly developed. Endodermis is reduced in wall thickness—Andropogoneae.
- Panicoid type—retention of an endodermis on large bundles in some species of *Panicum*. Typically, however, an endodermis is lacking. Cells of the parenchyma sheath do not have the specialized starch plastids typical of the tribe. Chlorenchyma is radially arranged around the bundles; cells were not very long and narrow or very obviously radially arranged. Air spaces are present in the mesophyll cells of *Panicum, Brachiaria, Tricholaena*, and so on.
- Arsitidoid type—no mesotome sheath, having double parenchyma sheath, containing specialized plastids. The genus *Aristida* is set apart from all other grasses by the unique sheath arrangement with no epidermis but is close to the Chloridoid type.
- Chloridoid type—characterized by an endodermis around the large bundles by a single parenchyma sheath, cells of which contain specialized plastids by a chlorenchyma. Long, narrow, radially arranged cells form one layer and contain few chloroplasts. Leaf has a dark-green parenchyma sheath, which darkens when stained with Iodine; radial cells are faintly green and contain no starch. Starch storage and photosynthesis take place in large cells of parenchymatous sheath. Examples of tribes are Chlorideae, Eragrosteae, Sporoboleae, Pappophoeae, and Zoysieae and of genera are *Bleopharoneuron, Muhlenbergia*, and *Lycurus*.

The chloridoid and aristidoid types are the most specialized in the grass family, and the species of both groups are typically arid land grasses. Each bundle sheath and chlorenchyma constitute a discrete unit of structure separated from the similar adjacent units by large, empty, bulliform cells.

The various types of leaf anatomy may be correlated with certain environmental conditions typical of the regions these grass groups occupy. Festucoid type is found in grasses grown in cool and cold regions. Panicoid type is found in tropical regions with medium to high temperatures. Chloridoid and aristidoid types are found in arid and semi-arid hot regions with very high light intensity. Chloridoid type of leaf anatomy may be related to drought resistance.

FRUIT ANATOMY OF GENUS *PIMPINELLA* (APIACEAE)

In Morison's (1672) *Plantarum umbelliferarum,* fruit morphology and anatomy have been regarded as essential to the taxonomy of Apiaceae. Koch (1824) divided the family into two principal groups, Multiiugatae and Pauciiugatae, on the basis of the number of mericarp ribs. He segregated these groups into 15 tribes according to the fruit shape and compression and characteristics of the mericarp ribs and vittae. De Candolle (1830) stressed the importance of endosperm shape, arranging the umbellifers into three groups: Orthospermae, Campylospermae, and Coelospermae. According to Koso-Poljanksy (1916), the essential features included the distribution of calcium oxalate crystals, vittae, aerenchyma, and sclerenchyma in the walls of the fruit. Some other research on the umbelliferous fruit and its importance in taxonomy of this family is based on the study of fruit anatomy of the genus *Thapsia* (Smitt, Jager, Adsersen, *et al.* 1995), anatomical research on fruits of the subfamily Saniculoideae and some related genera in Africa (Liu, van Wyk, and Tilney 2003), and carpological and molecular studies on the genus *Haussknechtia boirs* (Pimenov, Valiejo-Roman, Terentieva, *et al.* 2007). *Pimpinella* L. has 170–180 species all over the world, and it is one of the largest genera of the family Apiaceae (Pimenov and Leonov 1993). The genus includes 19 species in Iran; six of them are endemic. Khajepiri, Ghahremaninejad, and Mozaffariah (2010) studied the fruit anatomy of 17 Iranian species of *Pimpinella,* for the first time. The most important features were selected and an identification key was prepared based on them. Important anatomical fruit characteristics were listed: mericarp shape, homomorphism in mericarp, numbers of lateral and marginal ribs, ratio of mericarp width to its length, ratio of mericarp width to its thickness, presence or absence of hypodermal collenchyma, mesocarp and endocarp lignifications, number of vascular bundles, number of vallecular and commissural vittae, relative size of vittae and vascular bundles, ratio of thickness to width of vascular bundles, ratio of thickness to width of endosperm, and ratio of endosperm furrow depth to endosperm thickness. The 17 Iranian species of *Pimpinella* can be separated into two groups based on their fruit indumenta. The first group, including *P. anthriscoides, P. tragioides, P. peucedanifolia, P. rhodantha,* and *P. saxifrage,* has glabrous fruits. In this group, *P. athriscoides* can be clearly distinguished from other species by the number of vallecular vittae. Morphologically also, they have different appearances, that is, the absence of indumentum in all parts of the plant, size and shape of lower leaves, size of sheath, size of umbels, and rays and fruits indumentum. *P. tragioides* differs from the three other species in the ratio of width to thickness of marginal vascular bundles. Among other species, *P. peucedanifolia* has fruits with non-prominent marginal ribs and, therefore, it can be separated from *P. saxifragaa* and *P. rhodantha.* Finally, these two species are separated from each other on the basis of mericarp shape in transverse section.

The second group can be distinguished from the first one by the presence of an indumentum in their fruits. Species of this group fall into the main categories based on the number of vallecular vittae. Three species, *P. pastinacifolia, P. aurea,* and *P. deverroides,* belong to this category; fruits of these species have 5–10 vallecular vittae in each furrow and are anatomically very much similar to each other. They can be identified by the number of commissural vittae and length-to-width ratio of mericarp. In the second category, there are nine species, with three species (*P. olivieri, P. kotxhyana,* and *P. olivierioides*) being different from other species by means of mericarp shape. Most species have elliptic or half-round mericarps in transverse section. *P. affinis* can be separated for its different endosperm thickness-to-width ratio. *P. eriocarpa* is different due to its vallecular vittae (which are much larger than its vascular bundles) and absence of hypodermal collenchyma. *P. khayyami* can be recognized on the basis of puberulous mature fruits. Three species (*P. arisactis, P. khosasanica,* and *P. tragium*) are very similar to each other in fruit characteristics as well as morphologically.

SUCCESS AND LIMITATION OF PLANT ANATOMY IN SYSTEMATICS

Considering the number of publications on systematic wood anatomy listed in Gregory's (1994) comprehensive bibliography (as a sign of research productivity for systematic anatomy in general), there has been an explosive growth during the life span of the International Association for Plant Taxonomy (IAPT). There has been a decline in the number of publications after the 1980s because the attention has shifted from anatomical to molecular studies. Molecular systematics suffers from the same high levels of homoplasy in individual characters. Unconventional suggestions by anatomists and morphologists, which are vindicated by recent unequivocal molecular results, are then a good evidence of the phylogenetic information content of anatomical diversity patterns. The earliest, poorly resolved phylogeny proposed for the Oleaceae and entirely based on wood anatomical diversity (Bass, Esser, van der Westen, *et al.* 1988) was reproduced in detail but with better resolution by a recent molecular and total evidence cladistic analysis. Another strong anatomical support for unconventional molecular suggestions is the inclusion of Vochysiaceae in Myrtales (Conti, Litt, and Systma 1996) on account of the shared vestured pits and internal phloem (Metcalfe and Chalk 1950). Among the success, the placements of *Barbeya* by Dickison and Sweitzer (1970) near Utricales (now in Rosales) and Sphaerosepalaceae and Sarcolaenaceae in Malvales can be mentioned. A celastraceous affinity for the enigmatic genus *Plagiopteron* was first suggested by Baas (1975) on the basis of shared viscous latex in the leaves of some Celastraceae and *Plagiopteron*. It was found on the same clade as the Celastraceae in their molecular trees of the angiosperms. These successes are easily over-weighted by failures to find the closest phylogenetic relatives by vegetative anatomists and reproductive morphologists alike. All members of a recently identified robust clade in Asterales, Argophyclaceae, were misplaced in disparate orders by the anatomists (Karehed, Lundberg, Bremer, *et al.* 1999; Noshiro and Baas 1998). Similarly, *Medusagyne, Oncotheca, Oceanopapaver*, and Huaceae were misplaced by comparative anatomists. *Medusagyne*, credited by Dillenialean affinities by Dickison (1990) on the basis of vegetative anatomy and reproductive micromorphology, has been classified in Malpighiales by APG (1998). However, its molecular sister group relationship with Ochnaceae was predicted by Hickey and Wolfe (1975) on the basis of leaf venation and leaf margin characters. *Oncotheca,* a problem genus from New Caledonia, provisionally assigned the Theales by Baas (1975) and Dickison (1982), turned out to be a member of the Garryales (APG 1998). *Oceanopapaver,* confidently assigned to Capparaceae by Schmid, Carlquist, Hufford, *et al.* (1984) following a checkered taxonomic history, later found to belong to Malvaceae, in accordance with its suggested Tiliaceous affinities and typically malvalean combination of bark anatomy: stellate indumentum and conspicuous mucilage cells (Tirel, Jérémie, and Lobreau-Callen 1996).

A morphological skeleton phylogeny for the angiosperms as a whole has been published by Nandi, Chase, and Endress (1998) and compared with molecular trees based on rbcL sequences and combined morphological–molecular trees. With very rapid accumulation of more molecular data from nuclear, mitochondrial, and chloroplast DNA, the jury is still out as far as critical appraisal is concerned (of the phylogenetic signal contained in individual morphological and anatomical attributes) within the various clades of angiosperms. The genus *Dirachme* offers a good example (already subject to anatomical comparisons in the late 1960s); a proper wood and leaf anatomical match could only be found after molecular studies. Thulin, Bremer, Richardson, *et al.* (1998) have suggested *Barbeya* and Rhamnaceae as closest relatives, and ovule and seed anatomical study had confirmed this (Boesewinkel and Bouman 1997). Subsequent wood and leaf anatomical comparisons fully support the affinities of Dirachmaceae with Rhamnaceae. Baas, Wheeler, and Chase (2000) analysed all dicot orders of the APG system for increased wood anatomical support or conflict and found that the number of APG orders that showed increased wood anatomical cohesion far outnumbered the orders that showed greater wood anatomical heterogeneity. Combined study of anatomical and molecular systematics in monocots includes Asparagales (Rudall and Cutler 1995), Asteliaceae, and Hypoxidaceae (Rudall, Chase, Cutler, *et al.* 1998).

FUTURE PROSPECTS

Molecular systematics (having contributed better supported phylogenetic frameworks throughout the plant kingdom) is a tremendous stimulus for comparative morphology and anatomy. Combined molecular and structural analyses may give better resolution in phylogenetic trees. More importantly, once the branching topologies of phylogenetic trees are more firmly supported, the next step for systematics is to fill these trees with life, or to reconstruct the evolution of the biology of plants (Armbruster and Baldwin 1998). Major new plant forms are constantly being detected, and combined structural and molecular studies are necessary to explore them. This should also include linkage with field studies to elucidate the interdependence of ecological and organizational constraints on plant form. Thus, comparative evolutionary structural research should also play its role as a bond, a uniting force between centrifugal directions of new field in biology.

BIBLIOGRAPHY

APG (Angiosperm Phylogeny Group). 1998. **An ordinal classification for the families of flowering plants**. *Annals of the Missouri Botanical Garden* **85**: 531–553 (APG 2)

APG (Angiosperm Phylogeny Group). 2009. **An update of the angiosperm phylogeny group classification for the orders and families of plants**. *Botanical Journal of the Linnean Society* **161** (2): 105–121 (APG 3)

Armbruster W S and Baldwin B G. 1998. **Switch from specialized to generalized pollination**. *Nature* **394**: 632–637

Avdulov N P. 1931. **Kario-systematicheskoe issledovanie semeistva zlakov**. (Karyo systematic investigation of the grass family). *Bulletin of Applied Botany, Genetics, and Plant Breeding* **44** (Suppl.): 1–428

Baas P. 1975. **Vegetative anatomy and the affinities of Aquifoliaceae, *Sphonostemon*, *Phodlline*, and *Oncotheca***. *Blumea* **22**: 311–470

Baas P. 1982. **Systematic, phylogenetic, and ecological wood anatomy—history and perspectives**. In *New Perspectives in Wood Anatomy*, edited by P Baas, pp. 23–58. The Hague: M. Nijhoff/W. Junk

Baas P. 1986. **Ecological patterns of xylen anatomy**. In *Economy of Plant Form and Function*, edited by J Givnish, pp. 327–352. New York: Cambridge University Press

Baas P, Esser P M, van der Westen M E T, and Zandee M. 1988. **Wood anatomy of the Oleaceae**. *International Association of Wood Anatomists News Bulletin* **9**: 103–182

Baas P, Wheeler E A, and Chase M. 2000. **Dicotyledonous wood anatomy and the APG system of angiosperm classification**. *Botanical Journal of the Linnean Society* **134**: 3–17.

Barabe D, Bergeron Y, and Vincent G. 1982. **Etude quantitative de le classification des Hamemelididae**. *Taxon* **31**: 619–645

Boesewinkel F D and Bouman F. 1997. **Ovules and seeds of *Dirachma socotrane***. *Plant Systematics and Evolution* **205**: 195–204

Brown W V. 2011. **Leaf anatomy in grass systematics**. *Botanical Gazette* **119**: 170–178

Brown W V and Emery W H P. 1957. **Persistent nucleoli and grass systematics**. *American Journal of Botany* **44**: 585–590

Brown N V, Heimsch C, and Emery W H P. 1957. **The organization of the grass shot apex and systematics**. *American Journal of Botany* **44**: 590–595

Carlquist S. 1966. **Wood anatomy of compositae: a summary with comments on factors controlling wood evolution**. *Aliso* **6** (2): 25–44

Carlquist S. 1996. **Wood anatomy of primitive angiosperms: new perspectives and synthesis**. In *Flowering Plants Origin, Evolution and Phylogeny*, edited by D W Taylor and L J Hickey, pp 68–90. New York: Chapman and Hall

Conti E, Litt A, and Systma K J. 1996. **Circumscription of Myrtales and their relationships to other Rosids: evidence from rbcL sequence date**. *American Journal of Botany* **83**: 221–233

Covard H S. 1905. *The Waterlilies—a monograph of the genus Nymphaceae*, Vol. 5. Carnegie: Institution of Washington

Cronquist A. 1968. *The Evolution and Classification of Flowering Plants*. Boston: Houghton Miffin

Cronquist A. 1992. *An Integrated System of Classification of Flowering Plants*, 2nd edn. New York: Columbia University Press

Cutler D F and Gregory M (eds). 1998. *Anatomy of the Dicotyledons*, 2nd edn, Vol. IV. Saxifragales. Oxford: Clarendon Press. 324 pp.

Dahlgren G. 1989. **An updated angiosperm classification**. *Botanical Journal of the Linnean Society* **100**: 197–203

Dahlgren R M T and Rasmussen F N. 1983. **Monocotyledon evolution: characters and phylogenetic estimation**. *Evolutionary Biology* **16**: 255–395

Dahlgren R M T, Clifford H T, and Yeo P F. 1985. *The Families of the Monocotyledons*. Berlin: Springer

De Candolle A P. 1830. *Prodomus Systematics Naturalies Regni Vegetabiles*. Paris

Dickison W C. 1982. **Vegetative anatomy of *Oncotheca macrocarpa*, a newly described species of Oncothecaceae**. *Bulletin—Museum national d'histoire naturelle Andansonia* **4**: 177–181

Dickison W C. 1990. **The morphology and relationships of Medusagyne**. *Plant Systematics and Evolution*. **171**: 27–55

Dickison W C and Sweitzer E M. 1970. **The morphology and relationship of *Barbeya oleoides***. *American Journal of Botany* **57**: 468–476

Doweld A B. 1998. **Carpology, seed anatomy and taxonomic relationships of *Tetracentron* and *Trochodendron***. *Annals of Botany* **82**: 413–443

Doweld A B. 1998. **On the phylogenetic relationships of Medusagyne (Medusagynaceae) as evidenced by the structure of its fruits and seeds**. *Zurnal* **83** (2): 54–68

Duval-Jouve M J. 1875. **Histotaxie des feuiller de Graminees**. *Annales des Sciences Naturelles, Botanique, Series* **61**: 227–346

Endress P K. 1986. **Floral structure, systematics, and phylogeny in Trochodendrales**. *Annals of the Missouri Botanical Garden* **73**: 297–324

Endress P K, Baas P, and Gregory M. 2000. **Systematic plant morphology and anatomy—50 years of progress**. *Taxon* **49**: 401–434

Erwin D M and Stockey R A. 1989. **Permineralized monocotyledons from the middle Eocene Pricentron. Chert (Allen by formation) of British Columbia. Alismataceae**. *Canadian Journal of Botany* **67**: 2636–2645

Fabbri L T, Rua G H, and Bartoloni N. 2005. **Different patterns of aerenchyma formation in two hygrophytic species of *Paspalum* as response to flooding**. *Flora* **200**: 354–360

Fasset N C. 1955. **Echinodorus in the American tropics.** *Rhodora* **57**: 133–156, 174–188, 202–212

Gregory M. 1994. **Bibliography of systematic wood anatomy of dicotyledons.** *International Association of Wood Anatomists Journal* (Suppl. 1)

Hallier H. 1901. **Uber die verwand Schaftver halt nisse der ursprung der sympetalen etalen und die Anordnug der Angiospermen uberhaupt.** *Hambug* **16** (2): 1–112

Hallier H. 1912. **Lorigine et le system phyletique des Angiosperms exposes a l aide de leur arbre genealogique. Archives Neerlandaises des sciences Exactes et Naturelles Ser. 3B.** *Science Naturelles* **1**: 146–234

Hayens R R and Holm Nielsen L B. 1994. **The Alismataceae.** *Flora Neotropica* **64**: 1–112

Hickey L J and Wolfe J A. 1975. **The bases of angiosperm phylogeny: vegetative morphology.** *Annals of the Missouri Botanical Garden* **62**: 538–589

Howard R A. 1962. **The vascular structure of the petiole as a taxonomic character.** In *Advances in Horticultural Science and their Application*, edited by J C Garnand, Vol. 2, pp. 120–145. New York: Pergamon Press

IAWA Committee. 1989. **IAWA list of microscopic features for hardwood identification.** *International Association of Wood Anatomists Bulletin N.S.* **10**: 219–232

Janossonius H H. 1950. **Wood anatomy and relationship. Taxonomic notes in connection with the key to Javanese woods.** *Blumea* **6**: 407–461

Karehed J, Lundberg J, Bremer B, Bremer K. 1999. **Evolution of the Australian Asian families Alseuosmiaceae, Asgophyllaceae and Phellinaceae.** *Systematic Botany* **24**: 660–682

Khajepiri M, Ghahremaninejad F, and Mozaffariah V. 2010. **Fruit anatomy of the genus *Pimpinella* in Iran.** *Flora* **205**: 344–356

Koch W D J. 1824. **Generum tribhunque Plantarum umbelliferarum nova disposito. Nova Acta Acad. Caesaracae Leopoldive coralinae Germanicae.** *Naturae Curiosorum* **12**: 55–156

Koso-Poljanksy B M. 1916. **Sciadophytorum systematics lineamenta.** *Bulletin de la Société Imperiale des Naturalists de Moscou* **29**: 93–222

Les D H and Haynes R R. 1995. **Systematics of subclass Alismatideae a synthesis of approaches.** In *Monocotyledons: systematic and evolution*, edited by P J Rudall, P J Gibb, D F Culter, and C J Humphries, pp. 353–377. Kew: Royal Botanic Gardens

Liu M, van Wyk B E, and Tilney P M. 2003. **The taxonomic value of fruit structure in the subfamily Saniculoideae and related African genera (Apiaceae).** *Taxon* **52**: 261–270

Lohauss K. 1905. **Der anatomische Bau der Laubblatter der Festucaceen und dessen Bedeutung fur die systematic.** *Biblio. Bot.* **53**

Lot A and Novelo A. 1984. **Affinidades floristicas deda monocotyledoneas acuaticas mesoamericas**. In *Biogeograhy of Mesoamerica*, edited by S P Darwin, and A L Welden, pp. 147–153. *Proceedings of a Symposium*. New Orleans: Tulane University

Matias L Q, Soares A, and Scateva V L. 2007. **Systematic consideration of Petiole anatomy of species of *Echinodorus* from north eastern Brazil**. *Flora* **202**: 395–402

Melikian A P. 1973. **The types of seed coat in Hamamelidaceae and allied families relative to their systematic relationships**. *Botani ceskiy Zurnal* **58**: 350–359

Melikian A P. 1991. **Tetracenteraceae**. In *Anatomia seminu comparative*, edited by A L Takhata, Vol. 3. Leningrad: Nanka. 100 pp.

Metcalfe C R. 1960. ***Anatomy of monocotyledons***, Vol. I Gramineae. Oxford

Metcalfe C R. 1968. **Current developments in systematic plant anatomy**. In *Modern methods in Plant Taxonomy*, edited by V H Heywood, pp. 45–57

Metcalfe C R. 1971. ***Anatomy of the monocotyledens***, Vol. I Cyperaceae. Oxford

Metcalfe C R. 1987. ***Anatomy of the Dicotyledons***. Vol III. Magnoliales, Illiciales and Laurales, 2nd edn. Oxford: Clarendon Press

Metcalfe C R and Chalk L. 1950. ***Anatomy of Dicotyledons***. 2 vols. Oxford: Clarendon Press. 1500 pp.

Metcalfe C R and Chalk L. 1979. ***Anatomy of the dicotyledons***, 2nd edn, Vol. 1. Oxford: Clarendon Press

Metcalfe C R and Chalk L. 1983. ***Anatomy of the dicotyledons***, 2nd edn, Vol. 2. Oxford: Clarendon Press

Meyer F J. 1932. **Beitrage Zur anatomic der Alismataceen. I. Die Blaltanatomic Von Echonodorus**. *Beihefte zum Botanischen Centralblatt* **49**: 309–368

Meyer F J. 1935. **Untersuchungen an den leit bundelsystemen der Alismataceen blatter als Beitrage Zur Kewnntnis der Bedingtheit under Leistungen der Leit bundle – Verbindungen**. *Planta* **23**: 557–592

Mohana Rao P R. 1983. **Seed and fruit anatomy of *Trochodendron analiodes***. *Phytomorphology* **31**: 18–23

Morison R. 1672. **Plantarum Umbelliferarum distribution nova, per tabules congnations et affinitaties, ex libro Nature observate et delecta**. Oxford

Nandi O I, Chase M W, and Endress P K. 1998. **A combined cladistic analysis of angiosperms using rbcL and non-molecular data rebs**. *Annals of the Missouri Botanical Garden* **85**: 137–212

Nast C G and Bailey I W. 1945. **Morphology and relationships of *Trochodendron* and *Tetracentron* II. Infloresence, flower and fruit**. *Journal of the Arnold Arboretunn* **26**: 267–276

Nast C G and Bailey I W. 1946. **Morphology of *Euptelea* and comparison with *Trochodendron*.** *Journal of the Arnold Arboretum* **27**: 186–192

Noshiro S and Baas P. 1998. **Systematic wood anatomy of Cornaceae and allies.** *International Association of Wood Anatomists Journal* **19**: 43–97

Oliver D. 1889. ***Tetracentron Sinense Oliv Hooker's Icones Plantarum, or Figures with descriptive characters and remarks of new and rare plants selected from the Kew herbarium***, Series 3.9 (19). London: Williams and Norgate

Pee-Laby M E. 1898. **Etude anatomique de la femille des Graminees de la France.** *Annales Des Sciences Naturelles Series* **88**: 227–346

Pimenov M G and Leonov M V. 1993. ***The Genera of the Umbelliferae***. Kew: Royal Botanic Garden

Pimenov M G., Valiejo-Roman C M, Terentieva E I, Samigullin T H, and Mozaffarian V. 2007. **Enigmatic genus Haussknechtiq systematic relationships based on Molecular and carpological date.** *Nordic Journal of Botany* **24**: 555–564

Poole I and Francis J E. 2000. **The first record of fossil wood of Winteraceae from the upper cretaceous of Antarctica.** *Annals of Botany* **85**: 307–315

Prantl K. 1888. **Trochodendraceae.** In *Die naturlichen Pflanzen familien*, edited by A Engler and K Prantl, III Teil 2. Abteilung, pp 21–23. Leipzig: W Engelman

Prat H L. 1932. **Epiderme des Graminee. Etude anatomique et systematique.** *Annales Des Sciences Naturelles Series* 10 **14**: 117–324

Reeder J R. 1957. **The embryo in grass systematics.** *American Journal of Botany* **44**: 756–768

Rudall P J and Cutler D F. 1995. **Asparagales: a reappraisal.** In *Monocotyledons: Systematics and evolution*, edited P J Rudall, P J Cribb, D F Cutter, and C J Humphries, pp 157–168. Kew

Rudall P J, Chase M W, Cutler D F, Rusby J, de Burigin A Y. 1998. **Anatomical and molecular systematics of Asteliaceae and Hypoxidaceae.** *The Botanical Journal of the Linnean Society* **127**: 1–42

Schmid R K, Carlquist S, Hufford L D, Webster G L. 1984. **Systematic anatomy of Oceanopapaver a monotypic genus of the Capparaceae from New Caledonia.** *Botanical Journal of the Linnean Society* **89**(2): 119–152

Schwendener S. 1890. **Die Mestomscheiden der Gramineen blatter.** *Stizber. Akad. Beslin.* **1890**: 405–426

Seeman B. 1864. **Revision of the natural order of Hederaceae.** *Journal of Botany British and Foreign* **2**: 335–250, 289–309

Siebold P F Von. 1838. ***Flora Japonica: Give, Plantae quas in imperis Japonico Collegit, descripsit, exparie in ipsis locis Pingendas currant dr. Ph. Fr. De. Siebold, Sectio Prima Continue*** *plants ornatuvel usui inservientes*, Sect. 1. Apud Auctorem: Lug duni Botavorum

Smith A C. 1945. **A taxonomic review of *Trochodendron* and *Tetracentron*.** *Journal of the Arnold Arboretum* **26**: 123–142

Smith A C. 1972. **An appraisal of the orders and families of primitive extant angiosperms.** *Journal of the Indian Botanical Society* **50A**: 215–226

Smitt U W, Jager A K, Adsersen A, Gudicsen L. 1995. **Comparative studies in phytochemistry and fruit-anatomy of *Thapsie graganica* and *I. transtagane*. Apiaceae.** *Botanical Journal of the Linnean Society* **117**: 218–292

Solereder H. 1899. *Systematische Anatomic der Dicotyledonen*. Stullgart (English Translation by L A Boodle and F E Fritsch, 1908, Oxford: Clarendon Press)

Soltis D E. 1997. **Angiospherm phylogency inferred from 18S ribosomal DNA sequences.** *Annals of the Missouni Botanical Garden* **84**: 1–49

Soltis D E, Soltis P S, Chase M W, Mort M E, Albach D C, Zains M, Savolainen V, Fahn W H, Hoot S B, Fay M F, Axtell M, Swensen S M, Prince L M, Kress W J, Nixon K C, Farris J S. 2000. **Angiosperm phylogeny inferred from 18 S rdnA, rbcL, and atpB sequence.** *Botanical Journal of the Linnean Society* **133(4)**: 381–461

Stant M. 1964. **Anatomy of the Alisimataceae.** *Journal of the Linnean Society* (Botany) **59**: 1–42

Stevenson D W. 1973. **Phyllode theory in relation to leaf ontogeny in *Sansevierria trifasciate*.** *American Journal of Botany* **60**: 387–395

Takhtajan A L. 1966. **Cucmema u systema et aphytogenia Magnoliophytorum.** Leningrad: Nanka (in Russian)

Takhtajan A L. 1997. *Diversity and classification of flowering plants*. New York: Columbia University Press

Thulin M, Bremer B, Richardson J, Niklasson J, Fay M F, Chose M W. 1998. **Family relationships of the enigmatic Rosid genera *Barbeya Dirochma* from the horn of Africa region.** *Plant Systematics and Evolution* **213**: 103–119

Tieghem P H El Van. 1900. **Sur Leo dicotyledones du groupe des homoxylees.** *Journal de Botanique* **14**: 259–297, 330–361

Tirel C, Jérémie J, and Lobreau-Callen D. 1996. ***Corchorus neocaladonicus* (Tiliaceae), véritable identité de l'e'nigmatique *Oceanopapaver*.** *Adansonia* **18**: 35–43

Tomlinson P B. 1982. **VIII. Helobiae.** In *Anatomy of Monocotyledons*, edited by C R Metcalfee, Vol. 2, pp. 57–81. Oxford: Clarendon Press

Visser E J W, Voesenek L A C J, Vartapetian B B, Jackson M B. 2003. **Flooding and plant growth.** *Annals of Botany* **91**: 107–109

Wheeler E A and Baas P. 1991. **A survey of the fossil record for dicotyledonous wood and its significance for evolutionary

and ecological wood anatomy. *International Association of Wood Anatomists Bulletin Ns* **12**: 275–332

Williams W T and Barber D A. 1961. **The functional significance of paerenchyme in plants**. *Symposium of Society Experimental Biology 15, Mechanisms in Biological Completion*, Oxford, pp. 132–144

Wolfe J. 1973. **Fossil forms of Amentiferae**. *Brittonia* **25**: 334–335

Young D A. 1981. **Are the Angiosperms primitively vesselless?** *Systematic Botany* **6**: 313–330

11

Chemotaxonomy

Anand Sonkar and Sharda Mahilkar Sonkar

INTRODUCTION

Chemotaxonomy integrates the principles and procedures involved in the use of chemical evidences for classificatory purposes. Chemical systematics is the study of the chemical variation in a diversity of organisms and their relationships (Smith 1976). Extensive use of chemical information as an important taxonomic character has proved to be an extension of the range of recognized sources of taxonomic evidences. Chemical taxonomy provides evidences for and against schemes of relationship between the taxonomic groups.

The concept of utilizing chemical information in plant systematics investigations is ancient. For example, the recognition of the mint family or the umbellifers as distinct groups since the time of Theophrastos indicates that the early botanists recognized the chemical properties of essential oils for distinguishing different groups of plants. However, a genuine interest among the people to understand the possible relationship between plant constituents and classification has been relatively recent. The interest in this type of investigation has increased with the increase in data available from immunochemical, biochemical, and organic chemical research, as a result of the development of simple and relatively quick analytical techniques.

It is well known that a relationship exists between all plants that have descended from a common ancestor, which is often referred to as phylogenetic relationship. Chemical analysis has now revealed that certain plant species have similar kinds of chemicals or their variants stored in different parts of the plant. Many biologists, biochemists, and even chemists have now found it intriguing that chemical evidences from plants may be able to provide an improved account of classification.

One of the aims of plant systematics is to create a truly phylogenetic system of classification that reflects maximum possible natural relationships among plant taxa. However, to achieve this from the many classification systems of angiosperms that have been created in the past is not easy (Hutchinson 1959, 1973; Takhtajan 1959, 1969, 1980; Cronquist 1968, 1981; Dahlgren 1980; Thorne 1983, 1992; and others). Due to the more or less total absence of fossil data, morphological characters complemented those from other disciplines such as plant anatomy, palynology, embryology, and biochemistry. These have been used to find out and assess natural affinities.

A large step forward in the understanding of the phylogenetic relationships of plant families was the publication of the series *Chemotaxonomie der Pflanzen* by Hegnauer. In this monumental work (1962–73) and updates (1986–96), he reviewed the literature available on the distribution of secondary plant substances in the plant kingdom and some primary metabolites such as storage carbohydrates. Then he gave his views on the phylogenetic relationships of families with each other based on their chemical profiles (*Chemotaxonomische Betrachtungen*). These views were novel and often controversial when they were first published. For instance, the great systematist, Cronquist (1980) completely disagreed with Hegnauer's opinion that the Asteraceae and Apiaceae might be related to each other based on the fact that they shared many chemical characters, such as presence of polyacetylenes and sesquiterpene lactones and lack of iridoids. Cronquist said, "I have no patience with proposals…to associate the Asteraceae with the Apiaceae because of a degree of similarity in their secondary metabolites, when so many of their other features are very different". However, many other systematists provided evidence in favour of Hegnauer's opinions and accepted his views, such as Dahlgren and Thorne who based their systems partly on chemical characters (for example, Dahlgren 1975; Dahlgren, Rosendal-Jensen, and Nielsen 1981; Thorne 1981).

Much before Hegnauer, Greshoff (1909) suggested that natural classification should also include chemical characters and the description of every new plant discovered should be accompanied by a brief chemical description of the taxon. Baker and Smith (1920) were among the pioneers who successfully combined both morphological and chemical evidences in taxonomy of eucalypts. Species of eucalypts were divided into three groups based on their morphological structure and chemical constituents. McNair (1945) attempted to show that chemical ontogeny can be evidence for chemical phylogeny. Another classical example in higher plant taxonomy is provided by the work on tea (Roberts, Wight, and Wood 1958).

General emphasis has been placed on the mere presence or absence of a particular chemical compound, but with more sensitive techniques available now, one can elucidate the biosynthetic pathways as well. Initially, chemotaxonomy was restricted to certain groups of plants that contain secondary metabolites of medicinal use. But now with its application in systematics and the availability of easy-to-handle technologies, it has taken over almost the entire plant kingdom under its purview, whether extinct or extant. Chemotaxonomy has successfully been applied on bacteria (Rosch, Harz, Schmitt, et al. 2005), algae (Falshaw and Furneaux 2009), fungi (Frisvad, Andersen, and Thrane 2008), bryophytes (Irita, Hashimoto, Fukuyama, et al. 2000), pteridophytes (Huang, Sun, Xu, et al. 2003), gymnosperms (Bobrova Goremykin, and Troitskii 1995; Doyle 1998), and angiosperms. These data are useful as additional criteria in making taxonomic assessments even in palaeontology (Niklas and Chaloner 1976).

Chemicals present in plants can be broadly categorized into two major groups according to their molecule size. Compounds with relatively low molecular weight (1000 or less) are known as micromolecules. Those with higher molecular weight (more than 1000) shall fall in the category of macromolecules. Micromolecules can further be divided into two groups depending on their vitality in plant: primary metabolites and secondary metabolites. Primary metabolites are involved in the vital metabolic pathways such as formation of citric acid and protein amino acids. On the other hand, secondary metabolites are actually the by-products of metabolism. They constitute non-protein amino acids, alkaloids, glucosinolates (mustard oils), cyanogenic glycosides, terpenoids, fatty acids, flavonoid pigments, and other phenolic compounds (Figure 1). Macromolecules can also be further classified into two broad categories: semantide and non-semantide molecules. Semantides are the information-carrying molecules such as DNA, RNA, and proteins. DNA is often referred to as primary semantide, RNA as secondary, and proteins as tertiary semantides. Non-semantide molecules are not involved in transferring information, for example, starch and cellulose (Jones and Luchsinger 1986).

PRIMARY METABOLITES

Primary metabolites comprise compounds that are involved in the fundamental metabolic pathways. They are ubiquitous in nature and so are taxonomically insignificant. For example, various amino acids and sugar molecules that are involved in photosynthesis are present universally in plants. Although in some cases, quantitative differences of these metabolites may be taxonomically useful, but these are not the preferred metabolites.

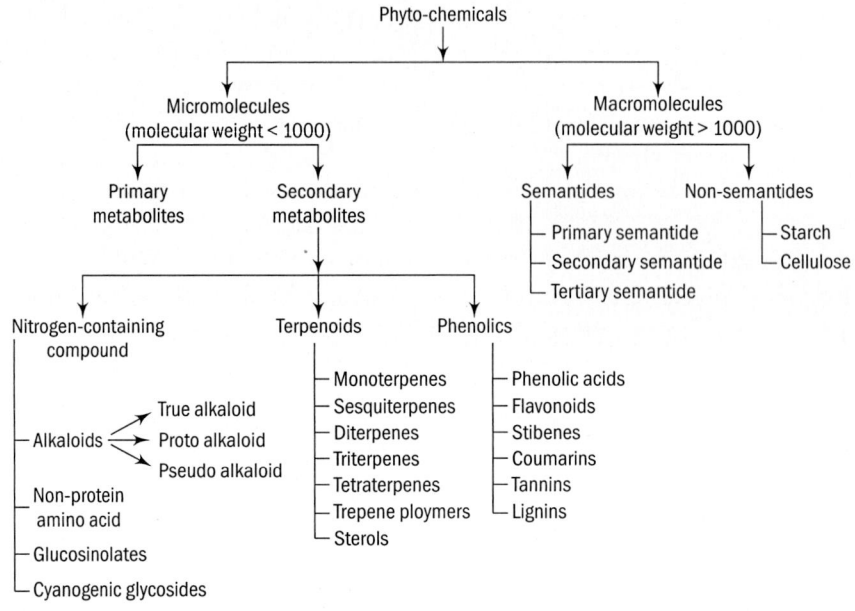

Figure 1 Classification of phyto-chemicals

SECONDARY METABOLITES

Plants produce a large array of diverse organic compounds that have no direct function in their growth and development. These are known as secondary products or natural products, which have very limited distribution in the plant kingdom. These compounds are specific in their occurrence, that is, they occur only in certain groups/species/families of plants, or in certain organ(s) of a particular plant. For many years, these compounds have been considered functionless or simply metabolic wastes. Only in late 19th and early 20th centuries, organic chemists became interested in them because of their importance as medicinal drugs and other industrial materials. Recently, it has been suggested that they may have important ecological functions as well. They keep a check on herbivory and also serve as attractants for pollinators. Secondary metabolites can further be divided into three major groups: nitrogen-containing compounds, terpenes, and phenolics.

Nitrogen-containing Compounds

Secondary metabolites in a large number of plants contain nitrogen in their structure. Alkaloids and cyanogenic glycosides are important and popular class of these compounds as they have been used in medicine owing to their important medicinal properties. Structure and properties of various nitrogen-containing secondary metabolites, including alkaloids, non-protein amino acids, glucosinolates, and cyanogenic glycosilates are discussed below.

Alkaloids

Alkaloids are organic, nitrogen-containing bases, usually with a heterocyclic ring. They are structurally diverse and are a heterogeneous group of compounds. Alkaloids can be synthesized from amino acids or mevalonic acid through various biosynthetic pathways. These are not essential for the growth of the plant, but physiologically active even at very low concentrations. The principal classes of alkaloids are pyrrolidines, pyridines, tropanes, pyrrolizidines, isoquinolines, indoles, and quinolines (Figure 2). Some well-known alkaloids include morphine, strychnine, quinine, ephedrine, and nicotine.

Three main categories of alkaloids are recognized according to their structure: true alkaloids, protoalkaloids, and pseudoalkaloids. *True alkaloids* have a nitrogen-containing heterocyclic nucleus derived from a biogenetic amine, which is formed by decarboxylation of an amino acid. Examples of this type are the phenylalanine, which after decarboxylation gives rise to isoquinoline. *Protoalkaloids* are similarly derived from amino acids but lack a heterocyclic ring. Natural protoalkaloids are usually simple amines with various additional groups; they may sometimes be precursors of true alkaloids. *Pseudoalkaloids* seem to be biogenetically unrelated to amino acids and thus considerably different from the other two categories. Most of them are derived from terpenes, sterols, aliphatic acids, nicotinic acid, or purines (Smith 1976).

Distribution

Although all families of plants do not accumulate alkaloids, some are known for it. Solanaceae, Ranunculaceae, Fabaceae, Papaveraceae, Apiaceae, and Salicaceae are among the most common families that contain alkaloids. Some families contain alkaloids of very specific type. For example, *Secologanin-type indole alkaloids* are restricted to very few families such as Apocynaceae, Gelsemiaceae, Loganiaceae, and Rubiaceae. *Tropane alkaloids* occur in a wide array of families, but similar ones are characteristic of Solanaceae, Erythroxylaceae, Proteaceae, Euphorbiaceae, Cruciferae, and Convolvulaceae (Griffin and Lin 2000). *Benzylisoquinoline alkaloids* occur in many members of Magnoliales, Laurales, Ranunculales, and Nelumbonaceae only. Other groups such as *isoprenoid alkaloids* and *pyrrolizidine alkaloids* show a more scattered distribution among angiosperms and are, therefore, of less systemic interest (Judd, Campbell, Kellogg, et al. 2008).

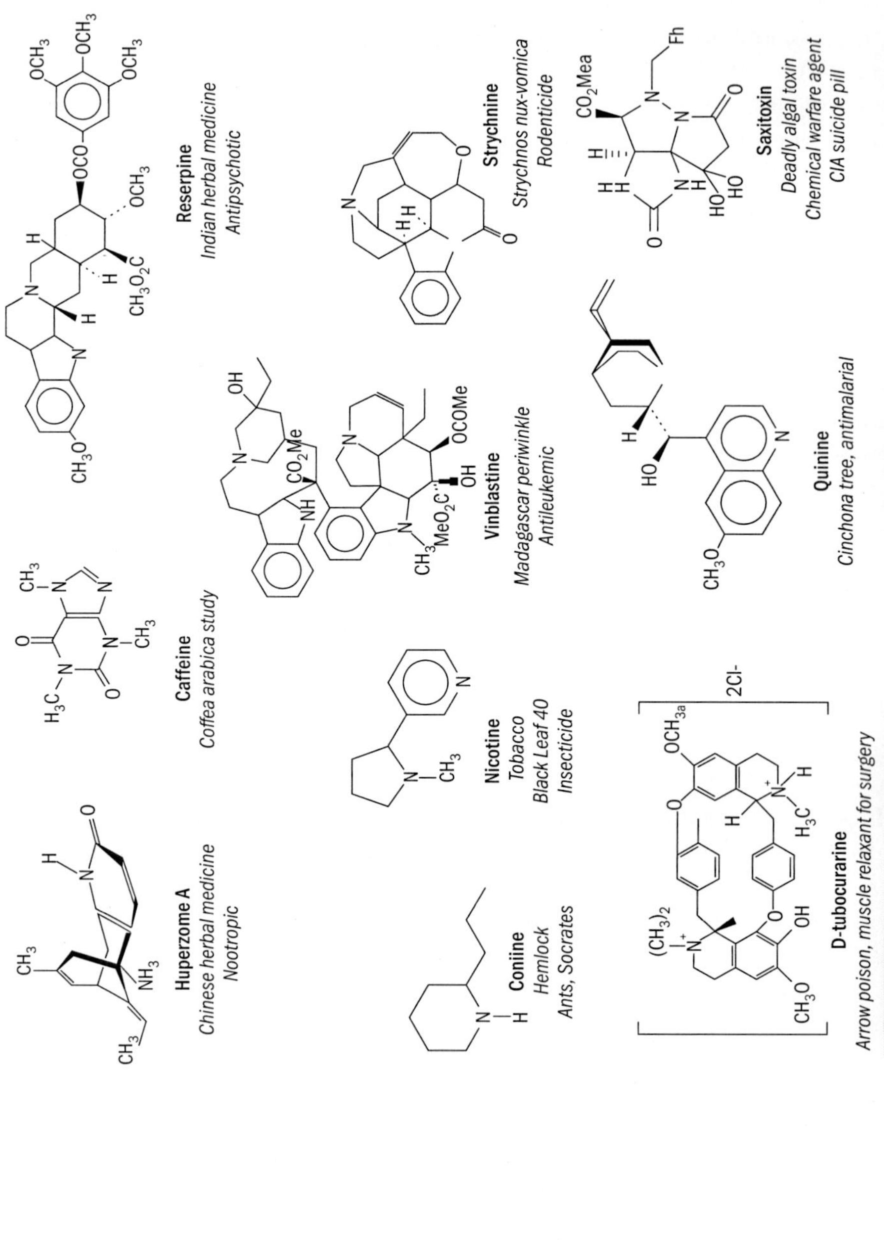

Figure 2 Select alkaloids and their uses

Alkaloids and Taxonomy

The commonly grown tea, *Camellia sinensis*, generally includes two taxonomically different subspecies: *sinensis* and *assamica*. Many hybrids of cultivated tea have come up during the many decades of long cultivation of these two subspecies. If the taxa are compared by their morphological features, it often leads to confusion due to past and present hybridization with cultivated forms. Adding to the problem has been the chance hybridization of wild tea with cultivated ones. The high commercial value of tea, owing to its alkaloid content, makes it very important to have a correct taxonomic understanding of it. Roberts, Wight, and Wood (1958), with the use of paper chromatography, demonstrated that China tea, obtained in as pure form as possible, with typical dull, microphyllous leaves contained glycosides of quercetin, kaempferol, and depsides. Plants of Assam tea that were entirely unaffected by past hybridization with China tea were not available; however, the purest form used contained only traces of triglycosides. It was assumed that these were introduced by hybridization and that typical Assam tea lacked triglycosides of this type. An unidentified phenolic, called "IC", was also present in trace form. A third kind of tea, with a southern distribution, completely lacked the triglycosides but contained large amounts of depsides and IC. This is possibly a discrete form of tea that has broad glossy leaves like Assam tea, but the red pigment may be related to other species of tea, for example, *Camellia taliensis* and *C. irrawadiensis*. Presence of IC in trace quantities in the Assam tea samples suggests that they contained genes from this southern tea, rather than typical China tea (Roberts, Wight, and Wood 1958).

Dahlgren (1983) classified Ancistrocladaceae, Dioncophyllaceae, and Nepenthaceae in his order Theales, which belongs to Theiflorae, and included Droseraceae in a rosiflorean order Droserales. Affinities of all four families mentioned are much disputed. In terms of secondary metabolism, all four families are characterized by the production of acetogenic plumbagin-type naphthoquinones and, in case of Ancistrocladaceae and Dioncophyllaceae, ancistrocladine-type acetogenic isoquinoline alkaloids. Probably all four families are phylogenetically related and should be included in the same superorder, as suggested by chemical evidences.

Didymelaceae represents a Madagascan taxon of obscure relationships. Dahlgren (1983) included the monogeneric family together with Buxaceae and Daphniphyllaceae in his rositlorean order Buxales. As far as Buxaceae is concerned, the proposed affinities are fully confirmed by its chemical characters. Buxaceae and Didymelaceae both produce aminopregnanes. Daphniphyllaceae, however, seems to lack Buxaceae-type alkaloids.

Instead, it accumulates iridoid glucosides and alkaloids that are synthesized via squalene and represents strongly modified triterpenes and octanortriterpnes. The classification of this taxon in Hamamelidae by Cronquist (1981) seems to be preferable.

Several species of the genus *Sophora* are highly heterogenous and are known to contain alkaloids of the matrine series, which are believed to be responsible for their pharmacological activity. Using alkaloid chemistry, the genus *Sophora* can be divided into four subgenera based on the presence of alkaloids of different series (Izaddoost 1975). Interestingly, the study of alkaloid distribution in some papilionaceous species in the tribes Sophoreae, Dalbergieae, Brongniartieae, and Bossiaeeae has revealed that tribe Loteae does not belong to the same lineage. Alkaloids have been identified in the tribes of Papilionoideae, whereas no alkaloids were detected in the seeds of several *Lotus* species (tribe Loteae) (Kinghorn, Balandrin, and Lin 1982).

For a long time, Aristolochiaceae included a taxon, *Incertae sedis*. Wettstein advocated its inclusion in Polycarpaceae near Annonaceae. This classification was fully confirmed by metabolic characters, especially accumulation of essential oils in idioblasts and of alkaloids produced from benzylisoquinoline pathway. The nearly family-specific nitrophenanthrene pigments, known as aristolochic acids, and corresponding aristololactams, which largely replace alkaloids in this family, have been proved to be strongly modified aporphine alkaloids (Hegnauer 1988).

Selected species of the tribe Antidesmeae (Euphorbiaceae, subfamily Phyllanthoideae) have been screened for antidesmone (a novel quinoline-type alkaloid) occurrence, and its content by quantitative high-performance liquid chromatograhy (HPLC) and qualitative liquid chromatography and mass spectrometry (LC–MS/MS) analysis (Buske Schmidt, and Hoffmann 2002). It has been proved beyond doubt that antidesmones are restricted to the members of tribe Antidesmeae. However, antidesmone-derived compounds are found in certain members of other tribes as well, but in very low concentrations.

Polyhydroxy alkaloids have been used for the classification of Convolvulaceae (Schimming, Jenett-Siems, Mann, *et al.* 2005). Similarly in Portuguese *Ulex,* the classification has been revised using quinolizidine alkaloids as taxonomic markers (Maximo, Lourenco, Tei, *et al.* 2006). Apparently, gas chromatography (GC) has been used in both the studies.

Non-protein Amino Acids

Amino acids are more frequently known as the building blocks from which proteins are synthesized. They are optically active compounds that are known to occur in both D- and L-enantiomorphs. They have broadly been classified into two categories: protein and non-protein amino acids.

Non-protein amino acids are not found in combination with proteins; in fact, they fall only in the category of secondary metabolites and not the protein ones. If the two categories are compared, non-protein amino acids are usually outnumbered. Non-protein amino acids exhibit scanty distribution with marked discontinuities; this discontinuous distribution has attracted much taxonomic attention. Information on these non-protein amino acids is not very clear; however, because they have been conserved in plants over the long period of evolution and even now plants are producing them in large quantities (Mothes 1966), they must be involved in certain process either directly or indirectly. It is likely that no single function exists for such a diverse group of molecules.

Paper chromatography has helped extensively in the identification of numerous non-protein amino acids (Fowden 1962). The non-protein amino acids of the legume genus *Bocoa* (Papilionoideae, Swartzieae) were investigated by LC–MS and GC–MS using extracts of herbarium leaf fragments. The chemical division of *Bocoa* concurs with the studies of other character types and recent molecular phylogenies (Kite and Ireland 2002).

Glucosinolates

Glucosinolates, also frequently known as mustard oil glycosides, are nitrogen- and sulphur-containing secondary metabolites (Figure 3). They are characteristic of Capparales and a few other dicotyledonous families and are ubiquitous and structurally diverse in the Cruciferae (Ettlinger and Kjaer 1968; Rodman 1978). Typically, the compounds occur in all parts of the plant but are most abundant in seeds (Josefsson 1970).

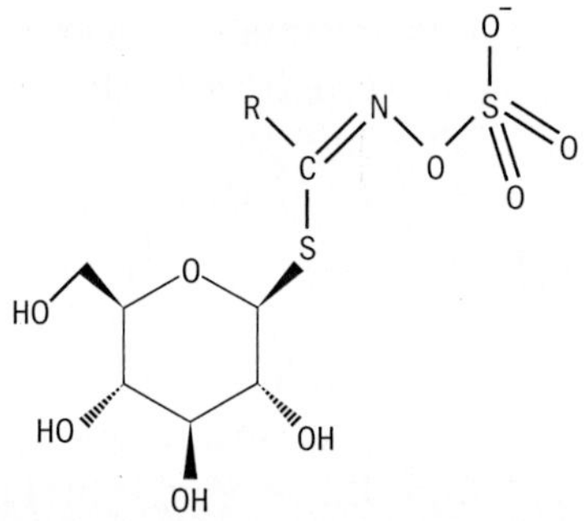

Figure 3 Glucosinolate

Glucosinolates have been very useful in the classification of the order Capparales, including Cruciferae and Capparaceae (Crawford and Giannasi 1982). Earlier Cruciferae, Capparaceae, Papaveraceae, and Fumariaceae were included in the single order Rhoeadales, but later chemical and other evidences supported the placement of Cruciferae and Capparaceae in the order Capparales (which produce glucosinolates) and of Papaveraceae and Fumariaceae in the order Papaverales (which contain alkaloids). Two families of uncertain affinities, Bataceae and Gyrostemonaceae, were once placed in Caryophyllales, but the presence of glucosinolates suggests that the two families be removed from this order (Jones and Luchsinger 1986). In addition, the compounds have been used as genetic markers in studies of plant migration (Rodman 1976) and hybridization (Rodman 1980).

Cyanogenic Glycosides

Cyanogenic glycosides are defensive compounds that are hydrolysed by various enzymes to release hydrogen cyanide (HCN) (Hegnauer 1977) through the process cyanogenesis. Breakdown of cyanogenic glycosides in plants is a two-step enzymatic process. Plants producing cyanogenic glycosides also produce the enzymes required to hydrolyse sugars and release HCN. In the first step, the sugar is cleaved by a glycosidase. In the second step, the resulting hydrolysis product, known as α-hydroxynitrile or cyanohydrin, can decompose spontaneously at a low rate to liberate HCN. Tubers of Cassava are known to contain high levels of cyanogenic glycosides.

Glycosides and Taxonomy

In Flacourtiaceae, cyclopentenoid cyanogenic glycosides showed the presence of only three compounds. These were found only in the tribes Berberidopsideae, Oncobeae, Pangieae, and Banaraeae. The family is thus divided into cyanogenic and acyanogenic members. The former group possesses compounds similar in structure to those of Passifloraceae and have been considered primitive morphologically. It seems improbable that Passifloraceae are derived from the acyanogenic, and putatively evolutionarily more advanced, group of Flacourtiaceae (Spencer and Seigler 1985).

Seigler, Pauli, Frohlich, *et al.* (2005) found that the major cyanogenic glycoside of *Guazuma ulmifolia* (Sterculiaceae) is (2R)-taxiphyllin, which co-occurs with (2S)-dhurrin. A few individuals of this species, but occasional other members of the family, have been reported to be cyanogenic. The cyanogenic glycosides of *Ostrya virginiana* (Betulaceae) are (2S)-dhurrin and (2R)-taxiphyllin in an approximate 2:1 ratio.

Based on nuclear magnetic resonance (NMR) spectroscopy and thin layer chromatography (TLC) data, major cyanogenic glycosides have been identified and found to be taxonomic markers.

Terpenes

Terpenes or terpenoids are the largest class of secondary products, usually derived from the union of 5-carbon isopentenyl diphosphate units. The basic structure is often referred to as isoprene unit and terpenes are referred to as isoprenoids. According to the number of carbons present in the terpenes, they have further been categorized: terpenes having 10 carbon atoms are known as monoterpenes, 15 carbon atoms as sesquiterpenes, 20 carbon atoms as diterpenes, 30 carbon atoms as triterpenes, 40 carbon atoms as tetraterpenes, and more than 40 carbon atoms as polyterpenes.

Steroids are triterpenes based on the cyclopentane perhydro-phenanthrene ring system. The triterpenoid derivatives, limonoids and quassinoids, are restricted to Rutaceae, Meliaceae, and Simaroubaceae. Cardenolide is a 23-carbon steroid known to be a poisonous glycoside found in Rannunculaceae, Euphorbiaceae, Apocyanaceae, Liliaceae, and Plantaginaceae. They are also known to occur in mints, umbellifers, citrus, and gymnosperms. However, they are not as much preferred as the flavonoids in systematics because of their restricted distribution among the vascular plants. Other shortcomings include the intricate instrumentation and practice needed for their analysis. They are formed by the union of 5-carbon isopentenyl diphosphate units formed in the mevalonic acid pathway.

Iridoids are 9- or 10-carbon monoterpenoid derivatives that usually occur as O-linked glycosides. Iridoid compounds are found in many families of the asterid clade (Jensen 1992). The iridoid types have been used in delimiting relationship within the group. *Seco-iridoids*, a derivative of iridoids, lack a carbocyclic ring and are known to occur in Gentianales, Dipsacales, and many other families of Cornales and Asterales. In contrast, *carbocyclic iridoids*, which have two rings (one composed entirely of carbon), are characteristic of Lamiales, except for Oleaceae, Tetrachondraceae, and Gesnariaceae. The presence of iridoids in Ericales and Cornales provides evidence that these taxa actually belong to asterid clade, even though they frequently have been excluded from that group (Cronquist 1981). Chemotaxonomy of *Plantago* using iridoid glucosides and caffeoyl phenylethanoid glycosides has been carried out by Ronsted, Gobel, Franzyk, *et al.* (2000).

Terpenes and Taxonomy

Several new constituents have been identified by analysing volatile oils from the foliage of individual plants from different populations of *Juniperus horizontalis, J. scopulorum,* and *J. virginiana* qualitatively and quantitatively. *J. horizontalis* can be differentiated from the other two species by the presence of relatively large percentages of cadinane-type sesquiterpenes and less of the elemol-eudesmol type. Although the oil of *J. scopulorum* is virtually devoid of cadinol-type sesquiterpenes, its differentiation from *J. virginiana* is difficult (Rudloff 1975).

The iridoid distribution and other studies in rubiaceous plants suggest that this family can be divided into three groups: (1) subfamily Ixoroideae, members of which contain gardenoside, geniposide; and ixoroside; (2) subfamily Rubioideae, all of which contain asperuloside and/or deacetylasperulosidic acid; and (3) subfamilies Cinchonoideae and Antirheoideae, which contain loganin, secoiridoids, and/or indole alkaloids biosynthesized through the latter two glucosides. This chemotaxonomic point of view throws some doubt on the taxonomic position of *Wendlandia formosana* and plants of *Mussaenda,* two taxa currently placed in the Cinchonoideae but chemically allied to Ixoroideae (Inouye, Takeda, Nishimura, *et al.* 1988).

Diospyros is an economically important genus of the family Ebenaceae. The uniqueness of the genus is the presence of a large number of pentacyclic triterpenes and juglone-based 1,4-naphthoquinone metabolites, because of which the leaves are exploited commercially. These metabolites have been used as chemical markers for taxonomic studies (Mallavadhani, Panda, and Rao 1999). In some *Portulaca* species (Portulacaceae), the diterpenoid constituents were isolated by solvent extraction, solvent partition, and various chromatographic methods. Their structural characterizations were performed on the basis of spectroscopic analyses, especially by NMR techniques, as already reported (Ohsaki, Shibata, Tokoroyama, *et al.* 1984; Ohsaki, Matsumoto, Shibata, *et al.* 1985; Ohsaki, Shibata, Tokoroyama, *et al.* 1986; Ohsaki, Ohno, Shibata, *et al.* 1986; Ohsaki, Shibata, Tokoroyama, *et al.* 1987; Ohsaki, Ohno, Shibata, *et al.* 1988; Ohsaki, Kasetani, Asaka, *et al.* 1991; Ohsaki, Kasetani, Asaka, *et al.* 1995; Ohsaki, Asaka, Kubota, *et al.* 1997). It was found that linear evolutionary relationship exists between the species that are studied on the basis of presence of similar kinds of diterpenes (Ohsaki, Shibata, Kubota, *et al.* 1999).

Cannabis possesses distinctive medicinal properties and hence is widely used for the extraction of many terpene based-drugs. Using gas chromatography, Hillig (2004) demonstrated how two economically important species (*C. sativa* and *C. indica*) can be segregated from each other.

Phytochemistry of *Bagassa guianensis* Aubl. (Moraceae) was studied using silica gel column chromatography. Moracin, stilbenes, and flavonones were found to be present. It has already been established that moracin is present in genus *Morus* (Takasugi, Nagao, and Masamune 1979; Hirakura, Fujimoto, Fukai, *et al.* 1986; Basnet, Kadota, Terashima, *et al.* 1993; Nguyen, Jin, Lee, *et al.* 2009) and was also isolated from *Artocarpus dadah* (Su, Cuendet, Hawthorne, *et al.* 2002). The presence of moracins justifies the Weiblen phylogenetic classification of *B. guianensis* as a member of the tribe Moreae *sensu stricto* rather than the tribe Artocarpeae. It was determined that the *Bagassa* genus is closely related to *Morus* and should be included in the tribe Moreae *sensu stricto* of the Moraceae family (Royer, Herbette, Eparvier, *et al.* 2010).

Gonosperminae is one of the 12 presently recognized subtribes of Anthemideae (Asteraceae) and one of the presumed examples of discontinuous distribution between the Canary Islands and South Africa. As compared with other members of the Anthemideae tribe, the taxonomy of these species has been controversial because the recognized morphological differences are little (Triana, Eiroa, Ortega, *et al.* 2010). Gonosperminae (with 15 species), comprising the three genera *Gonospermum* Less (Canary Islands), *Lugoa* (Canary Islands), and *Inulanthera* (South Africa), were considered a monophyletic group (Bremer and Humphries 1993). *Inulanthera* was initially included in the South African genus *Athanasia*. But later it was recognized as a separate genus based on chemical and morphological evidences, assuming a relationship between the Canarian and South African genera (Källersjö 1985). However, the phylogenetic analysis of internal transcribed spacer (ITS) sequences reveals that the Canarian genera are not connected to *Inulanthera* and do not support Gonosperminae monophyly. The Canarian Gonosperminae seems to be more closely related to the *Tanacetum* species endemic to the islands than to *Inulanthera* (Francisco-Ortega, Barber, Santos-Guerra, *et al.* 2001). *Inulanthera* has also been recommended to be assigned to the subtribe Ursiniinae (Oberprieler, Himmelreich, and Vogt 2007). Based on cytogenetic studies of the nine endemic taxa of *Gonospermum* Less, *Lugoa* DC, and *Tanacetum* L. in the Canary Islands, it has recently been proposed that all three taxa be included in the endemic genus *Gonospermum* (Febles 2008). Previous chemical studies of the Canarian endemics *Tanacetum ferulaceum* (González, Bermejo, Triana, *et al.* 1990); *T. ptarmiciflorum* (González, Bermejo, Triana, *et al.* 1992a); as well as *G. canariense* (Triana, López, Eiroa, *et al.* 2000), *G. elegans* (Triana, López, Rico, *et al.*, 2003), *G. gomerae*, and *G. fruticosum* collected in La Gomera (Triana, Eiroa, Ortega, *et al.* 2008); and *L. revoluta* (Triana, Eiroa, López, *et al.* 2001) showed a high content of similar sesquiterpene lactones. However, a chemical study of *G. fruticosum* collected on Tenerife (González, Bermejo, Triana, *et al.* 1992b) afforded sesquiterpene alcohols rather than lactones, and thus, in view of the wide distribution of this species, a new study was undertaken using samples from different locations. Using sesquiterpene lactones, Triana, Eiroa, Ortega, *et al.* (2010) have established that inclusion of *Gonospermum*, *Lugoa*, and species of *Tanacetum* endemic to the Canary Islands in a single genus does not support the monophyly of Gonosperminae based on chemotaxonomic markers.

Sesquiterpene dialdehyde variants have been used as species markers (Wayman, de Lange, Larsen, *et al.* 2010) in the taxonomy of *Pseudowintera* using HPLC, whereas essential oil composition markers have been used in *Salvia* species (Salimpour, Mazooji, and Darzikolaei 2011).

Phenolic Compounds

Plants produce a large variety of secondary products that contain a phenol group—a hydroxyl functional group on an aromatic ring. Nearly 10 000 individual compounds of this heterogeneous group have been identified. They are synthesized by two basic pathways in plants: the shikimic acid pathway and the malonic acid pathway. They can further be grouped into phenolic acids, flavonoids (betalains and anthocyanins), stilbenes, coumarins, lignins, and tannins.

On the basis of phenolic compounds analysed in 18 taxa of *Tephrosia*, it has been suggested that *T. calophylla* and *T. villosa* stand isolated from the rest. It also supports the maintenance of *T. hamiltonii* as distinct from *T. purpurea*, and the taxonomic distinction of *T. strigosa* (Rao and Rao 1993). Using similar techniques, phenol analysis of *Salvia* species has been carried out to find taxonomic similarities among the species (Qiao, Zhang, Ye, *et al.* 2009).

Flavonoids and Coumarins

Among the phenolic compounds, flavonoids are favoured micromolecules for chemotaxonomy. They have been most widely and effectively used for several reasons: their ubiquitous occurrence in plants, their easy separation and detection from small quantities of plant tissue, chemical stability that facilitates their use even after many years of collection, great structural diversity, and the fact that they can be used at all taxonomic levels in most groups of plants. They are primarily useful in assessing relationships among closely related species and also occasionally useful in assessing phylogenetic relationships at higher levels. Flavonoids are not only those substances that have the true flavonoid structure, but also closely related classes of compounds such as chalcones, isoflavones, aurones, stilbenes, cinnamic acids, and coumarins that are demonstrably associated with true flavonoids.

In the 1960s, simple but elegant use of the chromatographic patterns of flavonoids by Alston and Turner (1963) to document hybridization between species of *Baptisia* set the stage for the effective use of flavonoids data in plant systematics.

Thymbra capitata (L.) Cav. Lag. Griseb. [= *Thymus capitatus* (L.) Hoffmann and Link = *Corydothymus capitatus* (L.) Reichemb. fil.] is a Mediterranean plant that grows mainly in southern Spain and occasionally in eastern Spain (Alicante) and the Balearic islands. This plant constitutes a taxonomic problem that has been extensively discussed. Several authors have considered that there are two systematic groups sufficiently differentiated to sustain the existence of two separate genera, *Corydothymus* Reichemb. fil. and *Thymus* L. Other workers have considered that there is geobotanical, morphological, cytological, and chemotaxonomic evidence to include *Thymus capitatus* in *Thymus*, but separating it as a distinct subgenus *Corydothymus*. On the basis of flavonoid compounds studied and morphological and genetic similarities with *Thymbra spicata*, the plant has been renamed as *Thymbra capitata* (Barberan, Hernendez, and Tomas 1986).

Flavonoids have also been used as chemotaxonomic markers in the family Asteraceae (Emerenciano, Militao, Campos, *et al.* 2001) to establish the phylogeny of many controversial genera. Sea buckthorn berries comprise a mixture of different bioactive compounds belonging to various chemical groups, such as flavonoids, lipids, triterpenes, phenolic acids, and sterols. Most of them represent the chemical character and bioactivity of the sea buckthorn berries. Pharmacological and clinical studies indicated that these flavonoids have a wide spectrum of physiological activities. Using HPLC, a chemical fingerprint method was developed for investigating and demonstrating the variance of flavonoids among sea buckthorn berries. In the HPLC chromatograms, 12 compounds were identified as flavonoids in addition to some other compounds. Results revealed that the chromatographic fingerprint combining similarity evaluation could efficiently identify and distinguish sea buckthorn berries from different species. This method was considered suitable for fingerprint analysis to check the genuine origin and control the quality of sea buckthorn berries and extracts (Chen, Zhang, Xiao, *et al.* 2007).

TLC has also been used extensively for fingerprinting using flavonoids in important species of *Cyprus* (Zafar, Ahmad, Khan, *et al.* 2011).

Betalains and Anthocyanins

Betalains (formerly known as nitrogen-containing anthocyanins) are nitrogenous red and yellow pigments that are restricted to Caryophyllales, except for Caryophyllaceae and Molluginaceae. On the other hand, anthocyanins (a group of flavonoids) constitute red, yellow, blue, or purple pigments of the majority of other plants. These coloured materials are generally present in the showy parts of the plants, which primarily help in attracting the pollinators. Although they are also known to occur in other plant parts, their function in such places is totally different—probably a check on herbivory or UV absorption.

Betalains are of great taxonomic significance in higher plants. The presence of betalains in members of the order Caryophyllales has been an important criterion for their classification. The presence of betalains and anthocyanins is mutually exclusive in the angiosperms (Stafford 1994), that is, both of them have never been reported in the same plants. Betalains are water-soluble nitrogenous pigments. They can be divided into two major structural groups: the red to red-violet betacyanins (Latin *Beta*, beet and Greek *kyanos*, blue colour) and the yellow betaxanthins (Latin *Beta* and Greek *xanthos*, yellow). Betacyanins can further be classified, according to their chemical structures, into four types: betanin, amaranthin, gomphrenin, and bougainvillein (Strack, Steglich, and Wray 1993). In nature, betalains have so far been found to comprise approximately 50 red betacyanins and 20 yellow betaxanthins. Common beets usually contain both red betacyanins (consisting of 75%–95% betanin) and yellow betaxanthins (~95% vul-gaxanthin I), in various ratios depending on cultivar (Francis 1999; Piattelli 1981).

Affinities of Cactaceae were uncertain and much disputed. Wettstein included the family next to Aizoaceae in Centrospermae based on the fact that flower pigments were proved to be betalains. The same is valid for Didiereaceae, another family of long-disputed affinities (Hegnauer 1988).

Lignins and Tannins

Lignin, the most abundant organic substance after cellulose, is a highly branched polymer of phenylpropanoid. Lignin is difficult to extract from plants. So its exact structure is not known with precision. It is generally formed from three different phenyl propanoid alcohols: coniferyl, coumaryl, and sinapyl. Besides lignin, tannin is a second category plant phenolic polymer with defensive properties. Lignin is not frequently used in chemotaxonomic purpose, as it is difficult to be extracted from the plant tissue. Recently, hydrolysable tannins have been used in solving taxonomic problems in Rosaceae.

Fatty Acids

It has been found that the taxonomic position and lipid composition of octocorals are correlated. The fatty acid composition of coral specimens showed that total fatty acids are markers at the family level, and a good distinction can also be made at the genus level (Imbs and Dautova 2008). Using GC–MS, fatty acid composition has been used in the segregation of some Boraginaceae taxa (Ozcan 2008).

The chemotaxonomic relationship between *Coffea* (subgenus *Coffea*) species is not very well studied, and the compounds tested so far such as chlorogenic acids, diterpenoids, and purine alkaloids did not establish the phylogenetic relationships analogous to that revealed by chloroplast and nuclear DNA studies. Relationships between African Coffee species were assessed on the basis of their seed lipid composition. Fatty acids and sterols were determined in 59 genotypes. Groupings based on seed fatty acid composition showed remarkable ecological and geographical coherence; no phylogenetic explanation was found for the clusters retrieved from sterol data. When compared with previous phylogenetic studies, the groups deduced from seed fatty acid composition were remarkably congruent with the clades inferred from nuclear and plastid DNA sequences (Dussert, Laffargue, de Kochko, *et al.* 2008).

Malvales is an order of flowering plants with a contentious circumscription. The relationships between taxa, particularly Malvaceae, Bombacaceae, Sterculiaceae, and Tiliaceae, are not very well demarcated. The fatty acid composition of oilseeds from seven species of Malvaceae was determined by capillary GC–MS. The quantitative distribution of fatty acid was also analysed by a cluster analysis with Euclidean distance and unweighted pair group method with arithmetic mean (UPGMA). Based on the distributions of fatty acids, cluster similarities split the analysed species into two groups: (1) species rich in palmitic acid (*Herissantia tiubae, Sidastrum paniculatum,* and *Sida rhombifolia*) and (b) those rich in linoleic acid (*S. cordifolia, S. spinosa, S. salzmanii,* and *S. galheirensis*). Historically, it was believed that *Herissantia* have segregated from *Abutilon*, one of the largest Malvaceae genera, and *Sidastrum* from *Sida* (Fryxell 1997). In previous studies, seeds of *Abutilon indicum* L. and *A. muticum* DC. Sweet were characterized by high oleic acid content (Kashmiri, Yasmin, Ahmad, *et al.* 2009). In contrast, the seed oil of *A. pannosum* (G. Forst.) Schltdl. contained linoleic acid as the main fatty acid (Mariod and Matthäus 2008), as did the *Sida* species.

Sida is a large genus of the family Malvaceae and one of the more complex groups with reference to the species demarcation. The infrageneric classification of this genus exclusively on the basis of morphological characters has been proved to be complicated. Aguilar, Fryxell, and Jansen (2003) used sequences ITS of nuclear rDNA and placed *S. salzmanii*, *S. cordifolia*, *S. rhombifolia*, and *S. spinosa* inside the *Sida* core. But the preliminary data about the fatty acid profiles of seed oils do not clearly support the placement of *S. rhombifolia* within the *Sida* core. It should be noted, however, that *S. rhombifolia* has always been taxonomically troublesome (Verdcourt 2004). This author recognized six varieties of *S. rhombifolia* in east Africa and suggested further study to investigate the possibility that this assemblage contains more than one species.

Sidastrum Baker f. is another genus with unclear morphological limits that historically has been merged with *Sida* (Fryxell 1997). Although *Sidastrum* contains ca. eight species, no reports have been found on the seed oil chemical composition of these species. The ITS data of nrDNA placed *S. paniculatum* outside the *Sida* core (Aguilar, Fryxell, and Jansen 2003) and closer to some Australian *Sida* species and *Meximalva*. Although the data about fatty acid profiles support a distinction between *Sida* and *Sidastrum* (Figure 4), more species of both genera should be analysed to evaluate the real taxonomic value of fatty acid in Malvaceae (Da Silva, De Oliveira, Dos Santos, *et al.* 2010).

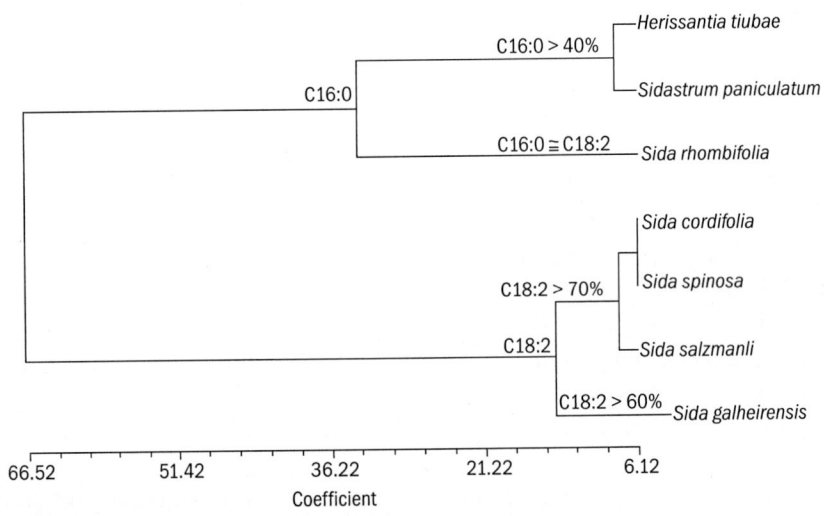

Figure 4 Dendrogram of similarity based on fatty acid distributions of oil from seeds of some species of Malvaceae

Notes 1. Analysed by Euclidean distances and UPGMA clustering methods.
2. C16:0 = palmitic acid; C18:0 = stearic acid; C18:1 = oleic acid; and C18:2 = linoleic acid

Source Da Silva, De Oliveira, Dos Santos, *et al.* (2010)

Salvia is one genus of family Labiatae where chemotaxonomy has been applied extensively. The plants are used in medicine owing to their secondary metabolite content; hence, the correct identity is an important requisite for increasing the commercial production, as species vary in their stored chemicals. Fatty acid composition of Salvia seed oils has been used as a chemotaxonomic marker by Kılıç, Dirmenci, and Gören (2007). Qiao, Zhang, Ye, et al. (2009) studied its chemical composition using the LC–MS method. Terpenes were used by Emboden and Lewis (1967), and recently by Zhou, Xu, Choi, et al. (2009). Essential oils obtained from the aerial parts have been used by Salimpour, Mazooji, and Darzikolaei (2011).

Polyacetylenes

Polyacetylenes are non-nitrogenous secondary metabolites in which fatty acids link acetate units. These compounds characterize a related group of asterid families, including Asteraceae, Apiaceae, Pittosporaceae, Campanulaceae, Goodeniaceae, and Caprifoliaceae. *Falcarinone polyacetylenes* are restricted to the Apiaceae, Araliaceae, and Pittosporaceae, which are similar in their essential oils, oleanene-and ursine-type saponins, caffeic acid esters, furanocoumarins, and flavonoid profiles (Judd, Campbell, Kellogg, et al. 2008). The analysis of polyacetylenes and a sesquiterpene lactone has confirmed Greuter's (1973) suggestion that *Cirsium strictum* (Ten.) Link should be placed in closely related genus *Ptilostemon* Cass. instead of *Cirsicum* Miller. Four species of the taxon under study were investigated and found to have similar type of chemical composition (by silica gel and TLC methods) (Janackovic, Tesevic, Marin, et al. 2002).

PROTEINS

Comparative protein analysis can be used as a fundamental taxonomic character, because proteins represent direct products of the DNA. They are often referred to as *semantides* along with nucleic acids. Some important techniques used in the study and comparison of proteins are serology, electrophoresis, and amino acid sequencing.

Serology

Serology is a branch of biology that is concerned with the nature and interactions of antigenic material and antibodies. Workers in the field of comparative serological systematics try to reveal similarities and/or dissimilarities detectable by serological procedures. Kraus (1897) first reported the precipitin reaction way back in 1897. The experiment is based on the feature of the precipitin reaction, which is the serological specificity of antibodies. This phenomenon is dependent on the ability of the antibodies to combine only with substances possessing certain antigenic determinants. Because of this antibody specificity, such reactions can be used as a powerful tool for the detection and serological identification of antigenic material (Hillebrand and Fairbrothers 1969). Recent works by a number of researchers have illustrated the importance of serological techniques in systematic studies. They also adequately illustrate the type of data that can be obtained by employing different immunochemical and serological methods.

Systemic serology techniques have been used to find out infrageneric relationships in *Viburnum* (Caprifoliaceae). Analyses of protein similarities among taxa of the genus *Viburnum* were performed using nephelometric and double-diffusion techniques. The results of serological studies were in correspondence with the sections of the genus derived from the analyses of gross morphological characters. Each section of the genus reacted as a serological unit, with representatives of the most primitive section showing the least serological correspondence with those of the most advanced section. A wide range of serological reactivity was observed among the species examined, suggesting the possibility of raising the sections to a taxonomic level above section. However, all taxa were included in a distinct serological group indicating that *Viburnum* is a distinct taxon (Hillebrand and Fairbrothers 1969).

In the *Ouchterlony double immunodiffusion* (ODI) technique, also known as *agar gel immunodiffusion* or *passive double immunodiffusion*, a series of wells are made in a gel, a sample extract of antigen is placed in one well, and sera or purified antibodies are placed in another well. The plate is left for 48 h; during this time the antigens in the sample extract and the antibodies diffuse out of their respective wells and make an immune complex, depending on the specificity, if they have any. This immune complex precipitates in the gel giving a thin white line, which is an illustration of antigen recognition. The method can also be carried out with multiple wells filled for the identification of multiple taxa simultaneously on the same gel.

Comparative serological studies on crude myrosinase extracts of *Sinapis alba* and *Brassica juncea* seed proteins have been made by immunodiffusion. The enzyme is serologically similar in the two species, but a considerable concentration difference exists between them, greater in S. *alba* (Vaughan and Gordon 1969).

Immunoelectrophoresis is a serological technique used in chemotaxonomy to compare protein extracts from different plant species. It is used for separation and characterization of proteins based on electrophoresis and reaction with antibodies. It is a two-step process: first, antigens are separated unidirectionally in gel by electrophoresis and then they are allowed to move towards an antiserum. But unlike the double-diffusion process, in this method only one sample can be treated at one time.

Relationships between seed protein patterns of 16 species of *Arachis* were studied by immunoelectrophoresis and double diffusion. Results support the recent schemes for the breakdown of genus into sections. It also suggests that A. *villosa* and A. *correntina* should probably be recognized as distinct species. The strong relationship indicated between A. *hypogaea* and A. *batizocoi* supports the hypothesis that the latter may be a source of one of the genomes of A. *hypogaea* (Klozova, Turkova, Smartt, et al. 1983).

Enzyme-linked immunosorbent assay (ELISA), also known as enzyme immunoassay (EIA), is used mainly to detect the presence of an antibody or an antigen in a sample. An unknown quantity of antigen is fastened to a surface, and then a specific antibody is applied over that surface to bind to the antigen. This antibody is linked to an enzyme, and a substrate is added so that the enzyme can be converted to some detectable signal, most commonly a colour change in a chemical substrate.

Using this technique, prolamin size heterogeneity and immunological cross-reactivities were examined in certain tribes of Poaceae, that is, the tribes Arundineae, Danthonieae, Cortaderieae, and Aristideae. Prolamins of the species examined were similar in size except for *Phragmites*. Structural similarities were measured using ELISA and immunoblotting techniques. These similarities were found to be very high among all genera except *Aristida* and *Phragmites*. Based on prolamin structure, *Aristida* was not similar to the core genera of the Arundinoideae. The remaining genera could not be distinguished as distinct tribes, and the Arundinoideae (as represented by the taxa tested) appeared monophyletic (Hilu and Esen 1990).

Radioimmunoassay (RIA) is another technique where a known quantity of radioactive antigen is taken. This radiolabelled antigen is then mixed with a known amount of antibody for that antigen; as a result, the two chemically bind to each other. This technique is advantageous because it detects even very minute quantities.

Interestingly, the levels of phytoecdysteroids (a type of terpene) have been quantified using RIA and bioassay in Ranunculaceae. Phytoecdysteroids are most prominently associated with the genus *Helleborus*. In this genus, species fall into two distinct classes: those with low or undetectable ecdysteroid levels and those with high ecdysteroid levels (Dinan, Savchenko, and Whiting 2002).

The relationship of gymnosperms with dicotyledons on the basis of immunochemical studies of seed proteins has been analysed. Twelve antisera were raised to proteins of taxa representing four classes of gymnosperm: Ginkgoopsida, Cycadopsida, Coniferopsida, and Gnetopsida. Seed proteins of eight dicotyledonous subclasses (after Takhtadzhyan, 1987) were used. The representatives of all dicotyledonous subclasses gave immunochemical reactions with those of all gymnospermous classes. The data obtained by Semikhov, Arefeva, Zolkin, et al. (2002) suggest sufficiently close immunochemical relations between gymnosperms and dicotyledons. Samples were found among the representatives of subclasses Dilleniidae, Hamamelididae, and Rosidae, which gave satisfactory reactions with eight to ten antisera to proteins of dicotyledonous seeds. Analysis of the data obtained suggests that gymnospermous and dicotyledonous plants originated from a common progymnospermous ancestor and later evolved independently, or that dicotyledons separated from gymnosperms at an early stage of their evolution before divergence of the latter into several phyletic lineages (Semikhov, Arefeva, Zolkin, et al. 2002).

Electrophoresis

In electrophoresis, molecules are separated on the basis of their charge when kept in an electric field. Molecules with positive charge move towards the cathode and those with the negative charge move towards the anode. The rate of separation depends on the net charge, which in turn depends on the pH of the medium. The size of the molecule and the strength of the voltage applied also affect the speed of the movement. Often the solvent is a gel of starch, agar, or polyacrylamide. This prevents the passage of small molecules as they become caught between the molecules of the gel, thus enabling clearer separation of the larger molecules. This technique allows separation and identification of various proteins.

The use of the seed protein profile obtained by electrophoresis, for resolving taxonomic and evolutionary problems, has significantly been increased in the 1960s. Seed proteins are preferred to other proteins as seed proteins are more stable. Electrophoresis is also being increasingly used as an additional approach for species identification and as a useful tool for tracing back the evolution of various groups of plants (Ladizinsky and Hymowitz 1979).

Use of Electrophoresis in Solving Taxonomic Problem

Numerous studies on the protein profile have been carried out using electrophoresis: on *Avena* by Murray, Craig, and Rajhathy (1970) and

Ladizinsky and Johnson (1972); *Hordeum* by McDaniel (1970); *Oryza* by Nakai (1977); *Poa* by Wilkinson and Beard (1972); *Sorghum* by Shechter (1975) and Shechter and de Wet (1975); *Triticum* by Johnson (1967a, 1967b, 1968); *Zea* by Paulis and Wall (1977); and so on. A few case studies are discussed below.

Chemotaxonomic characters in 12 species of the genus *Pennisetum* (Poaceae) have been worked out using the distribution pattern of three biochemical constituents, namely phenolic compounds, proteins, and esterase isozymes from the leaves, using paper chromatography and polyacrylamide gel electrophoresis. Data on these three chemotaxonomic characters were subjected to cluster analysis, which led to the grouping of the 12 species into five clusters (Rao, Sujatha, Rao, *et al.* 1988). As part of a chemotaxonomic revision of *Diplotaxis,* and two other related genera of the tribe *Brassiceae* (*Brassica* L. and *Erucastrum* Presl), the electrophoretic patterns of several seed isozymes have been studied in this group. A TLC of leaf and petal flavonoid aglycones has also been carried out. Electrophoretic zymograms showed the presence of many common bands in the species analysed, which confirms close infrageneric affinity. The presumed allopolyploid origin of *D. muralis* (with *D. tenuifolia* and *D. viminea* as parental species) is strongly supported (Sanchez-Yelamo and Martinez-Laborde 1991).

Species identification in *Festuca*, a native North American genus, has been a taxonomic puzzle. The banding patterns generated by sodium dodecylsulphate-polyacrylamide gel electrophoresis (SDS-PAGE) of extracts seed proteins have been a good taxonomic character. Study based on protein analysis supports (1) treating *F. dasyclada* Hackel ex. Beal as a *Festuca* and placing it in subgenus *Festuca*; (2) recognizing *F. brachyphylla* Schultes, *F. idahoensis* Elmer, and *F. saximontana* Rydb. as full species; and (3) recognizing F. *rubra* ssp. *densiuscula* (Hackel) Piper and *F. rubra* ssp. *richardsonii* (Hooker) Hulten as subspecies (Aiken, Ardinert, and Forde 1992).

Lectins are carbohydrate-binding proteins of nonimmune origin, which are widely distributed in nature and interact with sugar-containing substances. They are also capable of specific recognition and reversible binding to carbohydrates, without altering the covalent structures of any glycosyl ligands (Moreira, Ainouz, Oliveira, *et al.* 1991; Pusztai 1991). Seeds of eight species of *Canavalia* (that is, *C. dictyota, C. ensiformis, C. plagiosperma, C. brasiliensis, C. gladiata, C. maritima, C. bicarinata,* and *C. bonariensis*) were investigated with respect to their chemical composition. The seed extracts were investigated with respect to their behaviour in affinity chromatography on Sephadex column, polyacrylamide gel electrophoresis in the presence of SDS and 2-mercaptoethanol (SDS-PAGE), and ODI. All the species investigated have shown the presence of lectins, although their contents and extractability at different pH have some differences. Affinity chromatography reveals that every extracted lectin has been retained in the same column, but somehow their content in the seeds studied varied. Comparison of the SDS-PAGE protein bands showed that very close qualitative similarities are present, although important quantitative differences were observed. On the other hand, when the seed extracts were allowed to diffuse against IgG anti *C. brasiliensis* lectin, it obtained a distinct immunological identity among all the seeds. Cluster analyses were performed on the basis of lectin parameters (hemagglutinating activity and the lectin affinity peak), alone or grouped with the protein parameters. A statistical correlation was found with the six species belonging to the subgenus *Canavalia* grouped together but separated from the other two species that belong to the subgenera *Catadonia* and *Wenderothia*. A less significant correlation was found when the above parameters were analysed together with the seed morphological characteristics. In the phenogram (Figure 5), the quantitative data obtained clearly show that (as should be predicted by the systematic relationships) the species belonging to the subgenus *Canavalia* can be grouped together and separated from other two species belonging to the subgenera *Catadonia* and *Wenderothia*. In case of *Canavalia dictyota* (not clearly located in the subgenus *Canavalia* or *Maunaloa* [Aymard and Cuello 1991]), the results of cluster analysis

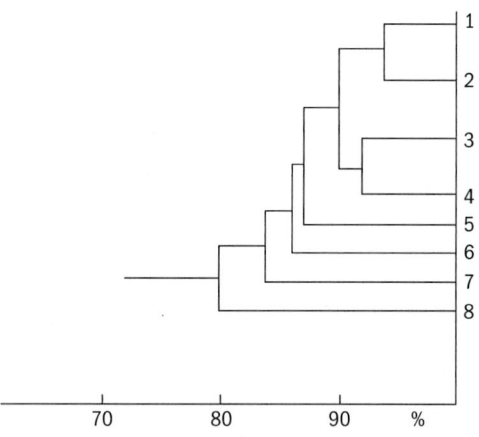

Figure 5 Phenetic relationships of the *Canavalia* species
(1) *C. dictyota*, (2) *C. ensiformis*, (3) *C. plagiosperma*, (4) *C. brasiliensis*, (5) *C. gladiata*, (6) *C. maritima*, (7) *C. bicarinata*, (8) *C. bonariensis*

Source Moreira, Cordeiro, Cavada, *et al.* (1993)

suggest its classification in *Canavalia* (as suggested by Sauer, 1964). On the other hand, when the (qualitative) results obtained from SDS-PAGE were compared, only small differences among the eight species were seen, with a very high similarity when the lectin bands were compared. The same was found when the pH 8.0 crude extracts were submitted to ODI, with IgG raised against the *C. brasiliensis* lectin. Every extract had an arc recognized by the IgG and, moreover, these arcs showed complete immunological identity. The study of Moreira, Cordeiro, Cavada, *et al.* (1993) suggested that the lectins can be used as chemotaxonomic marker within the genus *Canavalia*.

Disc Electrophoretic Method

In disc electrophoresis (the name is derived from the dependence of the method on discontinuities in the electrophoretic materials used; Ornstein and Davis 1962), a protein-containing extract is placed on the top of a polyacrylamide gel, and a voltage gradient is applied across it. Conditions are set in a way that the proteins are first concentrated into a very narrow starting zone, which increases subsequent resolution, and then separated from one another due to size differences and differential mobility in the applied voltage gradient. Proteins move in a voltage gradient depending on the pH at which the electrophoresis is carried out. This is due to the presence (in their make-up) of certain amino acids that carry positive or negative charges. The separation takes about 30 min; the gel is then removed and the separated proteins are located by treating the gel with a stain (Boulter, Thurman, and Turner 1966).

Capillary Zone Electrophoresis

The capillary zone electrophoresis technique has been used for the classification of oral treponemes (Brondz, Dahle, Greibrokk, *et al.* 1995) of bacteria, but it is yet to be used in the chemotaxonomy of higher plants.

Amino Acid Sequencing

The amino acid sequencing technique helps determine the amino acid sequence of a protein. The sequencing of homologous proteins from different species has been used for the determination of phylogenetic relationship. It is assumed that the number of differences in the sequence may be related to the length of time since different species diverged from a common ancestor. Phylogenetic trees have been constructed from data on the sequence of cytochrome c and plastocyanin.

Free protein amino acids have been used as chemotaxonomic markers in green algae belonging to the order Siphonales (Napoli, Fattorusso,

Mayol, et al. 1984). Amino acid composition of the gum exudates from *Acacia* species has also been a characterizing feature for segregation of species (Anderson and Mcdougall 1985).

Complete amino acid sequence of the respective [2Fe–2S] ferredoxins from two varieties of *Datura stramonium* (*D. stramonium* var. *stramonium* and *D. stramonium* var. *tatula*) has been determined. The ferredoxins from the two plants exhibited identical amino acid sequences, suggesting a very close taxonomic relationship between the two. This result supports the proposal by Blakeslee and others that these plants, initially identified by Linnaeus as two distinct species, that is, *D. stramonium* L. and *D. tatula* L., should be considered two varieties of a single species (Mino, Usami, Inoue, et al. 1993). Similar studies have been conducted on *Physalis*, another member of Solanaceae. Ferredoxin from *Physalis alkekengi* var. *francheti* was determined by automated Edman degradation. This ferredoxin exhibited 10, 10, and 9 differences, respectively, in the amino acid sequence, when compared with the ferredoxins of *D. stramonium*, *D. metel*, and *D. arborea*, but 21–28 differences for other angiosperms, and 34–37 differences for ferns and horsetails. These results are in absolute agreement with the taxonomic position for these plants (Mino and Yasuda 1998). Likewise, determination of the complete amino acid sequences of [2Fe–2S] ferredoxins showed four species of *Solanum* (*S. nigrum*, *S. lyratum*, *S. indicum*, and *S. abutiloides*) to be distantly related. Amino acid sequences of these four ferredoxins differed from each other by 12–19, whereas 0–4 differences have been observed among ferredoxins from plants in the same genus; 14–40 differences were seen between different families. These data suggest that these *Solanum* species are distantly related to each other taxonomically (Mino, Hazama, and Machida 2003).

Similar studies have been conducted for *Panax ginseng* (Araliaceae). It has been observed that ferredoxin had 18 differences in its amino acid sequence compared to that of *Petroselinum sativum* (Umbelliferae). In contrast, 23–33 differences were observed when compared to other dicotyledonous plants. This suggests that *P. ginseng* is taxonomically related to umbelliferous plants (Mino 2006).

Some Recent Techniques

With the progress in technology, techniques existing earlier have been refined as well as more sophisticated techniques are in use. Now, techniques such as infrared (IR) spectroscopy, Fourier transform infrared (FTIR) spectroscopy, electrospray ionization mass spectrometry (ESI-MS), and nuclear magnetic resonance (NMR) spectroscopy are being used.

Infrared Spectroscopy

IR spectroscopy is one of the most common spectroscopic techniques used by organic and inorganic chemists. It is simply the absorption measurement of different IR frequencies by a sample positioned in the path of an IR beam. The main goal of IR spectroscopic analysis is to determine the chemical functional groups in the sample. Functional groups are characterized by the absorption of different characteristic frequencies of IR radiation. Use of various sampling accessories allows IR spectrometers to accept a wide range of sample types such as solid, liquid, and gases. IR spectroscopy is emerging as an important and popular tool for structural elucidation and compound identification. A modified version of IR spectroscopy is known as FTIR spectroscopy.

Fourier Transform Infrared Spectroscopy

Fourier transform spectrometers have recently replaced dispersive instruments for most applications due to their superior speed and sensitivity. They have greatly extended the capabilities of IR spectroscopy and have been applied to areas that are very difficult or nearly impossible to analyse by dispersive instruments. Instead of viewing each component frequency sequentially, as in a dispersive IR spectrometer, all frequencies are examined simultaneously in FTIR spectroscopy.

Infrared Spectroscopy in Taxonomy

The taxonomy of the genus *Gigartina* (a red algae) based on morphology has been a problem in the past. However, these days it is well understood with the help of galactan structures present in the tetrasporophyte, which are determined using IR spectroscopy (Falshaw and Furneaux 2009). Medicinal plants in certain families, for example, Araliaceae, Campanulaceae, Magnoliaceae, Lauraceae, Leguminosae, Berberidaceae, and Pteridophyta, were studied with FTIR spectroscopy to find out similarities and differences within each family. The differences in spectra of samples from different parts, collected at different times, of the same plant are analysed. The characteristic radicals of the main effective components in plants were identified. It was considered that FTIR spectroscopy could become a rapid, reliable, impersonal, and effective method in chemotaxonomy as a supplement of morphologic plant taxonomy (Huang, Sun, Xu, *et al.* 2003).

Electrospray Ionization Mass Spectrometry

Mass spectrometry (MS) is an analytical technique that can provide both qualitative (structure) and quantitative (molecular mass or concentration) information on analyte (molecule to be studied) after their conversion to ions. The molecules of interest are first introduced into the ionization source of the mass spectrometer, where they are ionized to acquire either positive or negative charge. These ions then travel through the mass analyser and arrive at different parts of the detector according to their mass/charge (m/z) ratio. After the ions make contact with the detector, useable signals are generated and recorded by a computer system. The computer displays the signals graphically as a mass spectrum, showing the relative abundance of the signals according to their m/z ratio (Ho, Lam, Chan, et al. 2003).

ESI-MS in Taxonomy

Diterpene alkaloids were used to assess the chemical diversity in larkspur using ESI-MS. *Delphinium glaucum* samples were easily grouped and were significantly different from all other groups. *Delphinium barbeyi* and *D. occidentale* were found to be an out group, but more closely related. Samples from a hybrid between *D. barbeyi* and *D. occidentale* were found to be more closely related to *D. occidentale*, but were significantly different from all other groups. These data support the classification of *D. glaucum*, *D. barbeyi*, and *D. occidentale* as distinct species and suggest that the possible hybrid is more similar to *D. occidentale* than to *D. barbeyi* (Gardner, Ralphs, Turner, et al. 2002).

Organic, natural, or genetically modified (GM) soybeans were analysed by direct infusion ESI-MS. Free aglycones, monoglucosides, diglucosides, and esters, including isoflavones and flavones, provide characteristic fingerprinting mass spectra owing to the difference in their proportions. These polar components constitute a unique chemotaxonomic marker, which provides fast soybean typification (Santos, Catharino, Aguiar, et al. 2006).

Nuclear Magnetic Resonance Spectroscopy

NMR spectroscopy is a very powerful tool used for determining the structure of organic compounds. Atomic nuclei behave like small magnets and align themselves with an external magnetic field; NMR technique relies on this ability of atomic nuclei. When irradiated with a radiofrequency signal, the nuclei in a molecule can change from being aligned with the magnetic field to being opposed to it. The instrument works on stimulating the "nuclei" of the atoms to absorb radio waves, thus the name "nuclear". The energy frequency at which this occurs can be measured and is displayed as an NMR spectrum. The most common nuclei observed using this technique are ^{1}H and ^{13}C, but ^{31}P, ^{19}F, ^{29}Si, and ^{77}Se NMR are also available. It is used to identify and/or elucidate detailed structural information about chemical compounds.

Advantage of Using NMR Spectroscopy Over MS

MS is destructive, whereas NMR spectroscopy is not. However, the amount of material required to study is much smaller for MS. NMR spectroscopy and MS are often referred to as complementary techniques: while MS help determine the weight (and thus the molecular formula) of a molecule, NMR spectroscopy can even differentiate between structural isomers. It can also provide information about connectivity between atoms within a molecule.

NMR Spectroscopy in Taxonomy

Alkali-extractable and water-soluble cell-wall polysaccharides were purified from the cell walls of some species of *Fusarium* and *Gibberella*. Their structures were determined by chemical analysis and NMR spectroscopy. Interestingly, the polysaccharides demonstrated key differences from those of *Microdochium nivale*, *Plectosphaerella cucumerina*, *Fusarium ciliatum*, *F. aquaeductuum*, and *F. cavispermum*. Highly specific polyclonal antibodies were raised against this structure, which were used in immunocompetence and immunofluorescence experiments. This work emphasizes the fact that cell wall polysaccharide is also a good chemotaxonomic marker for *Fusarium* and its teleomorph *Gibberella* (Ahrazem, Gomez-Miranda, Prieto, et al. 2000).

The iridoid glucoside and the phenylethanoid glucoside, cornoside, have been isolated from species of *Veronica* (Plantaginaceae). Presence of these compounds has been screened by the NMR spectroscopy of crude extracts, a method more reliable than chromatography (Jensen, Albach, Ohno, et al. 2005).

NMR spectroscopy is a quite recent technique. It has predominantly been used in profiling because it is fast and simple. It is also a major analytical tool for many applications in plant metabolomics, chemotaxonomy, classification, and characterization (Le Gall, Colquhoun, and Defernez 2004). The analysis of the equivalence of GM plants (Le Gall, Colquhoun, Davis, et al. 2003; Colquhoun 2007) and interaction with other organisms (Abdel-Farid, Jahangir, van den Hondel, et al. 2009) and the environment (Jahangir, Kim, Choi, et al. 2008) also provides useful information. Although NMR spectroscopy has lower sensitivity than MS, it is particularly useful when high sensitivity is not required. Using NMR spectroscopy, Kim, Khan, Wilson, et al. (2010) have carried out metabolic classification of *Ilex* species.

CONCLUSION

In the past, chemical characters were undermined compared to molecular characters; however, now it has been proved beyond doubt that a comparison between chemical and molecular characters can help determine phylogenetic relationships among plant families. With the emergence of simple, rapid, and cheap techniques for analysis, chemotaxonomy has become a major growth area of systematic research. Time has come when taxonomists need to adapt to the changing scenario and equip themselves with chemical facts and techniques. Each plant/family has its own chemical characteristics, and one needs to detect and rely on chemicals and techniques carefully to avoid any further confusion. There can never be one answer to every taxonomic problem as one needs to be careful while choosing the phytochemical to be studied. The molecule needs to be chosen depending upon the type of classification required. Macromolecules were once thought to be of greatest value as they do not easily change with environmental modification, but they are often difficult to study.

The taxonomic validity of distribution and occurrence of small molecules is often doubted. To assess the extent of environmental variation, samples of sufficient size are chosen from different habitats at different times. Analysis of plants showing young, intermediate, and mature stages of growth is sufficient to have an idea about the degree of variation. Alkaloids, phenolics, glycosides, and other secondary metabolites help determine the palatability of foliage, thereby discouraging excessive grazing. Volatile constituents often serve as attractants to insects, while some may have phototoxic functions. Different plant taxa inhabit different ecological niches and interact with different pest and prey; this includes a lot of variation. This variation is likely to be a taxonomic marker.

BIBLIOGRAPHY

Abdel-Farid I B, Jahangir M, van den Hondel C A M J J, Kim H K, Choi Y H, Verpoorte R. 2009. **Fungal infection-induced metabolites in *Brassica rapa*.** *Plant Science* **176**(5): 608–615

Aguilar J F, Fryxell P A, and Jansen R K. 2003. **Phylogenetic relationships and classification of the *Sida* generic alliance (Malvaceae) based on nrDNA ITS evidence.** *Systematic Botany* **28**: 352–364

Ahrazem O, Gomez-Miranda B, Prieto A, Barasoain I, Bernabe M, Leal J A. 2000. **An acidic water-soluble cell wall polysaccharide: a chemotaxonomic marker for *Fusarium* and *Gibberella*.** *Mycological Research* **104**(5): 603–610

Aiken S G, Ardinert S E, and Forde M B. 1992. **Taxonomic implications of SDS-PAGE analyses of seed proteins in North American taxa of *Festuca* subgenus *Festuca* (Poaceae).** *Biochemical Systematics and Ecology* **20**(7): 615–629

Alston R E and Turner B L. 1963. ***Biochemical systematics***. Englewood Cliffs, New Jersey: Prentice-Hall

Anderson D M W and Mcdougall F J. 1985. **The proteinaceous components of the gum exudates from some phyllodinous *Acacia* species**. *Phytochemistry* **24**(6): 1237–1240

Aymard G C and Cuello N A. 1991. **Catalogo y adiciones a las especies neotropicales del genero *Canavalia* (Leguminosae-Papilionoideae-Diocleinae)**. *Seminario-Taller de Trabajo sobre Canavalia (Canavalia ensiformis (L) DC)*, Maracay, Venezuela.

Baker R T and Smith H G. 1920. ***A research on the eucalypts and their essential oils***, 2nd edn. Sydney: Government Printer

Barberan F A T, Hernendez L, and Tomas F. 1986. **A chemotaxonomic study of flavonoids in *Thymbra capitata***. *Phytochemistry* **25**: 561–562

Basnet P, Kadota S, Terashima S, Shimizu M, and Namba T. 1993. **Two new 2-arylbenzofuran derivatives from hypoglycemic activity-bearing fractions of *Morus insignis***. *Chemical and Pharmaceutical Bulletin* **41**: 1238–1243

Bobrova V K, Goremykin V V, and Troitskii A V. 1995. **Molecular-biology studies into origin of angiosperms**. *Zhurnal Obshchei Biologii* **56**(6): 645–661

Boulter D, Thurman D A, and Turner B L. 1966. **The use of disc electrophoresis of plant proteins in systematics**. *Taxon* **15**(4): 135–143

Bremer K and Humphries C J. 1993. **Generic monograph of the Asteraceae-Anthemideae**. *Bulletin of the Natural History Museum of London (Botany)* **23**: 71–177

Brondz I, Dahle U R, Greibrokk T, Olsen I. 1995. **Capillary zone electrophoresis as a new tool in the chemotaxonomy of oral treponemes**. *Journal of Chromatography B: biomedical applications* **667**: 161–165

Buske A, Schmidt J, and Hoffmann P. 2002. **Chemotaxonomy of the tribe Antidesmeae (Euphorbiaceae): antidesmone and related compounds**. *Phytochemistry* **60**: 489–496

Chen C, Zhang H, Xiao W, Yong Z, Bai N. 2007. **High-performance liquid chromatographic fingerprint analysis for different origins of sea buckthorn berries**. *Journal of Chromatography A* **1154**: 250–259

Colquhoun I J. 2007. **Use of NMR for metabolic profiling in plant systems**. *Journal of Pesticide Science* **32**: 200–212

Crawford D J and Giannasi D E. 1982. **Plant chemosystematics**. *Bioscience* **32**: 114–118, 123–124

Cronquist A. 1968. ***The Evolution and Classification of Flowering Plants***. London: T Nelson

Cronquist A. 1980. **Chemistry in plant taxonomy: an assessment where we stand**. In *Chemosystematics: principles and practice*, edited by F A Bisby, J G Vaughan, and C A Wright, pp. 1–27. London: Academic Press

Cronquist A. 1981. *An Integrated System of Classification of Flowering Plants*. New York: Columbia University Press

Da Silva A C O, De Oliveira A F M, Dos Santos D Y A C, Da Silva S I. 2010. **An approach to chemotaxonomy to the fatty acid content of some Malvaceae species**. *Biochemical Systematics and Ecology* **38**: 1035–1038

Dahlgren R M T. 1975. **A system of classification of the angiosperms to be used to demonstrate the distribution of characters**. *Botaniska Notiser* **128**: 119–147

Dahlgren R M T. 1980. **A revised system of classification of the angiosperms**. *The Botanical Journal of the Linnean Society* **80**: 91–124

Dahlgren R M T. 1983. **General aspects of angiosperm evolution and macrosystematics**. *Nordic I Bat* **3**: 119

Dahlgren R M T, Rosendal-Jensen S, and Nielsen B J. 1981. **A revised classification of the angiosperms with comments on correlation between chemical and other characters**. In *Phytochemistry and Angiosperm Phylogeny*, edited by D A Young and D S Seigler, pp. 149–204. New York: Praeger

Dinan L, Savchenko T, and Whiting P. 2002. **Chemotaxonomic significance of ecdysteroid agonists and antagonists in the Ranunculaceae: phytoecdysteroids in the genera *Helleborus* and *Hepatica*.** *Biochemical Systematics and Ecology* **30**: 171–182

Doyle J A. 1998. **Phylogeny vascular plants**. *Annual Review of Ecological Systems* **29**: 567–599

Dussert S, Laffargue A, de Kochko A, Joët T. 2008. **Effectiveness of the fatty acid and sterol composition of seeds for the chemotaxonomy of *Coffea* subgenus *Coffea***. *Phytochemistry* **69**: 2950–2960

Emboden W A and Lewis H. 1967. **Terpenes as taxonomic characters in *Salvia* section audibertia**. *Brittonia* **19**: 152–160

Emerenciano V P, Militao J S L T, Campos C C, Romoff P, Kaplan M A C, Zambon M, Brant A J C. 2001. **Flavonoids as chemotaxonomic markers for Asteraceae**. *Biochemical Systematics and Ecology* **29**: 947–957

Ettlinger M G and Kjaer A. 1968. **Sulfur compounds in plants**. In *Recent Advances in Phytochemistry*, edited by T J Mabry, R E Alston, and V C Runeckles, pp. 59–144. New York: Appleton-Century-Crofts

Falshaw R and Furneaux R H. 2009. **Chemotaxonomy of New Zealand red algae in the family Gigartinaceae (Rhodophyta) based on galactan structures from the tetrasporophyte life-stage**. *Carbohydrate Research* **344**: 210–216

Febles R. 2008. **Re-estructuración del Género Gonospermum Less. (Asteraceae: Anthemideae) en las Islas Canarias**. *Botanica Macaronésica* **27**: 101–105

Fowden L. 1962. **The non-protein amino acids of plants.** *Endeavour* **21**: 35–42

Francis F J. 1999. **Anthocyanins and betalains.** In *Colorants*, edited by F J Francis, pp. 55–66. St Paul, MN: Eagan Press

Francisco-Ortega J, Barber J C, Santos-Guerra A, Febles-Hernández R, Jansen R K. 2001. **Origin and evolution of the endemic genera of Gonosperminae (Asteraceae: Anthemideae) from the Canary Islands: evidence from nucleotide sequences of the internal transcribed spacers of the nuclear ribosomal DNA.** *American Journal of Botany* **88**: 161–169

Frisvad J C, Andersen B, and Thrane U. 2008. **The use of secondary metabolite profiling in chemotaxonomy of filamentous fungi.** *Mycological Research* **112**: 231–240

Fryxell P A. 1997. **The American genera of Malvaceae.** *Systematic Botany* **49**: 204–269

Gardner D R, Ralphs M H, Turner D L, Welsh S L. 2002. **Taxonomic implications of diterpene alkaloids in three toxic tall larkspur species (*Delphinium* spp.).** *Biochemical Systematics and Ecology* **30**: 77–90

González A G, Bermejo J, Triana J, López M, Eiroa J L. 1990. **Sesquiterpene lactones from *Tanacetum ferulaceum*.** *Phytochemistry* **29**: 2339–2341

González A G, Bermejo J, Triana J, López M, Eiroa J L. 1992a. **Sesquiterpene alcohols from *Gonospermum fruticosum*.** *Phytochemistry* **31**: 1816–1817

González A G, Bermejo J, Triana J, López M, Eiroa J L. 1992b. **Sesquiterpene lactones and other constituents of *Tanacetum* species.** *Phytochemistry* **31**: 1821–1822.

Greshoff M. 1909. **Phytochemical investigation at Kew.** *Kew Bulletin of Miscellaneous Information* **10**: 397–418

Greuter W. 1973. *Monographie der Gattung Ptilostemon (Compositae)*. Geneva: Boissiera. 22 pp.

Griffin W J and Lin G D. 2000. **Chemotaxonomy and geographical distribution of tropane alkaloids.** *Phytochemistry* **53**: 623–637

Hegnauer R. 1962–1996. *Chemotaxonomie der Planzen*, vol. 1–12. Basel: Birkhauser-Verlag

Hegnauer R. 1977. **Cyanogenic compounds as systematic markers in *Tracheophyta*.** *Plant Systematics and Evolution Supplement* **1**: 191–209

Hegnauer R. 1986. *Die Alkaloid-Familien der Dikotyledonen; Die Alkaloid-Familien der Gefdsspjlanzen; Chemotaxonomie der Pjanzen*, Bd. 3, pp. 18–28; Bd. 7, pp. 313–325. Basel: Birkhauser Verlag

Hegnauer R. 1988. **Biochemistry, distribution and taxonomic relevance of higher plant alkaloids.** *Phytochemistry* **21**(8): 2423–2427

Hillebrand G R and Fairbrothers D E. 1969. **A serological investigation of intrageneric relationships in *Viburnum* (Caprifoliaceae)**. *Bulletin of the Torrey Botanical Club* **96**(5): 556–567

Hillig K W. 2004. **A chemotaxonomic analysis of terpenoid variation in *Cannabis***. *Biochemical Systematics and Ecology* **32**: 875–891

Hilu K W and Esen A. 1990. **Prolamins in systematics of Poaceae subfam. Arundinoideae**. *Journal Plant Systematics and Evolution* **173**(1–2): 57–70

Hirakura K, Fujimoto Y, Fukai T, Nomura T. 1986. **Two phenolic glycosides from the root bark of the cultivated mulberry tree (*Morus ihou*)**. *Journal of Natural Products* **49**: 218–224

Ho C S, Lam C W K, Chan M H M, Cheung R C K, Law L K, Lit L C W, Ng K F, Suen M W M, Tai H L. 2003. **Electrospray ionisation mass spectrometry: principles and clinical applications**. *The Clinical Biochemists Review* **24**: 3–12

Huang H, Sun S Q, Xu J W, Wang Z. 2003. **Novel application of FTIR in medical herb chemotaxonomy**. *Guang Pu Xue Yu Guang Pu Fen Xi* **23**(2): 253–257

Hutchinson J. 1959. ***The Families of Flowering Plants***, 2nd edn, Vol. 1 and 2. Oxford: Clarendon Press

Hutchinson J. 1973. ***The Families of Flowering Plants Arranged According to a New System Based on their Probable Phylogeny***, 3rd edn. Oxford: Clarendon Press

Imbs A B and Dautova T N. 2008. **Use of lipids for chemotaxonomy of octocorals (Cnidaria: Alcyonaria)**. *Russian Journal of Marine Biology* **34**(3): 174–178

Inouye H, Takeda Y, Nishimura H, Kanomi A, Okuda T, Puff C. 1988. **Chemotaxonomic studies of rubiaceous plants containing iridoid glycosides**. *Phytochemistry* **27**(8): 2591–2598

Irita H, Hashimoto T, Fukuyama Y, Asakawa Y. 2000. **Herbertane-type sesquiterpenoids from the liverwort *Herbertus sakuraii***. *Phytochemistry* **55**: 247–253

Izaddoost R I. 1975. **Alkaloid chemotaxonomy of the genus *Sophora***. *Phytochemistry* **14**: 203–204

Jahangir M. 2008. **Metal ion-inducing metabolite accumulation in *Brassica rapa***. *Journal of Plant Physiology* **165**: 1429–1437

Jahangir M, Kim H K, Choi Y H, Verpoorte R. 2008. **Metabolomic response of *Brassica rapa* submitted to pre-harvest bacterial contamination**. *Food Chemistry* **107**: 362–368

Janackovic P, Tesevic V, Marin P D, Milosavljevic S M, Petkovic B, Sokovic M. 2002. **Polyacetylenes and a sesquiterpene lactone from *Ptilostemon strictus***. *Biochemical Systematics and Ecology* **30**: 69–71

Jensen S R. 1992. **Systematic implications of the distribution of the iridoids and other chemical compounds in the Loganiaceae and other families of the Asteridae. *Annual Report of Missouri Botanical Garden* **79**: 284–302

Jensen S R, Albach D C, Ohno T, Grayer R J. 2005. ***Veronica*: iridoids and cornoside as chemosystematic markers**. *Biochemical Systematics and Ecology* **33**: 1031–1047

Johnson B L. 1967a. **Confirmation of the genome donors of *Aegilops cylindrica***. *Nature* **216**: 859–862

Johnson B L. 1967b. **Tetraploid wheats: seed protein electrophoresis pattern of the emmer and timopheevi groups**. *Science* **158**: 131–132

Johnson B L. 1968. **Electrophoretic evidence on the origin of *Triticum zhukovskyi***. In *Third International Wheat Genetics Symposium*, edited by K W Finlay and K W Shepherd, pp. 105–110. London: Butterworth and Co. Ltd.

Jones S B and Luchsinger A E. 1986. ***Plant Systematics***, 2nd edn. New York: McGraw Hill Book Co.

Josefsson E. 1970. ***Pattern, Content, and Biosynthesis of Glucosinolates in Some Cultivated Cruciferae***. Lund: Swedish Seed Association.

Judd W S, Campbell C S, Kellogg E A, Stevens P F, Donoghur M J. 2008. ***Plant Systematics***, 3rd edn. Sunderland, MA, USA: Sinauer Associates, Inc.

Källersjö M. 1985. **Fruit structure and generic delimitation of *Athanasia* (Asteraceae-Anthemideae) and related South African genera**. *Nordic Journal of Botany* **5**: 527–542

Kashmiri M A, Yasmin S, Ahmad M, Mohy-ud-Din A. 2009. **Characterization, compositional studies, antioxid and antibacterial activities of seeds of *Abutilon indicum* and *Abutilon muticum* grown wild in Pakistan**. *Acta Chimica Slovenica* **56**: 345–352

Kılıç T, Dirmenci T, and Gören A C. 2007. **Chemotaxonomic evaluation of species of Turkish *Salvia*: fatty acid composition of seed oils. II**. *Records of Natural Products* **1**: 17–23

Kim H K, Choi Y H, and Verpoorte R. 2011. **NMR-based plant metabolomics: where do we stand, where do we go?** *Trends in Biotechnology* **29**: 6

Kim H K, Khan S, Wilson E G, Kricun S D, Meissner A, Goraler S, Deelder A M, Choi Y H, Verpoorte R. 2010. **Metabolic classification of South American *Ilex* species by NMR-based metabolomics**. *Phytochemistry* **71**(7): 773–784

Kinghorn A D, Balandrin M F, and Lin L J. 1982. **Alkaloid distribution in some species of the Papilionaceous tribes Sophoreae, Dalbergieae, Loteae, Brongniartieae and Bossiaeeae**. *Phytochemistry* **21**: 2269–2275

Kite G C and Ireland H. 2002. **Non-protein amino acids of *Bocoa* (Leguminosae; Papilionoideae)**. *Phytochemistry* **59**: 163–168

Klozova E, Turkova V, Smartt J, Pitterova K, Svachulova J. 1983. **Immunochemical characterization of seed proteins of some species of the *Arachis* L.** *Biologia Plantarum* **25** (3): 201–208

Kraus R. 1897. **Uber specifische Reactionen in keimfreien Filtration aus Cholera, Typhus und pestbouillon Culturen erzeugt durch homologes Serum.** *Wien klein Wschr* **10**: 726–738

Ladizinsky G and Hymowitz T. 1979. **Seed protein electrophoresis in taxonomic and evolutionary studies.** *Theoretical and Applied Genetics* **54**: 145–151

Ladizinsky G and Johnson B L. 1972. **Seed protein homologies and the evolution of polyploidy in *Arena*.** *Canadian Journal of Genetics and Cytology* **14**: 875–888

Le Gall G, Colquhoun I J, and Defernez M. 2004. **Metabolite profiling using ^{1}H NMR spectroscopy for quality assessment of green tea, *Camellia sinensis* (L.).** *Journal of Agricultural and Food Chemistry* **52** (4): 692–700

Le Gall G, Colquhoun I J, Davis A L, Collins G J, Verhoeyen M E. 2003. **Metabolite profiling of tomato (*Lycopersicon esculentum*) using ^{1}H NMR spectroscopy as a tool to detect potential unintended effects following a genetic modification.** *Journal of Agricultural and Food Chemistry* **51**(9): 2447–2456

Mallavadhani U V, Panda A K, and Rao Y R. 1999. **Pharmacology and chemotaxonomy of *Diospyros*.** *Phytochemistry* **38**(3): 840–890

Mariod A and Matthäus B. 2008. **Physico-chemical properties, fatty acid and tocopherol composition of oils from some Sudanese oil bearing sources.** *Grasas Aceites* **59**: 321–326

Maximo P, Lourenco A, Tei A, Wink M. 2006. **Chemotaxonomy of Portuguese *Ulex*: quinolizidine alkaloids as taxonomical markers.** *Phytochemistry* **67**: 1943–1949

McDaniel R G. 1970. **Electrophoretic characterization of proteins in *Hordeurn*.** *Journal of Heredity* **61**: 243–247

McNair J B. 1945. **Plant fats in relation to environment and evolution.** *The Botanical Review* **11**: 1–59

Mino Y. 2006. **Protein chemotaxonomy. XIII. Amino acid sequence of ferredoxin from *Panax ginseng*.** *Biological and Pharmaceutical Bulletin* **29**(8): 1771–1774

Mino Y and Yasuda K. 1998. **Amino acid sequence of ferredoxin from *Physalis alkekengi* var. *Francheti*.** *Phytochemistry* **49**(6): 1631–1636

Mino Y, Hazama T, and Machida Y. 2003. **Large differences in amino acid sequences among ferredoxins from several species of genus *Solanum*.** *Phytochemistry* **62**: 657–662.

Mino Y, Usami H, Inoue S, Ikeda K, Ota N. 1993. **Protein chemotaxonomy of genus *Datura*: identical amino acid sequence of ferredoxin from two varieties of *Datura stramonium*.** *Phytochemistry* **33**(3): 601–605

Moreira R A, Ainouz I L, Oliveira J T A, Cavada B S. 1991. **Plant lectins, chemical and biological aspects**. *Memórias do Instituto Oswarldo Cruz* **86**: 211–218

Moreira R A, Cordeiro E F, Cavada B S, Nunes E P, Fernandes A G, Oliveria J T A. 1993. **Plant seed lectins: a possible marker for chemotaxonomy of the genus *Canavalia***. *Revista Brasileira de Fisiologia Vegetal* 127–132

Mothes K. 1966. **Zur Problematik der metabolischen Exkretion bei Pflanzen**. *Naturwissenschaften* **53**: 317–323

Murray B E, Craig I L, and Rajhathy T. 1970. **A protein electrophoresis study of three amphidiploids and eight species in *Arena***. *Canadian Journal of Genetics and Cytology* **12**: 651–665

Nakai Y. 1977. **Variation of esterase isozymes and some soluble proteins in diploids and their autotetraploids in plants**. *The Japanese Journal of Genetics* **52**: 171–181

Napoli L D, Fattorusso E, Mayol L, Novellino E. 1984. **Free protein amino acids of some Mediterranean siphonales**. *Biochemical Systematics and Ecology* **12**(1): 19–21

Nguyen T D, Jin X, Lee K, Hog Y S, Young H K, Jung J L. 2009. **Hypoxiainducible factor-1 inhibitory benzofurans and chalcone-derived Diels–Alder adducts from *Morus* species**. *Journal of Natural Products* **72**: 39–43

Niklas K J and Chaloner W G. 1976. **Chemotaxonomy of some problematic palaeozoic plants**. *Review of Palaeobotany and Palynology* **22**: 81–104

Oberprieler C, Himmelreich S, and Vogt R. 2007. **A new subtribal classification of the tribe Anthemideae (Compositae)**. *Willdenowia* **37**: 89–114

Ohsaki A, Asaka Y, Kubota T, Shibata K, Tokoroyama T. 1997. **Portulene acetal, a novel minor constituent of *Portulaca grandiflora* with significance for biosynthesis of portulal**. *Journal of Natural Products* **60**: 912–914

Ohsaki A, Kasetani Y, Asaka Y, Shibata K, Tokoroyama T, Kubota T. 1991. **Clerodane diterpenoids from the roots of *Portulaca pilosa***. *Phytochemistry* **30**: 4075–4077

Ohsaki A, Kasetani Y, Asaka Y, Shibata K, Tokoroyama T, Kubota T. 1995. **A diterpenoid from *Portulaca pilosa***. *Phytochemistry* **40**: 205–207

Ohsaki A, Matsumoto K, Shibata K, Tokoroyama T. 1985. **Diterpenoid congeners in *Portulaca randiflora* hook**. *Chemical and Pharmaceutical Bulletin* **33**: 2171–2174

Ohsaki A, Ohno N, Shibata K, Tokoroyama T, Kubota T. 1986. **Clerodane diterpenoids from *Portulaca* cv. Jewel**. *Phytochemistry* **25**: 2414–2416

Ohsaki A, Ohno N, Shibata K, Tokoroyama T, Kubota T, Hirotsu K, Higuchi T. 1988. **Minor diterpenoids from *Portulaca* cv. Jewel.** *Phytochemistry* **27**: 2171–2173

Ohsaki A, Shibata K, Kubota T, Tokoroyama T. 1999. **Phylogenetic and chemotaxonomic significance of diterpenes in some *Portulaca* species (Portulacaceae).** *Biochemical Systematics and Ecology* **27**: 289–296

Ohsaki A, Shibata K, Tokoroyama T, Miura I. 1984. **Portulide, a clerodane diterpenoid from *Portulaca grandiflora* hook.** *Chemistry Letters* 1512–1522

Ohsaki A, Shibata K, Tokoroyama T, Kubota T, Naoki H. 1986. **Novel diterpenes with bicyclo[5.4.0]undecane skeleton from *Portulaca grandiflora* hook, possible linking intermediates in the biosynthesis of portulal.** *Chemistry Letters* 1585–1588

Ohsaki A, Shibata K, Tokoroyama T, Kubota T. 1987. **Structures of pilosanones A and B: novel diterpenoids with a bicyclo[5.4.0] undecane skeleton from *Portulaca pilosa* L.** *Journal of the Chemical Society, Chemical Communication* 151–152

Ornstein L and Davis B J. 1962. **Disc Electrophoresis**. Rochester, NY: Distillation Product Industries (Eastman Kodak Co.) [preprint]

Ozcan T. 2008. **Analysis of the total oil and fatty acid composition of seeds of some Boraginaceae taxa from Turkey.** *Plant Systematics and Evolution* **274**: 143–153

Paulis J W and Wall J S. 1977. **Comparison of the protein compositions of selected corns and their wild relatives teosinte and *Tripsacum*.** *Journal of Agriculture and Food Chemistry* **25**: 265–270

Piattelli M. 1981. **The betalains: structure, biosynthesis, and chemical taxonomy.** In *The Biochemistry of Plants: a comprehensive treatise*, edited by E E Conn, vol. 7, pp. 557–575. New York: Academic Press

Pusztai A. 1991. **Plant lectins**. Cambridge: Cambridge University Press. 263 pp.

Qiao X, Zhang Y T, Ye M, Wang B R, Han J, and Guo D. 2009. **Analysis of chemical constituents and taxonomic similarity of salvia species in China using LC/MS.** *Planta Medica* **75**: 1613–1617

Rao J V S and Rao R S S. 1993. **Phenolic compounds in the taxonomy of *Tephrosia* Pers. (Leguminosae).** *Feddes Repertorium* **104**(3–4): 245–250

Rao M V S, Sujatha D M, Rao Y S, Manga V. 1988. **Chemotaxonomic characters in twelve species of the genus *Pennisetum* (Poaceae).** *Proceedings of the Indian Academy of Science (Plant Science)* **98**(2): 111–120

Roberts E A H, Wight W, and Wood D J. 1958. **Paper chromatography as an aid to the taxonomy of the Thea Camellias.** *New Phytologist* **57**: 211–225

Rodman J E. 1976. **Differentiation and migration of Cakile (Cruciferae): seed glucosinolate evidence**. *Systematic Botany* **1**: 137–148

Rodman J E. 1978. **Glucosinolates: methods of analysis and some chemosystematic problems**. *Phytochemical Bulletin* **11**: 6–31

Rodman J E. 1980. **Population variation and hybridization in sea-rockets (Cakile, Cruciferae): seed glucosinolate characters**. *American Journal of Botany* **67**: 1145–1159

Ronsted N, Gobel E, Franzyk H, Jensen S R, Olsen C E. 2000. **Chemotaxonomy of *Plantago*. Iridoid glucosides and caffeoyl phenylethanoid glycosides**. *Phytochemistry* **55**: 337–348

Rosch P, Harz M, Schmitt M, Peschke K, Ronneberger O, Burkhardt H, Motzkus H, Lankers M, Hofer S, Thiele H, Popp J. 2005. **Chemotaxonomic identification of single bacteria by micro-Raman spectroscopy: application to clean-room-relevant biological contaminations**. *Applied and Environmental Microbiology* 1626–1637

Royer M, Herbette G, Eparvier V, Beauchêne J, Thibaut B, Stien D. 2010. **Secondary metabolites of *Bagassa guianensis* Aubl. wood: a study of the chemotaxonomy of the Moraceae family**. *Phytochemistry* **71**: 1708–1713

Rudloff E V. 1975. **Chemosystematic studies of the volatile oils of *Juniperus horizontalis*, *J. scopulorum* and *J. virginiana***. *Phytochemistry* **14**: 1319–1329

Salimpour F, Mazooji F, and Darzikolaei S A. 2011. **Chemotaxonomy of six *Salvia* species using essential oil composition markers**. *Journal of Medicinal Plants Research* **5**(9): 1795–1805

Sanchez-Yelamo M D and Martinez-Laborde J B. 1991. **Chemotaxonomic approach to *Diplotaxis muralis* (Cruciferae: Brassiceae) and related species**. *Biochemical Systematics and Ecology* **19**(6): 477–482

Santos L S, Catharino R R, Aguiar C L, Tsai S M, Eberlin M N. 2006. **Chemotaxonomic markers of organic, natural, and genetically modified soybeans detected by direct infusion electrospray ionization mass spectrometry**. *Journal of Radioanalytical and Nuclear Chemistry* **269**(2): 505–509

Sauer J. 1964. **Revision of *Canavalia***. *Brittonia* **16**: 106–181

Schimming T, Jenett-Siems K, Mann P, Tofern-Reblin B, Milson J, Johnson R W, Deroin T, Austin D F, Eich E. 2005. **Calystegines as chemotaxonomic markers in the Convolvulaceae**. *Phytochemistry* **66**: 469–480

Seigler D S, Pauli G F, Frohlich R, Wegelius E, Nahrstedt A, Glander K E, Ebinger J E. 2005. **Cyanogenic glycosides and menisdaurin from *Guazuma ulmifolia*, *Ostrya virginiana*, *Tiquilia plicata*, and *Tiquilia canescens***. *Phytochemistry* **66**: 1567–1580

Semikhov V F, Arefeva L P, Zolkin S Y, Novozhilova O A, Kostrikin D S. 2002. **Application of the serological method for evaluation of relations between gymnospermous and dicotyledonous plants**.

Biology Bulletin **29**(5): 437–446 [Translated from *Izvestiya Akademii Nauk, Seriya Biologicheskaya*. 2002. **5**: 541–551. Original Russian Text by Semikhov, Aref'eva, Zolkin, Novozhilova, Kostrikin.]

Shechter Y. 1975. **Biochemical systematic studies in *Sorghum bicolor*.** *Bulletin of the Torrey Botanical Club* **102**: 334–339

Shechter Y and de Wet J M J. 1975. **Comparative electrophoresis and isozyme analysis of seed protein from cultivated races of *Sorghum*.** *American Journal of Botany* **62**: 254–261

Smith P M. 1976. ***The chemotaxonomy of plants***. Bristol, England: J.W. Arrowsmith Ltd.

Spencer K C and Seigler D S. 1985. **Cyanogenic glycosides and the systematics of the Flacourtiaceae**. *Biochemical Systematics and Ecology* **13**(4): 421–431

Stafford H A. 1994. **Anthocyanins and betalains—evolution of the mutually exclusive pathways**. *Plant Science* **101**: 91–98

Strack D, Steglich W, and Wray V. 1993. **Betalains**. In *Methods in Plant Biochemistry: alkaloids and sulphur compounds*, edited by P M Dey and J B Harborne, vol. 8, pp. 421–450. London: Academic Press

Su B N, Cuendet M, Hawthorne M E, Kardono L B S, Riswan S F, Harry H S, Mehta R G, Pezzuto J M, Kinghorn A D. 2002. **Constituents of the bark and twigs of *Artocarpus dadah* with cyclooxygenase inhibitory activity**. *Journal of Natural Products* **65**: 163–169

Takasugi M, Nagao S, and Masamune T. 1979. **Structure of moracins E, F, G, and H, new phytoalexins from diseased mulberry**. *Tetrahedron Letters* **48**: 4675–4678

Takhtadzhyan A L. 1987. **Sistema magnoliofitov (System of magnoliophytes)**. Leningrad: Nauka

Takhtajan A. 1959. ***Die Evolution der Angiospermae***. Jena: Gustav Fischer

Takhtajan A. 1969. ***Flowering Plants, Origin and Dispersal***. Edinburgh: Oliver Boyd

Takhtajan A. 1980. **Outline of the classification of flowering plants (Magnoliophyta)**. *Botany Review* **46**: 225–359

Thorne R F. 1981. **Phytochemistry and angiosperm phylogeny. A summary statement**. In *Phytochemistry and Angiosperm Phylogeny*, edited by D A Young and D S Seigler, pp. 233–295. New York: Praeger

Thorne R F. 1983. **Proposed new realignments in the angiosperms**. *Nordic Journal of Botany* **3**: 85–117

Thorne R F. 1992. **An updated classification of the flowering plants**. *Aliso* **13**: 365–389

Triana J, Eiroa J L, López M, Ortega J J, González A, Bermejo J. 2001. **Sesquiterpene lactones from *Lugoa revoluta***. *Biochemical Systematics and Ecology* **29**: 869–871

Triana J, Eiroa J L, Ortega J J, León F, Brouard I, Hernández J C, Estévez F, Bermejo J. 2010. **Chemotaxonomy of *Gonospermum* and related genera**. *Phytochemistry* **71**: 627–634

Triana J, Eiroa J L, Ortega J J, León F, Brouard I, Torres F, Quintana J, Estévez F, Bermejo J. 2008. **Sesquiterpene lactones from *Gonospermum gomerae* and *G. fruticosum* and their cytotoxic activities**. *Journal of Natural Products* **71**: 2015–2020

Triana J, López M, Eiroa J L, González A, Bermejo J. 2000. **Sesquiterpene lactones and other constituents of *Gonospermum canariense***. *Biochemical Systematics and Ecology* **28**: 95–96

Triana J, López M, Rico M, González J F, Quintana J, Estévez F, León F, Bermejo J. 2003. **Sesquiterpenoid derivatives from *Gonospermum elegans* and their cytotoxic activity for HL-60 human promyelocytic cells**. *Journal of Natural Products* **66**: 943–948

Vaughan J G and Gordon E I. 1969. **Comparative serological studies of myrosinase from *Sinapis alba* and *Brassica juncea* seeds**. *Phytochemistry* **8**: 883–887

Verdcourt B. 2004. **The variation of *Sida rhombifolia* L. (Malvaceae) in East Africa**. *Kew Bulletin* **59**: 233–239

Wayman K A, de Lange P J, Larsen L, Sansom C E, Perry N B. 2010. **Chemotaxonomy of *Pseudowintera*: Sesquiterpene dialdehyde variants are species markers**. *Phytochemistry* **71**: 766–777

Wilkinson C F and Beard L B. 1972. **Electrophoretic identification of *Agrostis palustris* and *Poa pratensis* cultivars**. *Crop Science* **12**: 833–834

Zafar M, Ahmad M, Khan M A, Sultana S, Jan G, Ahmad F, Jabeen A, Shah G M, Shaheen S, Shah A, Nazir A, Marwat S K. 2011. **Chemotaxonomic clarification of pharmaceutically important species of *Cyperus* L**. *African Journal of Pharmacy and Pharmacology* **5**(1): 67–75

Zhou Y, Xu G, Choi F F K, Ding L, Han Q B, Song J Z, Qiao C F, Zhao Q, Xu H. 2009. **Qualitative and quantitative analysis of diterpenoids in *Salvia* species by liquid chromatography coupled with electrospray ionization quadrupole time-of-flight tandem mass spectrometry**. *Journal of Chromatography A* **1216**: 4847–4858

12

Cytotaxonomy and its Evolutionary Significance in the Evaluation of Orchidaceae and Cyperaceae

Prabha Sharma and P L Uniyal

INTRODUCTION

Plant cytology is a branch of botany that provides taxonomists with increasing number of facts for constructing a sound general classification. Chromosomal aspects have been important in evaluating relationship and deducing phylogenetic sequence in plants. Cytological research exposes the complexities and evolutionary potentialities of chromosomes. This leads to the discovery of new mechanisms for understanding how genes control chromosome and the patterns of evolution and variation of chromosomes. Cytological data also help in interpreting flora through analysis of the composition, endemism, conservation, and evolutionary status of plants. Further, the relationship with other floras and movement of flora, in relation to particular ecological, geographical, and geological compositions, can be understood. Cytological assessment enables botanists to fathom the level and pattern of genetic diversity. This is essential to (1) prevent further destruction of available genetic variability and potential existing in natural ecosystems owing to a lack of genuine understanding, (2) utilize the available genetic diversity for increased plant productivity, and (3) provide important characters for circumscription of species and distinction of closely related taxa.

Cytological Characters as a Tool in Taxonomy

Cytological studies are important for meeting the urgent need of scientific manipulation of the available genetic material containing novel genome complexes for meaningful conservation of floristic wealth and for solving taxonomic problems. Chromosome number, an important feature of cytotaxonomical analyses, can be a plesiomorphic characteristic of a large clade or a recurrent trait that arose independently in two or more clades. Chromosome number variation, base number, aneuploidy, paleopolyploidy, neopolyploidy, staining properties, m-chromosome, and secondary association are frequently used as taxonomic characters. The position of the centromere is a reliable feature of a chromosome structure and consequently makes a good taxonomic character. More detailed studies of meiotic behaviour can reveal, for example, the heterozygosity of some inversions. Such a feature may be consistent for a particular taxon, thus providing additional taxonomic evidence. Cytological data are sometimes considered to be of higher significance than other taxonomic evidence; therefore, use of such data in cytotaxonomy and karyotype evolution deserves much attention.

Cytotaxonomy is the application of cytological data to taxonomy. It is worth emphasizing that cytology is the study of animal and plant cells although nowadays it is considered the study of only chromosomes or at most of the nucleus and its structure. Size, shape, and behaviour of chromosomes throw more light on a taxonomic problem than their number alone. When all determinable characters derived from morphology, anatomy, cytology, genetics, phytogeography, ecology, and even pure physiology are used, these result in an increasing amount of synthetic taxonomy and a greatly improved classification. A large number of cytological records are based on the examination of single plants or plants from one locality, which is very often a botanical garden. A taxonomist suspects that there may often be cytological variation corresponding to the variation in other characteristics of species. It may also be noted that, to a systematist, the value of cytological research is greatly enhanced when it is combined with genetical experiments. There is another important matter that taxonomists use to check the determinations of the samples, rectify mistakes, and improve classifications. Unless cytologists preserve adequate specimens of the actual plants they have examined cytologically, it is often not possible to confirm (or correct with certainty) the determinations they have made or accepted. Mismatch of cytological findings with previously accepted taxonomy may lead to several possibilities: the old taxonomy may be wrong, the new cytology may be wrong, or the cytological findings may have been recorded under a wrong plant name. However, the taxonomists have been relieved by the declaration of a karyologist that the fictions, errors, and half-truths of taxonomy can be quickly remedied by a study of the chromosomes (Darlington 1965).

A REVIEW OF CYTOLOGICAL STUDIES ON AND EVOLUTIONARY PATTERN IN ORCHIDS

The pantropical orchid group is basically formed by the ancient subfamily Vandoideae (Dressler 1981) and is characterized by having two pollinia whose texture varies from firm to hard (Dressler 1993). Morphologically, the taxa show a great variation mainly in the subtribes Cyrtopodiinae (*Cymbidium*) and Oncidiinae (*Odontoglossum*, *Miltonia*, and *Oncidium*), which have more widely been studied cytologically (Sinotô 1962; Charanasri, Kamemoto, and Takashita 1973). The group displays the highest variation in chromosome number of all orchids: $2n = 10$ in *Psygmorchis pusilla* (Dodson 1957) to $2n = 168$ in a horticultural variety of *Oncidium varicosum* (Sinotô 1962). Chromosome number variation in orchids as a whole is intriguing because most of the genera have high ploidy levels and variable base numbers (Goldblatt 1980; Ehrendorfer 1980). The base number of the family is still uncertain, which makes it difficult to estimate species ploidy level and to understand the karyological evolution of the family. Raven (1975) reviewed the angiosperm's base number and suggested the need of more studies for consideration of a base number for Orchidaceae. Félix and Guerra (2000) investigated the chromosome number and interphase nuclear types in the 44 species of 20 genera of Cymbidioid orchids occurring in Brazil and reviewed the variability in chromosome number within the phylad along with its compatibility with the taxonomic treatment proposed by Dressler (1993).

Chromosome Numbers in Orchidaceae

The chromosome numbers reported by Félix and Guerra (2000) were as follows: $2n = 54$ (subtribe Eulophiinae), $2n = 44, 46, 92$ (subtribe Cyrtopodiinae), $2n = 54$, ca. 108 (subtribe Catasetinae), $2n = 52$, ca. 96 (subtribe Zygopetalinae), $2n = 40, 80$ (subtribe Lycastinae), $2n = 40, 42$ (subtribe Maxillariinae), $2n = 40$ (subtribe Stanhopeinae), $2n = 56$ (subtribe Ornithocephalinae), and $2n = 12, 20, 30, 36, 42, 44, 56, 112$, ca. 168 (subtribe Oncidiinae). They observed that the interphase nuclei varied widely from simple to complex chromocenter types, with no apparent cytotaxonomic value. The terrestrial and lithophytic species of *Catasetum* and *Oncidium* show higher ploidy levels than the epiphytic species, suggesting a higher adaptability of the polyploids to those habitats. They considered the primary base number $x = 7$, which seems to be associated with the haploid chromosome numbers of most Cymbidioid groups, although $n = 7$ was observed only in two extant genera of Oncidiinae.

In some of the taxa, analysis of the chromatin organization in interphase nuclei has helped understand the genomic diversification, independent of number and chromosome morphology (Morawetz 1986; Röser 1994). Generally, a single interphase nuclear type is conserved throughout a genus or a higher taxonomic category, as in Rutaceae, subfamily Aurantioideae (Guerra 1987). In orchids, Tanaka (1971) described five different types of interphase nuclei on the basis of his observations in 115 species of 52 genera. However, occurrence of more than one interphase nuclear type in a single genus has been described, for example, in *Habenaria* (Félix and Guerra 1998) and *Platanthera* (Yokota 1990). *Catasetum* and *Cyrtopodium* display uniformity in chromosome numbers and morphology; however, two different types of interphase nuclei are reported (Félix and Guerra 2000). Still, the presence of simple chromocenter nuclei in nearly all Oncidiinae species seems to reflect the uniformity of this group (Chase 1986).

To understand the chromosome numeric variation of the phylad, Félix and Guerra (1998) made a complete review of the recorded chromosome numbers, based on the review of Tanaka and Kamemoto (1984), followed by the chromosome number indexes published by Fedorov (1969), Moore (1973, 1974, 1977), Goldblatt (1984, 1985, 1988), and Goldblatt and Johnson (1990, 1991, 1994, 1996). The base number $x = 7$ is identified as one of the haploid numbers actually found in the genus, which explains the chromosome number variation found in the taxon and more related genera (Guerra 2000). It is possible to indicate the number that most probably represents the original haploid complement for each genus. The most frequent chromosome number has been accepted as an indicator of the base number only when it was well represented in the related genera.

Karyological Evolution in Orchids

An array of variable chromosome numbers is observed in orchids, and it is very difficult to relate it to a single base number. Cytotaxonomical analysis can be better understood in genera with great cytological diversity, which often correspond to the genera with the highest number of species in the tribe or family, such as *Boronia* in the tribe Boroniae, Rutaceae (Stace 1995), *Carex* in Cyperaceae (Luceño 1994), and *Passiflora* in Passifloraceae (Snow and MacDougal 1993). *Maxillaria*, one of the largest genera, is cytologically very poorly studied (six species), whereas *Oncidium* is most extensively studied (117 species). Chromosome number variability in *Oncidium* is also quite representative of the group. The known haploid numbers are $n = 13, 14, 15, 18, 19, 20, 21, 22, 25, 26, 27, 28, 29, 30, 36, 42, 56, 63, 70, 84$. This variation is clearly dominated by the polyploid series $n = 14, 21, 28, 42, 56, 63, 70, 84$. Most of the studied populations (64.8%) are reported to be ortoploid, with $n = 14, 21$, or 28, of which 46% display $n = 28$. These data strongly suggest $x_1 = 7$ as the primary base number for the genus; however, this number is hypothetical, because no species of the genus with $n = 7$ is known so far (Félix and Guerra 2000, 2005). The diploid *Oncidium* species either have not yet been reported or have probably become extinct, because the hexaploid $n = 21$ could only arise from a cross between tetraploids ($n = 14$) and putative diploids ($n = 7$) followed by polyploidization (Harlan and de Wet 1975). Therefore, if the genus originated from a tetraploid lineage, the hexaploid species could not belong to this same lineage and the genus would be artificial. The same might have occurred in *Rodriguezia*, with $2n = 28$ (Sinotô 1962) and $2n = 42$.

The variation of chromosome numbers in the subtribe Oncidiinae of Orchidaceae seems to be very similar to that of the genus *Oncidium*, with the numbers $n = 21$ and $n = 28$ prevailing. This suggests that the other genera have a common ancestor with *Oncidium*. The group also displays the smallest chromosome numbers of the family: $n = 7$ in *Lockartia* and $n = 5, 6$, and 7 in *Psygmorchis*. In three populations of *P. pusilla* studied by Félix and Guerra (1998), $2n = 12$ and $n = 6$ were most commonly found. The only *Lockartia* species analysed by Félix and Guerra (1998) exhibited $2n = 56$, which coincides with the previous reports of Charanasri and Kamemoto (1975) for *L. micrantha*. These data support the inclusion of *Lockartia* in Oncidiinae, in opposition to the assumption made by Freudenstein and Rasmussen (1999) based on the absence of leaf articulation in this genus.

In view of the presence of prominent polyploidy series reported in *Oncidium* and Oncidiinae, $x = 7$ is considered to be the primary base number, as suggested by Charanasri and Kamemoto (1975). Most Oncidiinae genera would have hexaploid (*Comparettia* and *Notylia*) or octoploid origin (*Aspasia*, *Gomesa*, *Miltonia*, *Sigmatostalix*, and *Trichopilia*). As polyploidy is quite a recurrent phenomenon in the evolution of angiosperms (Soltis and Soltis 1995; Leitch and Bennett 1997), it is very probable that higher polyploids arose de novo many times in a number of other genera.

Many authors (Chase 1986; Chase and Pippen 1988; Chase and Olmstead 1988; Chase and Palmer 1992) observed that the most primitive representatives of the subtribe had higher chromosome numbers, whereas *Psygmorchis* and *Lockartia*, with more derived morphological characters, displayed the lowest chromosome numbers. This led them to conclude that *Oncidium* and some Oncidiinae have the original chromosome numbers (x = 28, 30), which, through successive dysploidy, originated the low-numbered species with n = 7–5 (Félix and Guerra 2000, 2005). This conclusion was supported by isoenzymatic evidence from representatives of this group, which almost always exhibited a single locus for each isozyme (Chase and Olmstead 1988), like dysploids. However, the isoenzymatic analysis of several other definitely polyploid taxa also displayed a similar pattern (Haufler 1987), suggesting that it is not an accurate indicator of ploidy level (Soltis, Doyle, and Soltis 1992).

Polyploids often have very slow evolution rates, and they may conserve more primitive characters (Stebbins 1971), as observed in many present-day polyploids of Oncidiinae and other groups (Guerra 2000). The same interpretation can also be applied to other primitive and highly polyploid genera of orchids, such as *Neuwiedia* and *Apostasia* (Okada 1988). Diploids and recent polyploids exhibit more derived characters in different parallel evolutionary lines, as *Dipteranthus* in Ornithocephalinae (Williams, Toscano de Brito, Harborne, *et al.* 1994) and *Lockartia* in Oncidiinae (Chase 1986; Freudenstein and Rasmussen 1999).

Cytological analysis of Oncidiinae helped understand the unrelated chromosome numbers of the remaining taxa of tribe Maxillarieae. Lycastinae, Maxillariinae, and Stanhopeinae appear to be based on n = 20, which may have derived by descending dysploidy from a hexaploid lineage with n = 21. Ornithocephalinae, karyologically known only by two reports (Félix and Guerra 2000) for the genus *Dipteranthus* with $2n$ = 56, coincides with the base number of most Oncidiinae genera, supporting its affinity with that subtribe (Chase and Pippen 1988). Only the subtribe Zygopetalinae seems to be more diversified in the hexaploid–octoploid level (n = 26, 24/48, 23).

Relationship Among the Tribes of Orchidaceae

Félix and Guerra (2000) suggested the existence of three groups: a larger group (Oncidiinae and Ornithocephalinae) evolved from the base number x_1 = 7 and followed by successive cycles of polyploidy and secondary dysploidy; a second group (Lycastinae, Maxillariinae, and Stanhopeinae) made up of hexaploids with n = 21, which by dysploid reduction led to a secondary base number x_2 = 20; and a third group (Zygopetalinae) with a putative base number x_2 = 24 or 26 and no clear relationship with the polyploid series based on x_1 = 7. Morphologically, Stanhopeinae and Lycastinae share in common the presence of plicate leaves and elaborated pollination mechanisms (van der Pijl and Dodson 1966), whereas Oncidiinae and Ornithocephalinae have in common the absence of "unken glandular trichomes", found in Maxillariinae, Lycastinae, and Stanhopeinae (Toscano de Brito 1998).

In the other tribes of Orchidaceae, the predominant chromosome numbers are as follows: $n = 15, 21$ in Malaxideae, $n = 14, 21$ in Calypsoeae, and $n = 27$ in Cymbidieae. In Malaxideae, although $n = 15$ is a very common number, $n = 14$ has also been found at least in *Liparis* and *Malaxis*. The cytotaxonomic interpretation is difficult in *Liparis* because of the occurrence of secondary polyploid series based on $x = 10$ ($n = 10, 20, 40$). Existence of high frequency of $n = 15$ in the three genera of Malaxideae and a very rare haploid number in other tribes is noteworthy. In Calypsoeae, $n = 14$ has only been found in *Calypso*, with $n = 21$ prevailing in the other genera. These numbers probably have evolutionary history similar to that of *Oncidium*, and they are either lost or still awaiting discovery with $n = 7$.

High diversity of chromosome numbers is observed in the tribe Cymbidieae, which also displays a high morphological variability (Freudenstein and Rasmussen 1999). The main haploid numbers are $n = 27$ and 23 reported in the subtribe Eulophiinae, $n = 21$ and 20, $n = 28$ and 27 in the subtribe Cyrtopodiinae, $n = 20$ in a single species of Acriopsidinae, and $n = 27$ and 34 in Catasetinae. The subtribe Eulophiinae is known cytologically by *Eulophia*, which also displays high variation in chromosome numbers. In this genus, a polyploid series based on $x = 7$ ($n = 14, 21, 28, 35, 56$) is also reported, with the octoploid level ($n = 28, 27$) being strongly dominant. In *Oeceoclades*, the only two species analysed so far are also octoploids, where in *Dipodium* the only record ($n = 23$) is probably of a hexaploid. Poggio, Naranjo, and Jones (1986), while analysing the meiotic behaviour of several species of *Eulophia* with $n = 21$, observed the frequent secondary association of bivalent three-to-three, suggesting that it would be a remaining homoeology of the hexaploid condition with $x = 7$.

In comparison to other large families of angiosperms, such as Poaceae (Hunziker and Stebbins 1986) or Asteraceae (Watanabe, King, Yahara, *et al*. 1995), Orchidaceae stands out for the scarcity of representative diploids. These data suggest that the family Orchidaceae may be older than is generally considered (Garay 1972), thus there having been sufficient time for diploids to be widely substituted by polyploids.

Chromosome Numbers in Relation to Habitat

In plants, the conquest of new habitats is often related to the occurrence of polyploidy (Stebbins 1966). Frequently, polyploid races are associated with extreme environmental conditions (Ehrendorfer 1970; de Wet 1986; Uniyal 1998, 2007). In the orchid *Anacamptis pyramidalis* (L.) Rich., for example, the polyploid cytotypes are more adapted to regions with geologic formation different from those of diploid populations occurring in the same regions (Del Prete, Mazzola, and Miceli 1991).

Although the orchids constitute a paleopolyploid group (Jones 1974; Ehrendorfer 1980), the reversion to terrestrial habitat of typically epiphytic species is apparently acquired more easily when an increase in ploidy level occurs. In the genus *Pleione* (Orchidaceae), for instance, all the epiphytics have $2n = 40$, while about 50% of the terrestrial or lithophytic species are higher polyploids (Stergianou 1989). In the genus *Laelia*, subgenus *Cyrtolaelia*, the lithophytic species are generally allopolyploids (Blumenschein 1960). Félix and Guerra (2000, 2005) observed that *Catasetum* and *Oncidium* species, with lithophytic or terrestrial habitats, presented higher ploidy levels than epiphytic species. In *Oncidium*, *O.* aff. *flexuosum* with $2n = $ ca. 168, and lithophytic or terrestrial habitat is morphologically closely related to *O. flexuosum* with epiphytic habitat and chromosome number $2n = 56$. The same occurs in *O. blanchetii* and *O. varicosum* ($2n = 112$). Likewise, *Cyrtopodium blanchetii* ($2n = 92$), with underground pseudobulbs, is tetraploid in relation to the other species with aerial pseudobulbs. Similarly, *Catasetum discolor*, with terrestrial habitat, exhibited $2n = $ ca. 108, while the other species had $2n = 54$. On the other hand, the population of *Trigonidium acuminatum* collected in a lithophytic incidental habitat, under strong anthropic pressure, presented the same ploidy level as *T. obtusum* ($2n = 40$), with epiphytic habitat.

CYTOLOGY IN RELATION TO TAXONOMY OF CYPERACEAE

Cyperaceae is a widely distributed family, comprising nearly 5400 species in 103 genera. The family is known to show many cytological peculiarities, including psuedomonad pollen grain development (Tanaka 1941), diffuse kinetochores (Heilborn 1924), post-reductional meiosis (Wahl 1940), and chromosome fission and fusion, often referred to as agmatoploidy (Cayouette and Morisset 1986; Luceño, Vanzela, and Guerra 1998;). Roalson, McCubhin and Whitkus (2007) and Roalson (2008) reviewed the chromosome variation and potential processes of chromosome evolution in Cyperaceae. The most common count for the family is reported to be $2n = 30$, with high frequency of 28 and 29, as well as dominant chromosome numbers 10, 20, and 35. Counts have been made for all numbers between $n = 2$ and 60, and the highest count in the family is 114. The genus *Carex* has been studied widely. The aneuploid series peaks in frequency at $2n = 28$ and 30, with a second major peak at 35. *Cyperus* show an array of chromosome number distribution, and the numbers 8, 18, 32, 36, 40, 48, 52, 54, and 56 are most frequent. *Carex* shows low occurrence of polyploidy and it shows 18, 36, and 54 as most frequent numbers. *Fuirena, Isolepis, Lipocarpha, Bolboschoenus, Eriophorum, Scirpus*, and *Trichophorum* present an aneuploid series without clear polyploid chromosome numbers. *Schoenoplectus* has 19, 21, and 39 as dominant chromosome numbers. As the variation in the genus is largely distributed around 20 and 40, a case might be made for polyploidy with subsequent aneuploidy. *Eleocharis* (5, 10, 20), *Bulbostylis* (5, 10, 15, 30), and *Fimbristylis* (5, 10) show some aneuploid variation. Hakansson (1958) reported that at least within *Eleocharis*, non-localized centromeres are present, and qualitative aneuploidy such as agmatoploidy is contributing to the variation of chromosome number in some species.

In the family Cyperaceae, cytological studies of a number of species of *Carex, Cyperus, Eleocharis, Fimbristylis*, and *Scirpus* have revealed that while euploidy and possibly gene mutation have played the main role in the speciation in the genus *Fimbristylis* (Rath and Patnaik 1974a, 1974b), aneuploidy seems to be prevalent in genera such as *Cyperus* (Sanyal 1972), *Carex* (Tanaka 1948), and *Scirpus* (Sanyal and Sharma 1972). Meiotic analysis of the three species of *Fuirena* revealed haploid chromosome numbers ranging from $n = \sim18$ to 26 (Dujardin 1987). *Fuirena uncinata* Kunth showed $n = 18$. Dujardin (1987) observed a disparity in the size of the bivalents; he also observed that at diakinesis, very often two bivalents were found associated with the nucleolus in *F. uncinata*. The haploid chromosome number of *F. ciliaris* (L.) Roxb. was reported to be $n = 19$, and the cytological behaviour of this species was found to be similar to that of *F. uncinata* Kunth. Chromosome numbers of *F. umbellata* Rottb. revealed a haploid number of $n = \sim26$ with smaller size of the bivalents. Dujardin (1987) concluded that *F. uncinata* Kunth and *F. ciliaris* (L.) Roxb. are closely related in their sympatric growth as well as in general morphology (as suggested earlier by Hooker 1894), whereas *F. umbellata* Rottb. is distinct from them in its ecological preference and general morphology. Chromosome examination of these species also supported the closer cytotaxonomic relationship between *F. uncinata* Kunth and *F. ciliaris* (L.) Roxb. rather than with *F. umbellata* Rottb. The aneuploid variation in the chromosome numbers ($n = 18, 19, 26$) in the three species of *Fuirena* indicates that the evolutionary trend of the genus is similar to that of other members of the family. Summarizing the work of Heilborn (1939) and Lsve, Lsve, and Raymond (1957), Faulkner (1972) stated that comparing related species with close chromosome numbers, it has been found that the species with the lower number usually have more chromosomes of a larger size. It is generally suggested in polyploids or hybrid complexes that plants having low chromosome numbers are more primitive than plants having high chromosome numbers. The patterns of chromosome number distribution within lineages suggest that while aneuploidy may dominate the overall pattern of chromosome number distribution, there is evidence for both aneuploidy and polyploid chromosome number changes within many genera.

FUTURE PROSPECTS AND CONCLUSIONS

In the field of plant systematics and floristics, with the availability of more chromosome data, cytogenetic mechanism involved in the evolution and delimitation of taxa became clearer. Chromosomal features are regarded as decision-making characters in the study of phylogenetic affinities and evolutionary development and as indicators of appropriate classifications of plant groups. Analysis of chromosome numbers has been used to evaluate evolutionary and taxonomic relationships in diverse groups of plants. Chromosome numbers indicate the occurrence of polyploidy and reflect differences in the basic chromosome numbers among plants, which may be reflected in their treatments in floras. Studies of chromosomal variations in Orchidaceae have helped understand better the taxonomy of this family at many hierarchical levels. The basic number corresponds to the haploid number encountered in a given taxon that explains the variation in chromosome numbers seen in that taxon and related taxa (Guerra 2000). As is generally understood, speciation can be effected by polyploidy, hybridization, chromosome repatterning, and gene mutation.

Since changes in the number of chromosomes between species are not so prevalent in orchids, changes within the complement are assumed to be responsible for directing the course of evolution in the family. This could be brought about by hybridization, cryptic structural changes in the chromosomes, and gene mutations. Owing to certain morphological, genetical, and embryological peculiarities in orchids, hybridization is easily facilitated and the accompanying meiotic irregularities and consequent sterility of gametes are avoided to a considerable extent than in other plant families; this has resulted in hybridization becoming a continuing and major factor in the process of evolution in Orchidaceae, providing a very speedy form of evolving new species. Taken as a whole, the most prevalent form of speciation in orchids appears to be the one that has occurred without any change in the number of chromosomes. Euploidy appears to be the least favoured form of speciation in orchids, less than one-third of the genera investigated having produced species by this method. The incidence of aneuploidy, which seems to have played the second important role in speciation in Orchidaceae, is varied in the different tribes. Aneuploidy may result from occasional non-disjunction of chromosomes during anaphase or by the partial return to the original chromosome number in triploids and pentaploids. Dressler (1981) has pointed out that as a family undergoing rapid evolution, Orchidaceae provides excellent material for the study of evolution. The chromosome number 20 is more generalized and ancient of the two, being prevalent in both the hemispheres, and the number 19 is probably derived from it, which is frequently reported in advanced Orchid tribes.

Speciation can be accomplished through different mechanisms. Some of these imply an alteration of chromosome numbers between genera and closely related species, which can be due to two main processes: polyploidy and disploidy (the increase or decrease of one, or few, chromosomes). Polyploidy is one of the most frequent and important numeric chromosome alterations occurring in the evolution of plants (Stebbins 1971), affecting from 30% to 80% of plant species. Speciation mechanisms involve variation in chromosome number and may be investigated not only by counting the frequency of different chromosome numbers in a flora, but also by the analysis of the regularity of the microsporogenesis that occurs during the process of sexual reproduction. Interspecific hybrids and unbalanced polyploids, especially those with odd ploidy, generally present a high frequency of meiotic abnormalities (Stace 1995). This high frequency of meiotic regularity may indicate that most of the numeric alterations and hybridizations that could have occurred are already stabilized, implying that the process of sexual reproduction is effective. To confirm that one flora is derived from another flora by higher frequency of high chromosome numbers, a great amount of karyological and phylogenetic data on the species of different communities is required.

The experimental studies to elucidate genetical mechanisms involved in controlling the chromosomal system and speciation are lacking. The application of recently evolved techniques (chromosome banding, cytochemistry, gene ecology, and protoplast cultures) to the taxa of uncertain affinity shall answer many baffling questions of systematic treatments and interrelationships of the taxa of various ranks.

BIBLIOGRAPHY

Blumenschein A. 1960. Cayouette J and Morisset P. 1986. **Chromosome studies on Carex palacea Wahl., C nigra (L.) Reichard and C. aquatilis Wahl. in northeastern North America**. *Cytologia* **51**: 857–883 Estudo sobre a evolução no subgênero Cyrtolaelia **(Orchidaceae)**. *Livre-Docência thesis*, Escola Superior de Agricultura "Luis de Queiroz", Universidade de São Paulo, Piracicaba.

Cayouette J and Morisset P. 1986. **Chromosome studies on *Carex* palacea Wahl., C nigra (L.) Reichard and C. aquatilis Wahl. in northeastern North America**. *Cytologia* **51**: 857–883

Charanasri U and Kamemoto H. 1975. **Additional chromosome numbers in *Oncidium* and allied genera**. *American Orchid Society Bulletin* **44**: 686–691

Charanasri U, Kamemoto H, and Takashita M. 1973. **Chromosome numbers in the genus *Oncidium* and some allied genera**. *American Orchid Society Bulletin.* **42**: 518–524

Chase M W. 1986. **A reappraisal of the oncidioid orchids**. *Systematic Botany* **11**: 477–491

Chase M W and Olmstead R G. 1988. **Isoenzyme number in subtribe Oncidiinae (Orchidaceae): an evaluation of polyploidy**. *American Journal of Botany* **75**: 1080–1085

Chase M W and Palmer J D. 1992. **Floral morphology and chromosome number in subtribe Oncidiinae (Orchidaceae): evolutionary insights from a phylogenetic analysis of chloroplast DNA restriction site variation**. In *Molecular Systematics of Plants*, edited by P S Soltis, D E Soltis, and J J Doyle, pp. 324–332. New York: Chapman and Hall

Chase M W and Pippen J. 1988. **Seed morphology in the Oncidiinae and related subtribes (Orchidaceae)**. *Systematic Botany* **13**: 313–323

Darlington C D. 1965. ***Recent Advances in Cytology***. London: J A Churchill

de Wet J M J. 1986. **Hybridization and polyploidy in Poaceae**. In *Grass: systematics and evolution*, edited by T Soderstron, K W Hilu, C S Campbell, and M E Barkworth, pp. 188–194. London: Smithsonian Institution Press

Del Prete C, Mazzola P, and Miceli P. 1991. **Karyological differentiation and speciation in C. Mediterranean *Anacamptis* (Orchidaceae)**. *Plant Systematics and Evolution* **174**: 115–123

Dodson C H. 1957. **Chromosome number in *Oncidium* and allied genera**. *American Orchid Society Bulletin* **26**: 323–330

Dressler R L. 1981. *The orchids: natural history and classification*. Massachusetts: Harvard University Press

Dressler R L. 1993. *Phylogeny and Classification of the Orchid Family*. Portland: Dioscorides Press

Ehrendorfer F. 1970. **Evolutionary pattern and strategies in seed plants**. *Taxon* **19**: 185–195

Ehrendorfer F. 1980. **Polyploidy and distribution**. In *Polyploidy: biological relevance*, edited by W H Lewis, pp. 45–60. New York: Plenum Press

Faulkner J S. 1972. **Chromosome studies on *Carex* section *Acutae* in north-west Europe**. *Botanical Journal of the Linnean Society* **65**: 271–301

Fedorov A M A (ed). 1969. ***Chromosome Number of Flowering Plants***. Leningrad: Komarov Botanical Institute

Félix L P and Guerra M. 1998. **Cytological studies on species of *Habenaria* Willd. (Orchidaceae-Orchidoideae) occurring in the Northeast of Brazil**. *Lindleyana* **13**: 224–230

Félix L P and Guerra M. 2000. **Cytotogenetics and cytotaxonomy of some Brazilian species of *Cymbidiod* orchids**. *Genetics and Molecular Biology* **23**: 1–27

Félix L P and Guerra M. 2005. **Basic chromosome numbers of terrestrial orchids**. *Plant Systematics and Evolution* **254**: 131–148

Freudenstein J V and Rasmussen F N. 1999. **What does morphology tell us about orchid relationships?—a cladistic analysis**. *American Journal of Botany* **86**: 225–248

Garay L A. 1972. **On the origin of the Orchidaceae II**. *Journal of Arnold Arboretum* **53**: 202–215

Goldblatt P. 1980. **Polyploidy in angiosperms: monocotyledons**. In *Polyploidy: biological relevance*, edited by W H Lewis, pp. 219–232. New York: Plenum Press

Goldblatt P (ed). 1984. ***Index to Plant Chromosome Numbers 1979–1981***. St. Louis: Missouri Botanical Garden

Goldblatt P (ed). 1985. ***Index to Plant Chromosome Numbers 1982–1983***. St. Louis: Missouri Botanical Garden

Goldblatt P (ed). 1988. ***Index to Plant Chromosome Numbers 1984–1985***. St. Louis: Missouri Botanical Garden

Goldblatt P and Johnson D E (eds). 1990. ***Index to Plant Chromosome Numbers 1986–1987***. St Louis: Missouri Botanical Garden

Goldblatt P and Johnson D E (eds). 1991. ***Index to Plant Chromosome Numbers 1988–1989***. St Louis: Missouri Botanical Garden

Goldblatt P and Johnson D E (eds). 1994. ***Index to Plant Chromosome Numbers 1990–1991***. St Louis: Missouri Botanical Garden

Goldblatt P and Johnson D E (eds). 1996. ***Index to Plant Chromosome Numbers 1992–1993***. St Louis: Missouri Botanical Garden

Guerra M. 1987. **Cytogenetics of Rutaceae IV. Structure and systematic significance of interphase nuclei**. *Cytologia* **53**: 213–222

Guerra M. 2000. **Chromosome number variation and evolution in monocots**. In *Monocots II: systematics and evolution*, edited by K L Wilson and D A Morrison, pp. 127–136. Melbourne: CSIRO Publishing

Hakansson A. 1958. **Holocentric chromosomes in *Eleocharis***. *Hereditas* **44**: 531–540

Harlan J R and de Wet J M R. 1975. **On Ö Winge and a prayer: the origins of polyploidy**. *Botanical Review* **41**: 361–390

Haufler C H. 1987. **Electrophoresis is modifying our concepts of evolution in homosporous pteridophytes**. *American Journal of Botany* **74**: 953–966

Heilborn O. 1924. **Chromosome numbers and dimensions, species formation and phylogeny in the genus *Carex***. *Hereditas* **5**: 129–216

Heilborn O. 1939. **Chromosome studies in Cyperaceae III and IV**. *Hereditas* **25**: 224–240

Hooker J D. 1894. *Flora of British India,* vol 6, pp. 665–667. London: L. Reeve and Co. Ltd.

Hunziker J H and Stebbins G L. 1986. **Chromosomal evolution in the Gramineae**. In *Grass: systematics and evolution,* edited by T Soderstron, K W Hilu, C S Campbell, and M E Barkworth, pp. 179–187. London: Smithsonian Institution Press

Jones K. 1974. **Cytology and the study of orchids**. In *The orchids: scientific studies,* edited by C L Withner, pp. 383–389. New York: John Willey and Sons

Leitch I J and Bennett M D. 1997. **Polyploidy in angiosperms**. *Trends in Plant Science* **2**: 470–476

Lsve A, Lsve D, and Raymond M. 1957. **Cytotaxonomy of *Carex* section *Capillares***. *Canadian Journal of Botany* **35**: 715–761

Luceño M. 1994. **Cytotaxonomic studies in Iberian, Balearic, North African, and Macaronesian species of *Carex* Cyperaceae. II**. *Canadian Journal of Botany* **72**: 587–596

Luceño M, Vanzela A L L, and Guerra M. 1998. **Cytotaxonomic studies in Brazilian *Rhynchospora* (Cyperaceae), a genus exhibiting holocentric chromosomes**. *Canadian Journal of Botany* **76**: 440–449

Moore R J (ed). 1973.. **Index to plant chromosome number 1967–1971**. *Regnum Vegetabile* **90**: 1–539

Moore R J (ed). 1974. **Index to plant chromosome number 1972**. *Regnum Vegetabile* **91**: 1–108

Moore R J (ed). 1977. **Index to plant chromosome number 1973–1974**. *Regnum Vegetabile* **96**: 1–157

Morawetz W. 1986. **Remarks on karyological differentiation patterns in tropical wood plants**. *Plant Systematics and Evolution* **152**: 49–100

Nijalingappa B H M. 1977. **Cytological studies in *Fuirena* Rottb (Cyeraceae)**. *Current Science* **4**: 121–122

Okada H. 1988. **Karyomorphological observations of *Apostasia nuda* and *Neuwiedia veratifolia* (Apostasioideae-Orchidaceae)**. *Journal of Japanese Botany* **3**: 344–350

Poggio L, Naranjo C A, and Jones K. 1986. **The chromosomes of orchids IX. *Eulophia***. *Kew Bulletin* **41**: 45–49

Rath S P and Patnaik B N. 1974a. **A note on the cytotaxonomy of east Indian species of the genus *Fuirena* Rotth**. *The Journal of Plant Research* **87**: 333–336

Rath S P and Patnaik B N. 1974b. **Cytological studies in Cyperaceae with special reference to its taxonomy-I**. *Cytologia* **39**: 341–352

Raven P H. 1975. **The bases of angiosperm phylogeny: cytology**. *Annals of the Missouri Botanical Garden* **62**: 724–764

Roalson E H. 2008. **A synopsis on chromosome number variation in the Cyperaceae**. *Botanical Review* **74**: 209–393

Roalson E H, McCubbin A G, and Whitkus R. 2007. **Chromosome evolution in the Cyperales**. In *Monocots: comparative biology and evolution*

(Poales), edited by J T Columbus, E A Friar, J M Porter, L M Prince, and M G Simpson. *Aliso* **23**: 62–71

Röser M. 1994. **Pathways of karyological differentiation in palms (Arecaceae)**. *Plant Systematics and Evolution* **189**: 83–122

Sanyal B N. 1972. **Cytological studies on Indian Cyperaceae 11. Tribe Cyperae**. *Cytologia* **37**: 33–42

Sanyal B N and Sharma A. 1972. **Cytological studies in Indian Cyperaceae 1. Tribe Seirpeae**. *Cytologia* **37**: 13–32

Sinotô Y. 1962. **Chromosome numbers in *Oncidium* Alliance**. *Cytologia* **27**: 306–313

Snow N and MacDougal J M. 1993. **New chromosome reports in *Passiflora* (Passifloraceae)**. *Systematic Botany* **18**: 261–273

Soltis D E and Soltis P S. 1995. **The dynamic nature of polyploid genomes**. *Proceedings of the National Academy of Sciences* **92**: 8089–8091

Soltis P S, Doyle J J, and Soltis P E. 1992. **Molecular data and polyploid evolution in plants**. In *Molecular Systematics of Plants*, edited by P S Soltis, D E Soltis, and J J Doyle, pp. 177–201. New York: Chapman and Hall

Stace H M. 1995. **Primitive and advanced character states for chromosome number in Gondwanan angiosperm families of Australia, especially Rutaceae and Proteaceae**. In *Kew Chromosome Conference IV*, edited by P E Brandham and M D Bennett, pp. 223–232. Kew: Royal Botanical Gardens

Stebbins G L. 1966. **Chromosomal variation and evolution**. *Science* **152**: 1462–1469

Stebbins G L. 1971. *Chromosomal Evolution in Higher Plants*. London: Edward Arnold

Stergianou K K. 1989. **Habitat differentiation and chromosome evolution in *Pleione* (Orchidaceae)**. *Plant Systematics and Evolution* **166**: 253–264

Tanaka N. 1941. **Chromosome studies in the Cyperaceae. XII. Pollen development in five genera with special reference to *Rhynchospora***. *Botanical Magazine (Tokyo)* **55**: 55–67

Tanaka N. 1948. **The problem of aneuploidy (chromosome studies in Cyperaceae with special reference to the problem of aneuploidy)**. *Biological Contributions in Japan* **4**: 1–327

Tanaka R. 1971. **Types of nuclei in Orchidaceae**. *Botanical Magazine (Tokyo)* **84**: 118–122

Tanaka R and Kamemoto H. 1984. **Chromosomes in orchids: counting and numbers**. In *Orchid Biology: reviews and perspectives III*, edited by J Arditti, pp. 324–410. Ithaca: Cornell University Press

Toscano de Brito A L. 1998. **Leaf anatomy of Ornithocephalinae (Orchidaceae) and related subtribes**. *Lindleyana* **13**: 234–283

Uniyal P L. 1998. **Cytogenetics of bryophytes**. In *Topics in Bryology*, edited by R N Chopra, pp. 125–164. New Delhi: Allied Publishers

Uniyal P L. 2007. **Cytological evolution in mosses with special reference to west Himalayan Bryopsida**. In *Current Trends in Bryology*, edited by V Nath and A K Asthana, pp. 61–99. Dehra Dun, India: Bishen Singh Mahendrapal Singh

Van der Pijl L and Dodson C H. 1966. *Orchid Flowers: their pollination and evolution*. Coral Gabble: University of Miami Press

Wahl H. 1940. **Chromosome numbers and meiosis in the genus *Carex***. *American Journal of Botany* **27**: 458–470

Watanabe K, King R M, Yahara T, Ito M, Yokoyoama J, Suzuki T, Crawford D J. 1995. **Chromosomal cytology and evolution in Eupatorieae (Asteraceae)**. *Annals of the Missouri Botanical Garden* **82**: 581–592

Williams C A, Toscano de Brito A L, Harborne J B, Eagles J, Waterman P G. 1994. **Methylated C-glycosylflavones as taxonomic markers in orchids of the subtribe Ornithocephalinae**. *Phytochemistry* **37**: 1045–1053

Yokota M. 1990. **Karyomorphological studies on *Habenaria*, Orchidaceae and allied genera from Japan**. *Journal of Science of the Hiroshima University* **23**: 53–161

13
Palynology: Timeline

Meenakshi Prajneshu

INTRODUCTION

The term "palynology" was coined by Hyde and Williams (1944). It is a combination of two Greek words: "paluno" meaning "to sprinkle" and "pale" meaning "dust". It has familiarity with the Latin word "pollen". Formerly, palynology used to deal with walls of spores and pollen, but now even their living interiors are involved. Nehemiah Grew, an English botanist, observed pollen under the microscope in the 1640s. Robert Kidston examined spores in coal, Christian Gottfried Ehrenberg studied radiolarians and diatoms, Gideon Mantell examined desmids, and Henry Hopley White studied dinoflagellates. Lennart von Post started pollen analysis in 1916. Gunnar Erdtman did creditable work in pollen analysis and in correlating it to the studies of quaternary vegetation and climate change.

Palynology is either a basic or an applied science. Basic palynology has contacts with cytology, genetics, morphology, physics, chemistry, mathematics, and other branches of science. This branch of science deals with the study of present and fossil palynomorphs, including pollen, spores, orbicules, dinoflagellate cysts, acritarchs, chitinozoans, and scolecodonts, together with particulate organic matter and kerogen found in sedimentary rocks and sediments. Actually, it is an interdisciplinary subject and a branch of both geology and botany. Palynology encompasses the study of various microscopic organic structures called palynomorphs. The most notable examples of palynomorphs are spores and plant pollen. Most of the organic bodies that palynologists study are reproductive particles, much like pollen and spores, but some like algae are whole organisms themselves. Palynology is concerned with both living and fossilized specimens. However, many of the most important findings of the palynological community have come from fossils, which have given us new perspectives on everything from geological surveys to archaeological sites. Organic structures such as spores and pollen can help us understand all kinds of things in the world around us (archaeo-palynology). For example, fossilized pollen specimens found in the remains of ancient human civilizations can tell us about the dietary habits of past cultures. The study of pollen as presented within honey is called melissopalynology; hence, this science can be crucial to anyone dealing with honey. Many plants produce pollen and nectar that contain toxins; if these are turned into honey, it could be poisonous. Palynologists can examine honey and determine the pollen and nectar of the plants from which the honey is made, thus preventing any poisonous honey from reaching consumers. Palynologists have helped with the geological side of oil and natural gas exploration. Palynomorphs are resilient against decomposition, and they are also found embedded within sedimentary rocks in great supply. Therefore, by examining samples of ancient palynomorphs from various rocks, scientists are able to determine the age of different layers of rocks and also what plant life once existed in a now-buried area of the earth's crust. This information can help locate probable reserves of various fossil fuels. Palynology has wide applicability in the fields of botany, palaeontology, archaeology, pedology, geography, medicine, and allergy (Figure 1). Pollen can be used to study pollination mechanisms, foraging resources, migration routes (archaeo-palynology), source zone of insects, and other pollinators in entomology. Jones and Jones (2001) have effectively used rose pollen to study this science. Palynology is used by (1) geologists to help date rocks for petroleum, mining, and water exploration

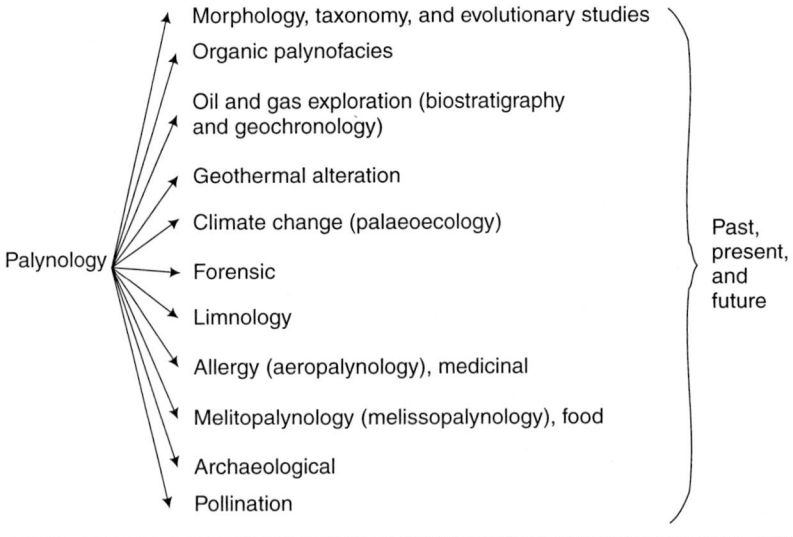

Figure 1 Areas of applications of palynology

and to help unravel the history of plants on earth; (2) geographers to investigate climatic and environmental change; (3) botanists for plant taxonomy and phylogeny; (4) immunologists to investigate allergenic pollen; (5) archaeologists to study the customs, rituals, and agricultural practices of ancient peoples; (6) zoologists to understand foraging habits of insects, birds, and mammals; and (7) environmental scientists to investigate past native vegetation and habitats to preserve and protect the present endangered species.

Pollen is a preferred research entity because it is distinctive, does not decay, and is a natural marker. It has microscopic size; most pollen grains are 10–70 µm in diameter. Moreover, it is abundant since pollen is produced in the anthers of flowers. Mostly wind or insects and small animals help transport pollen from the flower of a plant to the stigma of the flower of another plant of the same species. Therefore, the flowers must produce enormous amounts of pollen for at least some of it to reach its intended destination. Actually, most pollen ends up being particulate components of soil, dirt, dust, and rocks. Therefore, pollen is omnipresent.

Another character of pollen is that it is resistant to degradation. Pollen and spore walls are made up of sporopollenin, which is resistant to degradation; so if these are deposited in proper conditions, these can be preserved in rocks for millions of years. Complexity of pollen is another important factor. Mostly pollen or spores of one plant species differ from that of others. This property helps palynologists link dispersed pollen and spores found in rocks, soil, and dust with the plants that produced them.

BASIC PALYNOLOGY
Morphology, Systematics, and Evolution

Detailed cytological–palynological studies are numerous, and we have extensive knowledge of this field. Regarding chromosome numbers and meiotic disturbance, palynology has contacts with cytology as well as genetics. Occurrence of widely different pollen sizes within a single genus has attracted interest of cytologists, and so also the appearance of pollen with varying number of apertures (Garg 1980a). Polarity, symmetry, shape, size, sporoderm, and sratification belong more to morphology and microanatomy than to any other branch of science. Mathematical aspects of palynology, that is, formulae for calculation of grain surface areas as well as number, position, and character (NPC) of apertures, have not attracted much interest. Application of electron microscopy revealed a lot of new information. Using "scanning electron microscope" and from the observations of the sexine perforations, one can easily distinguish three species of *Hypericum* (Garg 1980a, Figure 2). Plant taxonomy is of huge utility in applied palynology. A cautious approach to this field may be through investigations showing whether taxonomical amendments on a megamorphological base can be confirmed by palynological evidence or not. For example, on the basis of pollen morphology, it was confirmed that genus *Coriaria* be retained in the family Coriariaceae. The pollen in the related genera is usually more or less of the same type, but there are striking exceptions. Occurrences of similar pollen in unrelated taxa have also been reported. The pollen of *Citrus medica*, *Hiptage madablota*, and *Pavetta tomentosa* have conspicuous cap-like bodies at the poral regions, an ornamentation similar to that present on the exine (Garg 1980b). On the other hand, more than one pollen type is found in certain genera, for example, dimorphic pollen grains in *Leptadenia parkeri* and *Berberis umbellate* (Garg 1980a), pollen heteromorphism in *Nicotiana tabacum* (Till-Bottraud, Mignot, De Paepe, *et al.* 1995), and polymorphism in pollen of *Salvia leucantha* (Gupta and Sharma 1990).

Vasanthy and Grard (2007) have developed a new tool for the identification and self-training in plant taxonomy (IDAO by Centre de cooperation internationale en recherché agronomique pour le developpement [CIRAD]). It is adapted to enable pollen identification by non-sequential choice of characters and without terminology.

Information on pollen is organized in a database accessed by the identification system. An electronic pollen flora, including digital and scanning electron micrographs and descriptions, distinguishes itself by the graphic identification system of pollen without the plethora of terms; it enables many levels of users (in south India as well as in the tropics, subtropics, and subtemperate regions in India, Asia, and other continents) to learn palynology with a "click of the mouse". The descriptive part furnishes palynological terms linked with the illustrated definitions and provides the bibliographic link for quick reference. A microtaxonomic research tool (adapted from IDAO by CIRAD) will aid users interested in climate-related past vegetational changes.

Applied palynology encompasses research into phytogeography. Detailed pollen morphology of 30 endemic plants was performed, since confinement of a plant to a certain ecological niche separates it as an entity very useful to geographers, geologists, and fossil botanists (palaeopalynologists) in interpreting their findings on Himalayas (Garg 1980a). Basic palynology has usefulness in systematics too; for example, with respect to anther and pollen characteristics, *Daphniphyllum* shows closer resemblance with Hamamelidaceae rather than with Magnoliaceae,

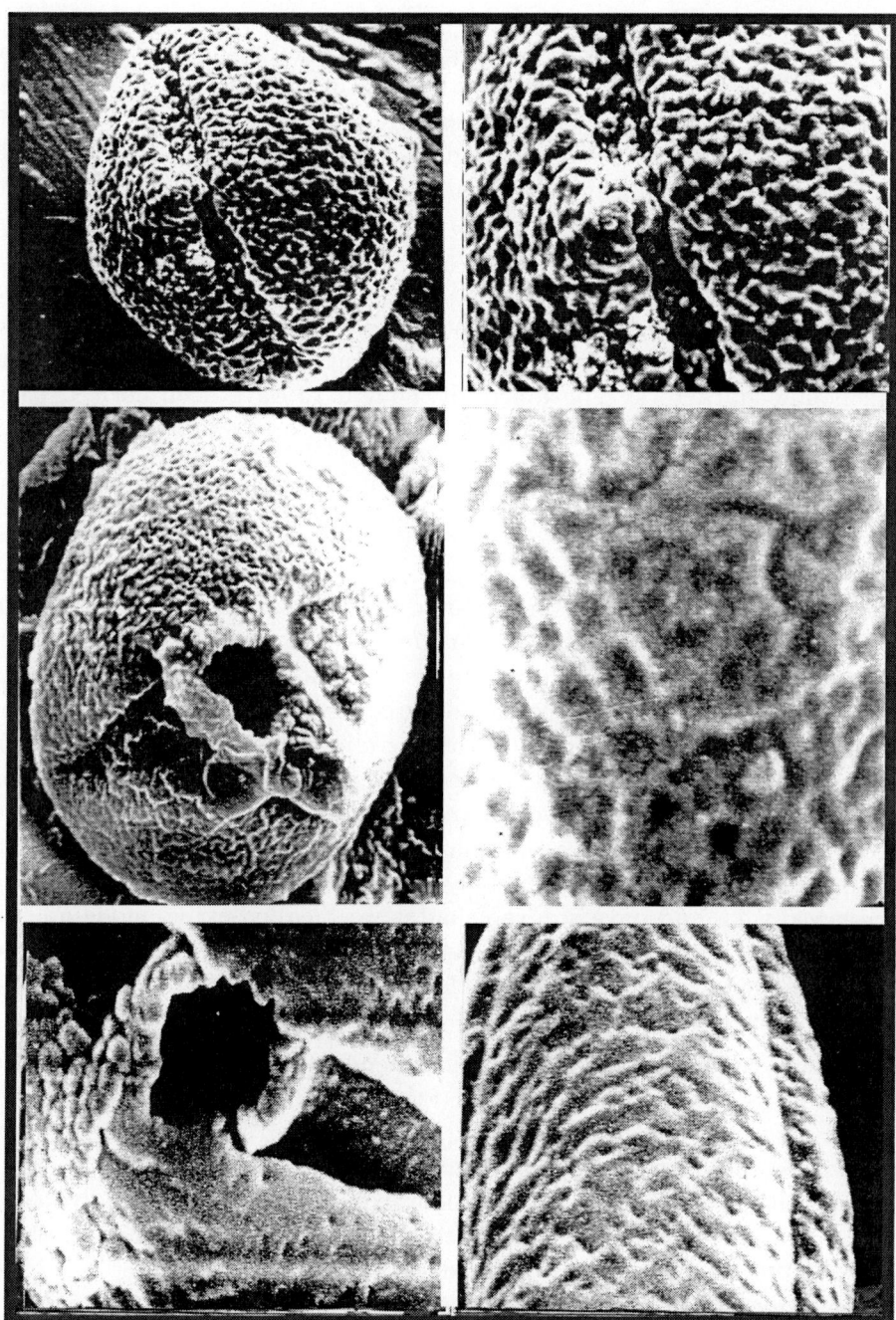

Figure 2 SEM pictures of sexine of *Hypericum* spp.

Euphorbiaceae, Pittosporales, or Geraniales, as envisaged by earlier investigators (Bhatnagar and Garg 1977). Another use of palynology is in evolutionary interpretations. For example, eudicots are a large, monophyletic assemblage of angiosperms, comprising roughly 190 000 described species, or 75% of all angiosperms. The monophyly of eudicots is well supported from molecular data and is delimited by at least one palynological apomorphy: tricolpates are derived pollen grains. Tricolpate pollen evolved from a monosulcate type (having a single distal aperture; Figure 3), which is considered to be ancestral in angiosperms as well as in many seed plants. Many eudicots have more than three apertures, varying greatly in numbers, shapes, and position (constituting important taxonomic characters). These are all thought to have been derived from tricolpate type (Simpson 2006).

Pollination and Entomology

Plants disperse their pollen or spores by a number of different methods. Many aquatic angiosperms live completely submerged and release their pollen underwater, relying on water currents to transport pollen from the male reproductive part to the female stigma of a neighbouring flower. Water pollination is a chance method. Pollen of such plants has only a single-layered cellulose wall. This pollen is almost never preserved in sediments and generally oxidizes rapidly if removed from water. Therefore, these pollen types are of little potential value for sedimentary and forensic work. Another small group of plants, called autogamous, self-pollinates and needs little quantity of pollen. Pollen grains from these plants are rarely dispersed into the atmosphere, are preserved well, and have a durable outer wall (called exine) made up of a stable chemical compound called sporopollenin. Like the pollen produced by submerged plants, the pollen of autogamous plants is also of little value in forensics.

In zoogamous plants, pollination is dependent on the transport of pollen by some types of insects or animals (for example, hummingbirds, lizards, nectar-feeding marsupials and bats, or other small mammals). Although their pollen productivity is low, it is more than that of autogamous plants. The potential value of zoogamous pollen in forensics and palaeopalynology is excellent because they have most durable exines. Therefore, their pollen often remains preserved in deposits for long periods of time and is generally less susceptible to destruction. Zoogamous pollen is produced in low amounts; thus, these are generally not the contaminants present in the pollen rain of an area, which may be both good and bad. It is

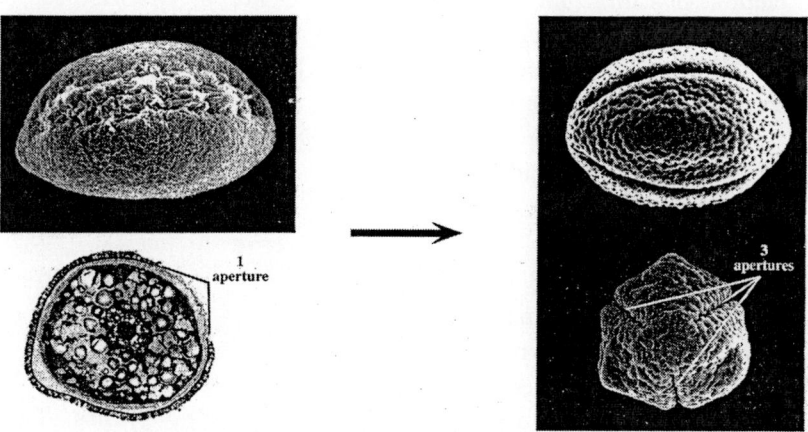

Figure 3 Monophyly of eudicots

good because if the pollen of a zoogamous plant is found in a forensic sample, it almost certainly belongs to the forensic sample and is not an atmospheric contaminant. It is bad because so little pollen is produced by each plant that the chances of its pollen getting into a forensic sample are reduced. The time of production of viable pollen, stigmatic exudates, and nectar along with prevailing environmental conditions play important roles in attracting thrips to chilly flowers for food. Foraging activities of thrips within the same flowers cause self-pollination. Larger population of thrips creates competition for food, leading to their migration to other flowers, thus ensuring cross-pollination. In particular, *Scirtothrips dorsalis* helps in pollen transfer between two varieties of chillies (Saxena, Vijayaraghavan, Sarbhoy, et al. 1996).

Wind-pollinated (anemophilous) plants include the majority of gymnosperms, a significant number of angiosperms, and also spore-producing plants such as fungi, ferns, and mosses. Anemophilous plants produce vast quantities of light-weight grains that travel easily in air currents. When large fields of these anemophilous plants grow together, their flowers can produce millions of pollen grains that are dispersed daily during the flowering season. This abundance becomes a disadvantage in forensics to catch culprits of drug trade. For example, *Cannabis* (marijuana) pollen may be found in trace amounts on the shoes of people involved in this trade. Nevertheless, when such evidence is found, a palynologist cannot state in court that "traces of *Cannabis* pollen could only have come from direct association with or use of the actual plant". This is because traces of marijuana pollen on a suspect's shoes could have come from almost anywhere as a result of "random air dispersal". To cite an example, in 1995, European newspapers reported that "clouds" of *Cannabis* pollen were drifting across the Mediterranean from source areas. Residents were warned not to breathe "too much" of the *Cannabis* pollen as these could cause hallucinations (it is false as *Cannabis* pollen do not contain any of hallucinogenic cannabinoids). As such, the high atmospheric counts of *Cannabis* pollen illustrate how some of these grains might accidentally occur in some forensic samples.

Palaeoecology

Ecology is the study of environment and its impact and feedback relationships with the organisms, which in turn are part of the environment. It is well known that the vegetation that lived on earth during the geological past responded to the physical and ecological variations in the same manner as the present-day plants. Thus, to reconstruct the palaeoenvironment, it is essential to know the existing characteristic elements of flora and climatic conditions that are prevalent now, because contemporaneous fossil assemblage is indicative of their closeness to ecological conditions (concept of "uniformitarianism", that is, "present is the key to the past"). Palynomorphs have considerably helped in the reconstitution of palaeoenvironment and vegetational history of the sub-recent and quaternary deposits of India. Chanda and Mukherjee's (1969) work on Calcutta peat revealed that ca. 5000 years ago, this city and its suburbs consisted of a marshy place full of mangrove vegetation, which is now restricted only to Sunderbans and the estuarine parts of the Ganges. According to Vishnu-Mittre (1976), important members of oak woods, for example, *Quercus* spp., *Alnus*, Lauraceae, *Larix*, and *Carpinus*, do not occur in the Kashmir valley today, although these, along with *Corylus* and *Picea* (not in dominance now), were most significant constituents of the early quaternary forests. After second glaciation in the valley, these forests showed fluctuations between oak-mixed woods and conifers, and non-tree pollen far exceeded tree pollen in their remains. An analysis of the pollen of quaternary sediments of Kashmir shows the predominance of *Typha* in lower Karewa, followed by abundance of *Alnus,* which is now replaced by *Plantago* and chenopods (Nair 1960). Singh (1963) and Vishnu-Mittre and Sharma (1966) have attributed the occurrence of certain oaks, alders, and elms in the pollen diagram of Kashmir valley and their absence in the present flora mainly to the change in climatic conditions and human interference. It is speculated that in Ootacamund and Nilgiri mountains, bushes appeared 32 000 years BP; bush-tree Savannah was formed 26 000 years BP, which later declined. Here shola forests were established between 15 000 and 11 400 years BP, and now the forests are of shrub-tree Savannah type. Almost negligible information is available on the palaeopalynology from the

Siwalik sediments of Punjab basin. Nandi and Bandopadhyay's (1970) account dealt only with the upper and middle Siwaliks and was based on individual grab samples. The palaeoenvironmental interpretations based on this observations are, therefore, not conclusive.

The microfloral analysis of middle and lower Siwalik formations from the Bhakra Nangal area of Punjab discloses that during the uplift of lower Siwaliks, the vegetation was dominated by Palmae, Gramineae, Compositae, and Polypodiaceae, indicating near-shore environment of deposition and a moist subtropical climate (Banerjee 1968). On the contrary, the upper Siwalik sediments are dominated by pollen grains of *Pinus* and bisaccate pollen-bearing gymnosperms, whereas those of palms and other subtropical elements are totally missing. Grass pollen considerably decreased in its frequency. The assemblage thus points towards a cool-dry climate and a distinct change from subtropical to temperate environment. The pollen-rich quaternary sediments occur at a few places such as Kashmir and other regions of north-west Himalayas, Ootacamund, and possibly in Assam in eastern Himalayas. Their studies, if pursued, can prove rewarding. Studies on altitudinal belts of Sumatran and Javan sites indicate abundance of gymnosperm pollen and forest altitude boundaries to be much lower than today's (Stuijts, Newsome, and Flenley 1988), thereby imparting ample evidence for late-quaternary vegetational change in the highlands.

Basic palynology also includes pollen and spore dispersal (aeropalynology), and pollen and spore content of peat and sediment under formation (palynostratigraphy). Palynological study of Kasaragod Formation in deep-water well KKD-AA, Kerala–Konkan Basin, has enabled dating of the sediments between 3520 m and 1925 m and interpreting environment of deposition based on dinoflagellate cysts, spores, and pollen. Early part of late Palaeocene (3205–3180 m) and early Eocene (2035–1925 m) suggest prevalence of inner shelf environment. The record of early Palaeocene (Danian) dinoflagellate cysts immediately above the trap wash (3524 m) suggests that the rifting of Kerala–Konkan Basin had already started during the Danian, prior to the initiation of Mumbai Offshore Basin during late Palaeocene.

Limnology Studies

Limnology is a science related to aquatic ecology and hydrobiology. It involves the study of aquatic organisms in relation to their hydrological environment. Past lake levels and long-term climate change are studied using freshwater palynomorphs, animals, and plant fragments, including the prasinophytes and desmids (green algae).

APPLIED PALYNOLOGY

Biostratigraphy, Geochronology, Geothermal Alteration

The significant role of palynology in hydrocarbon exploration is in the dating of sediments and high-resolution biostratigraphy for finer zonation and marine sediments, sequence biostratigraphy, evaluation of hydrocarbon source rock potential, kerogene analysis, and palaeogeographic reconstruction. The study of organic contents of sediments from both geochemical and particulate standpoints provides maturity, that is, thermal alteration index, total organic content, and also the subsidence history of sedimentary basins with generative window that lead to investigation of genesis and pooling of liquid and gaseous hydrocarbons.

Towards the end of the 1980s came a thrust in the application aspects of palynomorphs and palynofacies. There is an increasing emphasis on hydrocarbon exploration in different on-shore and off-shore sedimentary basins. With the development of sequence stratigraphy in sedimentary geology, palynology and palynofacies have emerged as important components in integrated multidisciplinary studies. Significance of palynomorphs has been amply demonstrated for characterizing various sequence components, correlation of terrestrial and marine deposits, and relative sea-level fluctuations. Studies on different palynofossil groups have helped in precise dating and correlation of sediments. For dating, they are used both individually and in combination to achieve fine-time slicing. Correlation is made on the basis of spores–pollen (for terrestrial) and dinoflagellate cysts, algae, and nannoplankton (for marine sequence of sediments). Such studies help in the identification of "onlapofflap sequences" in sequence stratigraphy. Statistical analyses based on the ratios of spores–pollen versus dinocysts are utilized for deciphering marine transgressive/regressive (T/R) cycles. Hydrocarbon is generated from organic-rich sediments. During burial of the sediments, temperature increases, resulting in a series of geochemical reactions. These reactions lead to the transformation of organic matter to kerogen, which is the precursor of hydrocarbons. The amount, type, and composition of hydrocarbons generated can be reflected by the palynological expressions of the source rocks and geological history of the basins. Basin analysis is an integral part of any petroleum exploration programme that use all modern geoscientific tools for its geological history. The geochemical techniques have emerged as potential tools in hydrocarbon research, exploration, and exploitation. The amount of organic matter, its type, and maturity as evaluated are used for the source rock characterization (immature or mature), which helps in determining prospects of the basin. Correlations between oil and gas-source rock are used to understand migration pathways of hydrocarbons. Ultimately, geochemical basin modelling is employed to determine the amount of oil and gas generated, the time of hydrocarbon generation, and so on to evaluate ultimate oil and gas reserves of a sedimentary basin. Modern geochemical techniques, tools, and applications of computer basin modelling are also used for hydrocarbon prospect evaluation.

Palynofacies

Palynofacies studies examine preservation of particulate organic matter and palynomorphs to provide information on the depositional environment of sediments and depositional palaeoenvironments of sedimentary rocks. The term palynofacies was introduced by French geologist Combaz in 1964. Palynofacies studies are often linked to investigations of the palynology and organic geochemistry of sedimentary rocks. Palynofacies can be used in two ways.

1. Organic palynofacies considers all the acid-insoluble particulate organic matter, including kerogen and palynomorphs, in sediments and palynological preparations of sedimentary rocks. The sieved or unsieved preparations may be examined using light microscopy or ultraviolet fluorescence microscopy. The abundance, composition, and preservation of various components, together with thermal alteration of the organic matter, are to be considered for this type of palynofacies.

2. Palynomorph palynofacies considers the abundance, composition, and diversity of palynomorphs in palynological preparation of sedimentary rocks. The ratio of marine fossil phytoplankton (acritarchs and dinoflagellate cysts) together with chitinozoans to terrestrial palymomorphs (pollen and spores) is used to derive a terrestrial input index in marine sediments.

Both types of palynofacies studies are used for the geological interpretation of sedimentary basins in exploration geology, often in conjunction with palynological analysis and vitrinite reflectance. The organic matter types, palynofacies, and organic maturity play important roles in the origin of hydrocarbons. Hydrocarbon source capabilities of sedimentary rocks can be determined by the qualitative and quantitative assessments of organic matter types and maturation. These require the study of organic matter entities, such as structured terrestrial matter, spores and pollen, biodegraded terrestrial matter, charcoal, amorphous organic matter, biodegraded aqueous organic matter, and structured aqueous organic matter. The data obtained are integrated with various geochemical parameters for the precise assessment of source potential. Litho and biostratigraphic parameters are considered along with the environment to understand the spatiotemporal distribution of source potential facies. The Palakollu Shale (Palaeocene) in the Bhimanapalli and Mori structures has good source potential to generate gas. The Pasarlapudi Formation (Eocene) has good potential to generate hydrocarbons. The Ravva Formation (Mio-Pliocene) in GS-3 structure has poor-to-marginal source potential to generate gas, whereas the Early Eocene sequence in GS-21 structure has good source potential for gas. Genesis of methane in coal beds mainly occurs during biochemical and geochemical stages of coal formation (throughout the entire coalification series) and also during post-depositional phases at the time of magmatic intrusions. Geothermal gradient also plays a significant role in coalification, for example, Damodar Valley coalfields have high-rank coals with intense magnetic ramification and huge coal bed methane (CBM) reserves. In early phases of coal deposition (peat to lignite), biochemical and, further onwards, geochemical stages are responsible for generation of CBM, coal mine methane, and abandoned mine methane in Indian coalfields. During the biochemical stage of coal formation, a lot of gases, including methane, evolved owing to transformation of plant parts, particularly transformation of cellulose and lignin into humic acid and humus (together called ulmin). Methane is no more considered only a hazardous gas, but it is a source of alternate energy in several countries.

Allergy

It is a hypersensitivity disorder of the immune system. Allergic reactions, which can be acquired, predictable, or rapid, occur to normally harmless environmental substances known as allergens. Many allergens such as dust or pollen are airborne particles. In these cases, symptoms arise in areas in contact with air, such as eyes, nose, and lungs. For instance, allergic rhinitis, also known as hay fever, causes irritation of the nose, sneezing, and itching and redness of the eyes (Bope and Rakel 2005). Inhaled allergens can also lead to asthmatic symptoms, caused by narrowing of the airways (bronchoconstriction) and increased production of mucus in the lungs, shortness of breath (dyspnoea), coughing, and wheezing (Holgate 1998).

Forensic Palynology

Forensic palynology is the study of pollen and powdered minerals, their identification, and where and when they occur, to ascertain that a body or other object was in a certain place at a certain time. Pollen grains are indicative of where a person or object has been because specific regions of the world, or even more particular locations such as certain set of bushes, will have a distinctive collection of pollen species. Pollen evidence can also reveal the season when a particular object picked up the pollen, for example, activity of a burglar at mass graves in Bosnia were traced using pollen. The burglar accidentally brushed against a *Hypericum* bush during a crime (Mildenhall 2006). Pollen can be used as an additive for bullets to enable tracking them. Time of pollen dehiscence can be helpful in tracking a crime. For instance, a dead body may be found in a wood and the clothes may contain pollen that were released after death (the time of death can be determined by forensic entomology) but in a place other than where it was found. This clearly indicates that the body was moved. Studies of palynomorphs trapped in materials associated with criminal or civil investigations are slowly gaining recognition as valuable forensic techniques. Today, New Zealand leads the world in the use of forensic palynology and the acceptance of this type of evidence in courts of law.

The term forensic palynology refers to the use of pollen and spore evidence in legal cases (Mildenhall 1992). In its broader application, it also includes legal information derived from the analysis of a broad range of microscopic organisms, such as dinoflagellates, acritarchs, and chitinozoans that can be found in both fresh and marine environments (Faegri and Johs 1964). The inception of forensic palynology is difficult to establish. Attempts might have been made prior to the 1950s, yielding successful or unsuccessful results; these probably did not gain much public attention and, therefore, were not reported. If earlier attempts were made, the results might have been purposely hidden from the media in order not to alert criminals about this new technique. The earliest case was reported by Erdtman (1969). In Sweden, during the court hearing of a case regarding a woman killed in May, a number of experts, including a palynologist, were asked to examine the dirt attached to her clothing. The objective of those studies was to determine whether or not the woman was killed where she was found. Preliminary studies of the pollen in the dirt samples suggested that she had been killed elsewhere because the dirt lacked pollen from plants common in the area where the body was found (that is, *Plantago, Rumex,* and grasses). However, a later reinterpretation of the forensic pollen samples noted that the murder could have occurred in May because that was before the grasses and herbs in the region had pollinated. Both the opinions were entered as evidence in the court proceedings. The importance of this case is that it is one of the earliest records in which pollen data were considered important forensic evidence in a court case. Pollen grains have been used as evidence in a fictional murder mystery novel called *Probable Cause* (Pearson 1991). According to the author, the identification of specific pollen grains found in the earwax of a murdered person was significant enough evidence to link the suspect to the murder and to swing the jury to win a conviction.

Grains are very important in forensics because of their production and dispersion. If one knows the expected production and dispersal patterns of spores and pollen for the plants in a given region, one expects certain type of pollen fingerprint in samples from that area (Bryant 1989). Therefore, the first task of the forensic palynologist is to try to find a match between the pollen in a known geographical region. Pollen of woody perennials in various altitudinal belts of western Himalayas were studied by Garg (1980a; Figure 4). For forensic examination, samples of sediment such as soil, dirt, and dust are taken. These samples should be collected carefully because they often contain abundant pollen and spores. Samples of dirt collected from the clothing, skin, hair, shoes, or vehicle of a victim may prove useful in linking the victim with the location of the crime (Mildenhall and Deer 1988). The same will be true of any suspects thought to be associated with a crime. Mud found on a stolen vehicle or a vehicle used in a crime can link the vehicle with the scene of a crime or the place from where it was stolen, that is, specific geographical locale (Brown and Llewellyn 1991). For example, a thief robbed a store and escaped on a motorcycle. Police gave chase and almost caught the thief, but at the last minute the thief abandoned his motorcycle and ran up a muddy hill and escaped into a wooded area. The next day a man comes to the local police headquarters and reports that his motorcycle had been stolen the day before. Because the motorcycle claimant was about the same physique as the thief, police considered him to be their prime suspect. However, he could not be linked with the crime. The police searched his house but could not locate any stolen item; however, they collected a pair of muddy boots owned by the suspect from there. When asked about the mud, the suspect said that the mud came from the farm where he worked and denied that it could have come from the hill where his motorcycle had been abandoned. Dirt samples from the suspect's boots were sent to the forensic laboratory for analysis, along with soil samples of mud collected from the hillside where the motorcycle was abandoned and from the farm where the suspect worked. The pollen types recovered from the mud on the boots closely matched those recovered from the mud on the hillside but was quite different from the pollen types found in the soil samples collected from the farm. In spite of convincing pollen evidence, the suspect was released because the court remained unconvinced that forensic pollen studies were truly accurate enough to link the suspect to the hillside where his motorcycle had been abandoned (Bryant, Mildenhall, and Jones 1990). Forensic palynology is in its infancy; it remains untried in many regions of the world and is yet to be accepted or recognized as a valuable evidence in most court systems. There are still misconceptions about what type of information forensic pollen samples can provide. Often, forensic data regarding pollen results do not actually convict a suspect. It only serves as a useful tool that can point investigators in the right direction, or narrow the number of suspects or perhaps even eliminate a person as a prime suspect. Nevertheless, even in this type of supporting role, forensic palynology can become a powerful tool of the forensic scientist.

Melitopalynology or Melissopalynology

It is the study of pollen and spores contained in honey and, in particular, the pollen's source. By studying the pollen in a sample of honey, it is possible to gain evidence of the geographical location and genus of the plants that the honey bees visited; however, it should be kept in mind that honey may also contain airborne pollens from anemophilous plants, spores, and dust due to attraction by the electrostatic charge of bees. Generally, melissopalynology is used to combat fraud and inaccurate labelling of honey. Information gained from the study of a given sample of honey (and pollen) is useful when substantiating claims of a particular source. Monofloral honey derived from one particular source plant may be more valuable than honey derived from many types of plants. The price of honey also varies according to the region from where it originates.

Achromatic analysis by tristimulus colorimetry and a pollen analysis of pollen grains contained in 33 *Eucalyptus* unifloral honey samples were carried out along with colour of the pollen grains. It allowed the prediction of ultimate colour of the honey. The results obtained show that lightness is significantly related to the pollen type—*Olea europaea*. On the other hand, the variable that better relates to the chroma is the *Zea mays* pollen type (Anass, Escudero, González-Miret, *et al.* 2004).

Archaeology

It is the study of human society, primarily through the recovery and analysis of the material culture and environmental data that they have left behind. This includes artefacts, architecture, biofacts (including pollen), and cultural landscapes (the archaeological record). Because archaeology employs a wide range of different procedures, it can be considered to be both a science and a subject of humanities. In the USA, it is thought of as a branch of anthropology, although in Europe it is viewed as a separate discipline. Archaeology studies human history from the development of first stone tools in eastern Africa 3.4 million

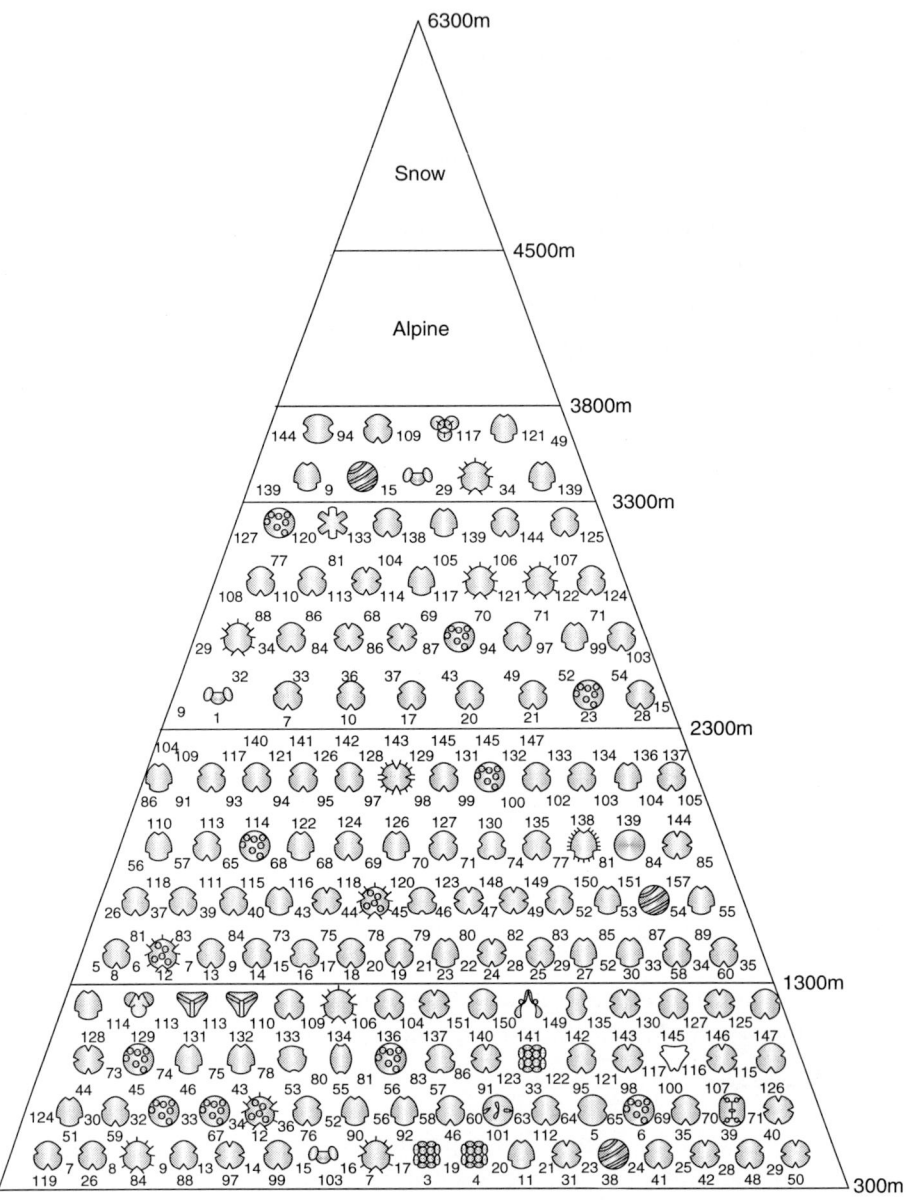

Figure 4 Pollen in various altitudinal belts of Western Himalayas

years ago to recent decades. It is most important for learning about prehistoric societies, when there are no written records for historians to study, making up over 99% of total human history, from the Palaeolithic until the advent of literacy in any given society. Archaeology has various different goals, which range from studying human evolution to cultural evolution and understanding cultural history. The discipline involves survey, excavation, and eventually analysis of data collected to learn more about the past. In broad scope, archaeology relies on cross-disciplinary research, involving anthropology, history, art history, classics, ethnology, geography, geology, linguistics, physics, information sciences, chemistry, statistics, palaeoecology, palaeontology, palaeozoology, palaeoethnobotany, and palaeobotany.

Medicine

Pollen have various types of therapeutic uses: rye pollen is known to cure congestion and gastric ulcers; pollen mixed with honey improve appetite; they cure high-altitude sickness and lower oxygen tension (Hong-fu, Zheng-sheng, Fang, *et al.* 1990).

Food

Beetles, bees, mites, chickens, and calves are known to feed on pollen probably in the want of protein. The polynomorphs contain ascorbic acid, vitamins, proteins, carbohydrates, phosphorus, and calcium in much higher proportion than most of the common eatables. In addition, they supply high energy to the body, thereby suggesting it to be an excellent food for human and livestock. Ninety grams of pollen per day is a good protein supplement. In vivo and in vitro studies in hazelnut were carried out (Franchi 1987) to gather information regarding the digestibility of pollen. Surface enzymes are digested first, then the apertural region, followed by the internal part. Such experiments must be carried out in other plant species. The presence of vitamins in pollen also makes them a preferred food for warriors. It prevents brain damage, induces slimming, and cures indigestion. It is an alternative food for famine times. Bee-bread (pollen stored by bees in hives) is also regarded as a choicest food.

IMPORTANT JOURNALS ON PALYNOLOGY

- *Pollen Analysis Circular* (first journal on palynology, from North America [1944])
- *Journal of Palynology* (Indian)
- *Grana: International Journal of Palynology*
- *Palynology*
- *The Journal of Palynology*
- *Review of Palaeobotany and Palynology*

MAIN PALYNOLOGY LABORATORIES OF INDIA

- Birbal Sahni Institute of Palaeobotany, Lucknow, Uttar Pradesh
- French Institute of Pondicherry, Puducherry

FUTURE OF PALYNOLOGY

Pioneering discoveries in palynology were made up to 1916; then came a slag (up to 1944) and then again an acceleration until the 1970s. With the use of electron microscopy, especially the scanning electron microscope, there was a boost in basic palynology and also in its application to applied aspects. In the last 40 years, it has become an accepted discipline in oil companies, both in research and in exploration. Micropalentology is also very useful in oil exploration, but as pollen is produced in abundance, palynology is more successful. This way palynology moved out of academics and government circles and became an important tool for oil exploration. As a result, the number of palynologists also increased exponentially. It is still used by a few government organizations and a handful of academicians. Palynology is another palaeontological discipline with versatility and a wide range of applicability; fossil group is also available to it. Therefore, it can focus on environmental concerns, that is, climatic fluctuations, acid rain, and pollution. It plays an important role in basin maturation studies, sequence stratigraphy, and development of hydrocarbon charged models. Modern palynologists can also pursue a carrier in forensics and archaeology; however, if they stick only to morphological and pollen analysis, the scope of growth is narrow. Palynology is multidimensional, helping the entire field of botany and quite a few branches of geology. Pollen, apart from its immense positivity, has some negativity too, for example, allergy to human beings. Study of pollen grains in air in different flowering seasons, coupled with their surface and inside chemicals, makes the science of pathology very useful to patients world over. Pollen contains lots of proteins and other beneficial ingredients; more research efforts are required in this direction. Studies on pollen also find ground in incompatibility studies in plants; so physiological and biotechnological findings have ample academic and applied utility. Pollen is important in molecular biology, DNA fingerprinting (as pollen is a pool of genetic variability), plant biodiversity, tissue culture, and hybridization experiments. Pollen has a haploid set of chromosomes containing all characters of a plant, which makes it a preferred material for applied biology.

BIBLIOGRAPHY

Anass T, Escudero M L, González-Miret M L, Heredia F J. 2004. **Colour characteristics of honeys as influenced by pollen grain content: a multivariate study.** *Journal of the Science of Food and Agriculture* **84**: 380–386

Banerjee D. 1968. **Siwalik microflora from Punjab, India.** *Review of Palaeobotany and Palynology* **6**: 171–176

Bhatnagar A K and Garg M. 1977. **Affinities of *Daphniphyllum*: palynological approach.** *Phytomorphology* **27**: 92–97

Bope E T and Rakel R E. 2005. ***Conn's Current Therapy 2005***. Philadelphia: W B Saunders Co.

Brown G and Llewellyn P. 1991. ***Traces of Guilt: science fights crime in New Zealand***. Auckland, New Zealand: Collins Publishers

Bryant V M Jr. 1989. ***Pollen: nature's fingerprints of plants. 1990 Yearbook of Science and the Future. Encyclopedia Britannica***, pp. 92–111. Chicago, Illinois: Encyclopedia Britannica Inc.

Bryant V M Jr, Mildenhall D C, and Jones J G. 1990. **Forensic palynology in the United States of America.** *Palynology* **14**: 193–208

Chanda S and Mukherjee B B. 1969. **Radio carbon dating of two microfossiliferous quaternary deposits in and around Calcutta.** *Science and Culture* **35**: 275–276

Erdtman G. 1969. ***Handbook of Palynology***. New York: Hafner Publishing Co.

Faegri K and Johs I. 1964. ***Textbook of Pollen Analysis***. Oxford: Blackwell Scientific Publications

Franchi G G. 1987. **Researches on pollen digestibility.** *Atti Societa di sciennze Naturali Memorie Series* **4**: 43–52

Garg M. 1980a. **Palynological studies on some woody perennials of western Himalaya.** *PhD thesis*, University of Delhi (unpublished)

Garg M. 1980b. **Pollen morphology and systematic position of *Coriaria*.** *Phytomorphology* **30**: 234–239

Gupta A and Sharma C. 1990. **Polymorphism in pollen of *Salvia leucantha*.** *Lamiaceae* **29**: 277–284

Holgate S T. 1998. **Asthma and allergy: disorders of civilization?** *QJM: An International Journal of Medicine* **91**: 171–184

Hong-fu P, Zheng-sheng X, Fang M, Dung-pi P, Zi-ming L, Zhong-wen L, Shau-rong L, Shung-zing T. 1990. **The effect of pollen enhancing tolerance to hypoxia and promoting adaptation to highlands** (in Chinese). *Journal of Chinese Medicine* **70**: 77–81

Hyde H A and Williams D A. 1944. **The right word.** *Pollen Analysis Circular* **8**: 6

Jones G D and Jones S D. 2001. **The uses of pollen and its implications for entomology.** *Neotropical Entomology* **30**: 341–350

Mildenhall D C. 1992. **Pollen plays part in crime-busting.** *Forensic Focus* **11**: 1–4

Mildenhall D C. 2006. **Hypericum pollen determines the presence of burglars at the scene of a crime: an example of forensic palynology.** *Forensic Science International* **163**: 231–235

Mildenhall D C and Deer V. 1988. **Palynology: an example of the use of forensic palynology in New Zealand.** *Tuatara* **30**: 1–11

Nair P K K. 1960. **Palynology in India: a review.** *Journal of Scientific and Industrial Research* **19**: 253–260

Nandi B and Bandopadhyay N N. 1970. **Preliminary observations of the microfossils and microstructures of Siwalik lignite from Himachal Pradesh, India.** *Science and Culture* **36**: 240–242

Pearson R. 1991. *Probable Cause*. New York: St Martins Paperbacks

Saxena P, Vijayaraghavan M R, Sarbhoy R K, Raizada U. 1996. **Pollination and gene flow in chillies with *Scirtothrips dorsalis* as pollen vectors.** *Phytomorphology* **46**: 317–327

Simpson M G. 2006. *Plant Systematics*. USA: Academic Press

Singh G. 1963. **A preliminary survey of the post-glacial vegetational history of Kashmir valley.** *Palaeobotanist* **12**: 73–108

Stuijts I, Newsome J C, and Flenley J R. 1988. **Evidence for late quaternary vegetational change in the Sumatran and Javan highlands.** *Review of Palaeobotany and Palynology* **55**: 207–216

Till-Bottraud I, Mignot A, De Paepe R, Dajoz I. 1995. **Pollen heteromorphism in *Nicotiana tabacum* (Solanaceae).** *American Journal of Botany* **82**:1040–1048

Vasanthy G and Grard P. 2007. **Pollen grains of South Indian trees: a user-friendly multimedia identification software.** *Collection Ecologie* (45)

Vishnu-Mittre. 1976. **Quaternary vegetation of India (late abstract).** In *4th International Palynology Conference*, pp. 38–40. Lucknow, India: Birbal Sahni Institute of Palaeobotany

Vishnu-Mittre and Sharma B D. 1966. **Studies of post-glacial vegetational history from the Kashmir Valley-1, Haigam lake.** *Palaeobotanist* **15**: 185–212

14

Role of Molecular Markers in Evaluation of Plant Diversity

Mandeep Kaur, Gurveen Kaur, Rajneet Kour Soodan, Jatinder Kaur Katnoria, and Avinash Nagpal

INTRODUCTION

Plants have always been considered a key source of energy for the survival and evolution of the animal kingdom, thus forming a base for every ecological pyramid. Plant taxonomy involves the analysis of genetic diversity and relatedness between or within different species, populations, and individuals. Identification of plants and knowledge of differences among plant species are important in agricultural production and use of plants in herbal medicines, pharmaceuticals, food, cosmetics, and other industries. Traditional approaches to measure the diversity of plants relied upon comparative morphology, anatomy, physiology, embryology, electron microscopy, and biochemical and phytochemical assays. Over the last few decades, these techniques have largely been complemented by molecular methods, which provide valuable tools for the reliable and precise identification of plants (Henry 1997). Molecular markers offer numerous advantages over conventional phenotype-based methods because of their stability and detectability in all tissues regardless of growth, differentiation or development (Agarwal, Shrivastava, and Padh 2008). Furthermore, defence status of the cell is not confounded by the environment, pleiotrophic, and epistatic effects.

Molecular markers generally refer to biochemical constituents, including primary and secondary metabolites and other macromolecules such as proteins and DNA (Joshi, Ranjekar, and Gupta 1999). Secondary metabolites have extensively been used as markers in quality control and standardization of botanical drugs. The analysis of genetic diversity within a species may involve storage proteins such as glutenins, gliadins, and hordeins (Vapa and Radovic 1998; Metakovsky 1991; Shariflou, Hassani, and Sharp 2001; Kraic, Horvath, Gregova, et al. 1995; Cerny and Sasek 1996a, 1996b). A piece of DNA or a protein can be used as a marker (Ovesna, Polakova, and Leisova 2002). During the late 1950s, development of protein electrophoresis techniques provided selectively neutral markers spanning several genes, thus contributing significantly to the current understanding of plant genetic diversity (Hunter and Markert 1957; Smithies 1955).

PROTEIN-BASED MOLECULAR MARKERS

Owing to their low cost and user friendliness, allozymes have been considered the molecular markers of choice to discern genetic variations within and among plant populations for quite a long period of time (Brown 1979; Hamrick and Godt 1990; Loveless and Hamrick 1984). Allozymes are distinct forms of the same enzyme encoded by different alleles on a single locus. One of the advantages of allozymes as useful genetic markers in the studies of plant genetic diversity is the possibility of efficient processing of a large number of samples (Karp, Isaac, and Ingram 1998; Parker, Snow, Schug, et al. 1998). Besides, allozymes are usually expressed codominantly so as to distinguish homozygous and heterozygous genotypes. Allozymes also have some limitations in studying diversity in plants. The genes encoding allozymes represent a small biased sample of the genome (Karp, Isaac, and Ingram 1998; Parker, Snow, Schug, et al. 1998). Many allelic variants remain undetected because of the redundancy in the genetic code (Jasieniuk and Maxwell 2001). Another limitation is that allozymes may differ in metabolic function and thus be exposed to natural selection. Furthermore, allozyme markers are unable to resolve extremely small genetic differences.

Allozymes have extensively been used for the estimation of genetic variations in weed populations. Low genetic diversity was observed in populations of velvet leaf (*Abutilon theophrasti* Medicus) (Warwick and Black 1986a; Warwick 1990), Jimson weed (*Datura stramonium* L.) (Warwick 1990), barnyard grass [*Echinochloa crusgalli* (L.) Beauv.] (Barrett and Shore 1990), wild-prosomillet (*Panicum miliaceum* L.) (Warwick 1987, 1990), giant foxtail (*Setaria faberi* Herrm.) (Warwick, Thompson, and Black 1987a; Warwick 1990), and Johnson grass [*Sorghum halepense* (L.) Pers.] (Warwick, Thompson, and Black 1984; Warwick 1990). In contrast, high levels of allozyme variations were observed in *Apera spicaventi* (L.) Beauv. (Warwick, Thompson, and Black 1987b), broom snakeweed [*Gutierrezia sarothrae* (Pursh) Britt. et Rusby], and thread leaf snakeweed [*Gutierrezia microcephala* (D C) Gray] (Sterling and Hou 1997). In a comprehensive review of allozyme diversity in plants by Hamrick and Godt (1990), it was found that plant species are polymorphic at approximately 50% of their loci, with an average of two alleles per locus. Some studies on allozyme variations in herbicide susceptibility/tolerance of weed species have also been carried out (Warwick and Black 1986b; Warwick and Black 1993). Allozymes have also been used to study the levels of outcrossing, hybridization, and introgression in weeds (Warwick and Thompson 1989; Warwick, Bain, Wheatcroft, *et al.* 1989). These limitations have led the scientists to focus on DNA molecule as a source of informative polymorphism. Each individual DNA sequence is unique; hence, it can be exploited for any study of genetic diversity and relatedness between organisms.

DNA-BASED MARKERS

Following the discovery of restriction endonucleases and polymerase chain reaction (PCR), DNA markers were developed, which offered new possibilities in studying genetic diversity within and between plant species. According to Agarwal, Shrivastava, and Padh (2008), an ideal DNA marker technique should (1) be polymorphic and evenly distributed throughout the genome; (2) provide adequate resolution of genetic differences; (3) generate multiple, independent, and reliable markers; (4) be simple, quick, and inexpensive; (5) need small amounts of tissue and DNA samples; (6) have linkage to distinct phenotypes; and (7) require no prior information about the genome of an organism. A wide variety of techniques (known as DNA fingerprinting) have been developed over the last two decades to visualize the DNA sequence polymorphism. DNA fingerprinting techniques can be grouped into different categories depending on whether or not the assays are PCR based, and if they are PCR based, whether primers used are arbitrary or semi-arbitrary for unknown sequences or specifically designed for known sequences. Although DNA sequence is a straightforward approach for identifying variations at a locus, it is expensive and laborious. Therefore, a wide variety of techniques have been developed to visualize DNA sequence polymorphism. The term DNA fingerprinting was coined by Alec Jeffreys in 1985 to describe the bar-code-like DNA fragment patterns that were obtained by hybridization of multilocus DNA probes to electrophoretically separated genomic DNA fragments (Jeffreys, Wilson, and Thein 1985a). The emerging patterns are unique to the analysed individual and are currently considered to be the ultimate tool for biological individualization.

Non-PCR-based Techniques

Restriction Fragment Length Polymorphism

Restriction fragment length polymorphism (RFLP) is a non-PCR-based method that involves digestion of genomic DNA with a restriction endonuclease, electrophoretic separation of resultant DNA fragments according to size, transfer of restriction fragments to a filter by southern blotting, and hybridization with a labelled complementary DNA sequence (the so-called probe) (Karp, Seberg, and Buiatti 1996). Historically, the preferred method of labelling probes was to use radioisotopes (Brettschneider 1998). Hybridization of genomic DNA with probes for hypervariable regions composed of tandem repeats (known as "microsatellites" or SSRs, having basic repeat unit of approximately 16–100 bs) gives multilocus patterns that can resolve variations at the level of populations and individuals (Jeffreys, Wilson, and Thein 1985a; Bruford and Wayne 1993). The latter technique is often referred to as the variable number of tandom repeats (VNTRs) or oligonucleotide fingerprinting. The differential profile obtained is a result of nucleotide substitutions, DNA rearrangements such as insertion or deletion, or single nucleotide polymorphism.

The RFLP markers are relatively highly polymorphic, codominantly inherited, and highly reproducible (Karp, Seberg, and Buiatti 1996; Parker, Snow, Schug, et al. 1998), and also help screen numerous samples simultaneously. DNA blots can be analysed by repeatedly stripping and reprobing (usually 8–10 times) with different RFLP probes. RFLPs were the first DNA markers to be used by population biologists (Parker, Snow, Schug, et al. 1998). Differences in restriction patterns among individuals or species can arise from base substitutions within restriction sites or from insertions, deletions, or sequence rearrangements between two restrictions sites (Burr, Evola, Burr, et al. 1983; Karp, Seberg, and Buiatti 1996). These differences can be used to evaluate plant genetic variation. Nuclear DNA RFLP markers have extensively been used to create linkage maps, assess the diversity of germplasm collections, and analyse the segregation of progenies by plant breeders (Brettschneider 1998; Tanksley, Young, Paterson, et al. 1989). Organellar DNA RFLPs are extremely well suited to be used for the studies of population genetic divergence over large geographic areas (Schaal, Hayworth, Olsen, et al. 1998; Newton, Allnutt, Gillies, et al. 1999). Despite its advantages over earlier allozyme techniques, RFLP analysis also has limitations. The procedures involved, in particular the southern blotting and hybridization steps, are very labour intensive and expensive for the quantity of information obtained (Karp, Seberg, and Buiatti 1996; Karp, Isaac, and Ingram 1998; Parker, Snow, Schug, et al. 1998). Also, RFLP methods require relatively large amounts of high-quality DNA and thus are not useful when very limited amounts of plant material or preserved tissue are available.

Dot/slot Blots

Direct hybridization in dot or slot blots may also be used for the detection of specific sequences for the identification of a specific plant variety or species. The procedure involves the application of test DNA samples directly to the membrane, denaturation of the DNA by heating, addition of labelled probes, washing of the membrane to remove unhybridized probes, and observation of hybridized dots. In this case, the probe is designed on species-specific repeated DNA sequences. According to Guidet, Rogowsky, Taylor, et al. (1991), this method has limited applications and is useful for detecting viruses.

Restriction Landmark Genomic Scanning

Restriction landmark genomic scanning (RLGS) was introduced by Hatada, Hayashizaki, Hirotsune, et al. in 1991 for genomic DNA analysis of higher organisms. This method is based on the principle that restriction enzyme sites can be used as landmarks. It involves direct labelling of genomic DNA at the restriction site and identification of these landmarks. The technique has been proved to be useful in the genetic analysis of closely related cultivars and for obtaining polymorphic markers that can be cloned by spot target method (Hirotsune, Shibata, Okazaki, et al. 1993). The RLGS method has been used as a new fingerprinting technique for rice cultivars (Kawase 1994). This technique is time consuming, involves expensive and radioactive/toxic reagents, and requires large quantity of high-quality genomic DNA; hence, it is not used extensively. The requirement of prior sequence information for probe generation increases the complexity of the methodology. These limitations led to the conceptualization of a new set of less technically complex methods known as PCR-based techniques.

PCR-BASED TECHNIQUES

The invention of PCR by Karl Mullis and colleagues revolutionized the applicability of molecular methods (Saiki, Scharf, Faloona, et al. 1985; Saiki, Gelfand, and Stoffel 1988; Mullis and Faloona 1987). This technique allows the amplification of any DNA sequence of interest to high copy numbers, thereby circumventing the need of molecular cloning. For the amplification of a particular DNA sequence, two single-stranded oligonucleotide primers complementary to the motifs on the template DNA are designed. The primer sequences are chosen so as to allow base-specific binding to the template in reverse orientation. Addition of a thermostable DNA polymerase to a suitable buffer system and cyclic programming of primer annealing, polymerization, and denaturation steps result in exponential amplification of sequences between primer sites. Shortly after the introduction of PCR, a range of new technologies that could overcome many technical limitations of RFLPs for the detection of DNA polymorphisms were developed. Initially, specific primers complementary to known sequences were used to reveal polymorphism. Highly polymorphic amplification products were obtained by using primers complementary to flanking regions of minisatellite (Jeffreys, Wilson, and Thein 1985a, 1985b) and microsatellite loci (Litt and Luty 1989; Weber and May 1989). In another set of experiments, semispecific primers complementary to repetitive DNA elements were used. For human genome analysis, considerable level of polymorphism was observed using an abundant class of randomly interspersed DNA, called "Alu repeats", as primers (Ledbetter 1992). Later on, the principle of Alu-PCR (that is, use of primers complementary to interspersed repeats) was adopted for other species as well (Kaukinen and Varvio 1992). In 1990, several laboratories started using one or two short GC-rich primers of arbitrary sequence to generate PCR amplification products for genomic DNA.

Arbitrarily Primed PCR-based Techniques

Three arbitrarily primed PCR-based techniques, which did not require any sequence information, were developed independently: randomly amplified polymorphic DNA (RAPD) analysis (Williams, Kubelik, Livark, et al. 1990), arbitrarily primed polymerase chain reaction (AP-PCR) (Welsh and McClelland 1990), or DNA amplification fingerprinting (DAF) (Caetano-Anolles, Bassam, and Gresshoff 1991a, 1991b). In RAPD analysis, the amplification products are separated on agarose gels in the presence of ethidium bromide and visualized under UV light (Williams, Kubelik, Livark, et al. 1990). AP-PCR and DAF differ from RAPDs principally in primer length, primer-to-template ratio, gel matrix used, and the visualization procedure. A common term, multiple arbitrary amplicon profiling (MAAP), has been suggested to describe the common characteristic of all three techniques (Caetano-Anolles, Bassam, and Gresshoff 1994). The amplification of anonymous DNA sequences using arbitrary primers is common to all three techniques. Although nothing is known about the identity and sequence of a particular amplification product, its presence or absence in different organisms can serve as a highly informative character for the evaluation of genetic diversity and relatedness. These arbitrarily primed techniques are popular owing to the following advantages:

- DNA sequence information is not required for the design of specific primers.
- Because the procedure involves no blotting or hybridization steps, it is quick, simple, and automatable.
- A very small amount of DNA (10 mg per reaction) is required.

The data obtained from RAPDs (or AP-PCR or DAF) can very well be used to distinguish individuals, cultivars or accessions. However, the difficulty of achieving robust profiles makes their reliability for "typing" questionable. The presence or absence of bands can be scored, and the data can be converted into similarity matrices for the calculation of genetic distances (Ellsworth, Rittenhouse, and Honeycutt 1993). The following points should be realized while using arbitrarily priming procedures.

- The markers are dominant and heterozygotes cannot be detected.
- In the absence of pedigree analysis, the identity of individual bands in the multiband profiles is not known and there can be uncertainty in assigning markers to specific loci.
- The presence of a band of apparently identical molecular weight in different individuals cannot confirm that the two individuals share the homologous fragment, although the assumption is commonly made.
- Single bands on the gel can sometimes consist of several comigrating amplification products.

Randomly Amplified Polymorphic DNA

RAPD analysis is the technically simplest variation of the AP-PCR methods. Primers with 10 nucleotides and a GC content of at least 50% are used. Amplification products are separated on an agrose gel and detected by staining with ethidium bromide. This method involved autoradiography and was not very popular. The RAPD technique requires the presence of only a single "randomly chosen" oligonucleotide. Individual RAPD primers are able to hybridize to several hundred sites on complete strands of the target DNA; however, not all of these hybridizations lead to the production of PCR fragments. A discrete DNA product is formed through thermocyclic amplification only when two primer binding sites are within an amplifiable range of each other. On an average, each primer directs the amplification of several discrete loci in the genome, making the assay useful for efficient screening of nucleotide sequence polymorphism between individuals (Tingey, Rafalski, and Williams 1993). However, because of the stochastic nature of DNA amplification with random sequence primers, it is important to optimize and maintain consistent reaction conditions for reproducible DNA amplification. They are dominant markers and hence have limitations in their use as markers for mapping, which can be overcome to some extent by selecting those markers that are linked in coupling (Williams, Hanafey, Rafalski, et al. 1993). Several groups have used RAPD assay for the identification of markers linked to agronomically important traits, which are introgressed during the development of near-isogenic lines. RAPDs are able to produce multiple bands using a single primer; hence, a relatively small number of primers are sufficient to generate a very large number of fragments. Generally, these fragments are generated from different regions of the genome and hence multiple loci may be examined very quickly (Edwards 1998).

RAPD is a fast technique, easy to perform, and comparatively cheap. It is applicable to the analysis of most organisms because universal sets of primers can be used without any prior sequence information (Hallden, Hansen, Nilsson, et al. 1996). This marker system was used in many different applications involving the detection of DNA sequence polymorphisms, mapping in different types of populations (Carlson, Tulsieram, Glubitz, et al. 1991; Reiter, Williams, Feldmann, et al. 1992), isolation of markers linked to various traits or specific targeted intervals (Giovannoni, Wing, Ganal, et al. 1991; Michelmore, Paran, and Kesseli 1991), and applications such as variety identification and analysis of parentage (Tinker, Fortin, and Mather 1993; Mailer, Scarth, and Fristenski 1994). The RAPD technology, however, has some limitations. RAPD markers are in general dominant, thereby having lower information content than codominant markers in the linkage analysis of F2 populations (Williams, Kubelik, Livark, et al. 1990). Penner, Bush, Wise, et al. (1993) reported difficulties in obtaining identical band patterns from the same set of primers and materials among different laboratories. In their study, the type of thermocycler used for RAPD analysis seemed to be a key determinant of the reproducibility of band patterns. Another problem associated with these markers is the occurrence of RAPD bands in progeny but not in their parental DNAs, a phenomenon explained as heteroduplex formation (Riedy, Hamilton, and Aquadro 1992; Hunt and Page 1992; Ayliffe, Lawrence, Ellis, et al. 1994). It has been suggested that the outcome of RAPD reaction is, in part, determined by a competition for priming sites in the genome (Williams, Hanafey, Rafalski, et al. 1993). In several mapping projects, non-Mendelian inheritance for a significant fraction of all polymorphic bands was detected, possibly indicating problems with reproducibility and competition (Reiter, Williams, Feldmann, et al. 1992; Echt, Erdah, and Mccoy 1992; Giese, Holm-Jensen, Mathiassen, et al. 1994). On the other hand, Obara-Okeyo and Kako (1998) reported that the amplifications were generally reproducible and examples of successful application of the methods are known.

The basis of RAPD technique is differential PCR amplification of genomic DNA. It deduces DNA polymorphisms produced by "rearrangements or deletions at or between oligonucleotide primer binding sites in the genome" using short, random oligonucleotide sequences (mostly 10 bases long) (Williams, Kubelik, Livark, et al. 1990). The approach requires no prior knowledge of the genome that is being analysed; hence, it can be employed across species using universal primers. The major drawback of this method is that the profiling is dependent on the reaction conditions. Thus, the results of one laboratory may vary from those of another. Also, as several discrete loci in the genome are amplified by each primer, profiles are not able to distinguish between heterozygous and homozygous individuals (Bardakci 2001). The high speed and efficiency of RAPD analysis helped develop high-density genetic mapping in many plant species such as alfalfa (Kiss, Csanadi, Kalman, et al. 1993), faba bean (Torress, Weeden, and Martin 1993), and apple (Hemmat, Weeden, Manganaris, et al. 1994) in a relatively short time. The RAPD analysis of non-isogenic lines (NILs) has been successful in identifying markers linked to disease-resistant genes in tomato (*Lycopersicon* sp.) (Martin, Williams, and Tanksley 1991), lettuce (*Lactuca* sp.) (Paran, Kesseli, and Michelmore 1991), and common bean (*Phaseolus vulgaris*) (Adam-Blondon, Sevignac, Bannerot, et al. 1994).

Arbitrarily Primed Polymerase Chain Reaction

This is a special case of RAPD, wherein discrete amplification patterns are generated by employing single primers of 10–50 bases in length in the PCR of genomic DNA (Welsh and McClelland 1991). Annealing is performed under non-stringent conditions in the first two cycles, followed by 30–40 cycles of high stringency. In the last 20–30 cycles, radioactively labelled nucleotides are added. Radiolabelled produces are separated by polyacrylamide gel electrophoresis (PAGE) and visualized by autoradiography. This variant of arbitrarily primed techniques is not very popular as it involves autoradiography. Recently, however, it has been simplified by separating the fragments on agarose gels and using ethidium bromide staining for visualization. The final products are structurally similar to the RAPD products.

DNA Amplification Fingerprinting

Caetano-Anolles, Bassam, and Gresshoff (1991a, 1991b) employed single arbitrary primers (as short as five bases in length) to amplify DNA using PCR. In a spectrum of products obtained, simple patterns are useful as genetic markers for mapping, while more complex patterns are useful for DNA fingerprinting. Band patterns are reproducible and can be analysed using PAGE and silver staining. DAF requires careful optimization of parameters; however, it is extremely amenable to automation and fluorescent tagging of primers for early and easy determination of amplified products. DAF profiles can be tailored by employing various modifications such as predigestion of template. This technique has been useful in genetic typing and mapping.

Other PCR-based Techniques

Amplified Fragment Length Polymorphism

Vos, Hogers, Bleeker, *et al.* (1995) developed a new PCR-based technique, termed amplified fragment length polymorphism (AFLP), which is essentially intermediate between RFLP and RAPD. This technique involves restriction digestion of genomic DNA, followed by selective PCR amplification of the restricted fragments. Amplified products are radioactively or fluorescently labelled and separated on sequencing gels. AFLPs appear to be as reproducible as RFLPs, but are technically more demanding and require more DNA (1 µg per reaction) than RAPDs. On average, they give 100 bands per gel as compared to 20 bands by RAPDs. The AFLP procedure mainly involves three steps.

1. Restriction of DNA using simultaneously a rare cutting and a commonly cutting restriction enzyme (such as *Mse*I and *Eco*RI), followed by ligation of oligonucleotide adapters, of defined sequences, including the respective restriction enzyme sites.
2. Selective amplification of sets of restriction fragments using specifically designed primers. This can be achieved if the 5′ region of the primer contains both the restriction enzyme sites on either side of the fragment complementary to the respective adapters, while the 3′ ends extend for a few arbitrarily chosen nucleotides into the restriction fragments.
3. Gel analysis of the amplified fragments.

AFLPs are good for mapping, fingerprinting, and calculating genetic distances between the genotypes. However, they also share many of the limitations of RAPDs. To generate diversity data from specific sequences such as genes, it is necessary to have knowledge of the sequence surrounding the target to design specific primer pairs. The amplified products separated on an agarose gel can be compared with the corresponding products from another individual; however, only those differences in length that result from many base pair changes will be detected. A number of gel systems, such as thermal gradient gel electrophoresis (TGGE) (Riesner, Steger, Zimmat, *et al.* 1989), denaturing gradient gel electrophoresis (DGGE), single-strand conformational polymorphism (SSCP) (Hayashi 1992), and heteroduplex formation (White, Carvalho, Derse, *et al.* 1992), provide sensitive detection assays of sequence differences without the need for complete sequencing, but they require highly controllable conditions.

AFLP analysis depicts unique fingerprints regardless of the origin and complexity of the genome. Most AFLP fragments correspond to unique positions on the genome and hence can be exploited as landmarks in genetic and physical mapping (Vos, Hogers, Bleeker, *et al.* 1995). AFLPs are extremely useful for DNA fingerprinting (Hongtrakul, Huestis, and Knapp 1997) and also for cloning and mapping of variety-specific genomic DNA sequences (Yong, Glenn, Buss, *et al.* 1996; Paglia, Olivieri, and Morgante 1998). Similar to RAPDs, the bands of interest obtained by AFLP can be converted into SCARs. Thus, AFLP is a newly developed, important tool that is used for a variety of applications.

AFLP analysis involves selective amplification of a subset of restriction fragments from a complex mixture of DNA fragments obtained after the digestion of genomic DNA with restriction endonucleases. Polymorphisms are detected from differences in the length of the amplified fragments using PAGE (Matthes, Daly, and Edwards 1998) or capillary electrophoresis methods.

AFLP technology is used for the detection and evaluation of genetic variation in germplasm collections and in the screening of biodiversity, as well as for fingerprinting studies (Werner, Pellio, Ordon, *et al.* 2000). Using the above-mentioned tools, many molecular markers have been developed throughout the world. They can be successfully used for marker-assisted selection. A few examples that document the usefulness of the approach are as follows: molecular markers linked to the Rfo restorer gene used for the Ogu-INRA cytoplasmic male-sterility system in rapeseed (Delourme, Foisset, Horvais, *et al.* 1998), markers developed for linolenic acid content in rapeseed (Ho, Li, Struss, *et al.* 1999), markers allowing selection of BaYMV resistance in barley (Ordon, Schiemann, Pellio, *et al.* 1999), markers used to identify quantitative trait loci for grain yield and grain-related traits in maize (Ajmone-Marsan, Monfredini, Brandolini, *et al.* 1996), markers closely linked to the Rph7.g resistance gene of barley (Graner, Streng, Drescher, *et al.* 2000), markers allowing marker-assisted breeding for *Fusarium* head blight resistance (Lin, Aken, Kaul, *et al.* 2000), and markers discriminating mutant and normal alleles at the Wx-D1 locus in wheat (Shariflou, Hassani, and Sharp 2001). Application of markers helps speed up the breeding process and change some paradigms in plant breeding (Gupta, Varshney, Sharma, et al. 1999; Ranade, Farooqui, Bhattacharya, et al. 2001; Koornneef and Stam 2001). The above-mentioned techniques have been used to develop linkage maps of many plant species and DNA markers (Klein, Klein, Cartinhour, et al. 2000; Li-Wei, Tang, Wu, et al. 2000). However, the technical background makes it possible to employ still more effective approaches to genome characterization (Cai, Chen, Gibbs, et al. 2001).

Cleaved Amplified Polymorphic Sequences

In another method, termed PCR-RFLP or cleaved amplified polymorphic sequence (CAPS), the amplified product is digested with a specific restriction enzyme and the product visualized on an agarose gel after ethidium bromide staining (Akopyanz, Bukanov, Westblom, et al. 1992). This approach is most informative when restriction sites are mapped, rather than simply detected as RFLPs, and will not, of course, resolve all differences. Potentially, only sequencing the fragments will resolve all possible differences between the samples. These polymorphic patterns are generated by the restriction enzyme digestion of PCR products. Such digests are compared for their differential migration during electrophoresis (Koniecyzn and Ausubel 1993; Jarvis, Lister, Szabo, et al. 1994). For this process, PCR primer can be synthesized on the basis of the sequence information available in databank of genomic or cDNA sequences or cloned RAPD bands. These markers are codominant in nature.

Randomly Amplified Microsatellite Polymorphisms

Randomly amplified microsatellite polymorphisms (RAMPO) involve amplification of genomic DNA using arbitrary (RAPD) primers. The amplified products are then electrophoretically separated, and the dried gel is hybridized with microsatellite oligonucleotide probes. This method combines several advantages of oligonucleotide fingerprinting (Epplen, 1992), RAPD (Williams, Kubelik, Livark, et al. 1990), and microsatellite-primed PCR (Weising, Nybom, Wolff, et al. 1995; Gupta, Chyi, Romero-Severson, et al. 1994), including speed of the assay, high sensitivity, high level of variability detected, and non-requirement of prior DNA sequence information (Richardson, Cato, Ramser, et al. 1995). This technique has successfully been employed in the genetic fingerprinting of tomato, kiwi fruit, and closely related genotypes of *Dioscorea bulbifera* (Richardson, Cato, Ramser, et al. 1995; Joshi, Ranjekar, and Gupta 1999).

Although molecular marker techniques are largely available for application to biodiversity evaluation and conservation, it should be recognized that it is an emerging, rapidly evolving field in which the technology has advanced faster than scientific understanding. There is a trade-off between different types of markers with regard to their use for diversity assessment. Techniques that generate multilocus profiles provide information on numerous dispersed loci, whereas those based on random markers have been proved useful in restricted and specific applications such as analysis of closely related species or cultivar/strain identification. Importantly, the use of molecular markers has helped in understanding the processes and dynamics of biodiversity in evolution and natural preservation.

BIBLIOGRAPHY

Adam-Blondon A F, Sevignac M, Bannerot H, Dron M. 1994. **SCAR, RAPD, and RFLP markers linked to a dominant gene (Are) conferring resistance to anthracnose in common bean**. *Theoretical and Applied Genetics* **88**: 865–870

Agarwal M, Shrivastava N, and Padh H. 2008. **Advances in molecular marker techniques and their applications in plant sciences**. *Plant Cell Reports* **27**: 617–631

Ajmone-Marsan P, Monfredini G, Brandolini A, Melchinger A E, Garay G, Motto M. 1996. **Identification of QTL for grain yield in an elite hybrid of maize: repeatability of map position and effects in independent samples derived from the same population**. *Maydica* **41**: 49–57

Akopyanz N, Bukanov N O, Westblom T U, Kresovich S, Berg D E. 1992. **DNA diversity among clinical isolates of Helicobacter pylori detected by PCR-based RAPD fingerprinting**. *Nucleic Acids Research* **20**: 5137–5142

Ayliffe A M, Lawrence G J, Ellis J G, Pryor A J. 1994. Heteroduplex molecules formed between allelic sequences cause nonparental RAPD bands. Nucleic Acids Research **22**: 1632–1636

Bardakci F. 2001. **Random amplified polymorphic DNA (RAPD) markers**. *Turk Journal of Biology* **25**: 185–196

Barrett S C H and Shore J S. 1990. **Isozyme variation in colonizing plants**. In *Isozymes in Plant Biology*, edited by D E Soltis and P S Soltis, pp. 106–126. Porteland: Dioscorides Press

Brettschneider R. 1998. **RFLP analysis**. In *Molecular Tools for Screening Biodiversity*, edited by A Karp, P G Isaac, and D S Ingram, pp. 85–96. London: Chapman and Hall

Brown A H D. 1979. **Enzyme polymorphisms in plant populations**. *Theoretical Population Biology* **15**: 1–42

Bruford M W and Wayne R K. 1993. Microsatellites and their application to population genetic studies. Current Opinion *in* Genetics and Development **3**: 939–943

Burr B, Evola S V, Burr F A, Beckmann J S. 1983. The application of restriction fragment length polymorphisms to plant breeding. In Genetic

Engineering Principles and Methods, edited by J K Setlow and A Hollaender, vol. 5, pp. 45–59. New York: Plenum Press

Caetano-Anolles G, Bassam B J, and Gresshoff P M. 1991a. **DNA amplification fingerprinting using very short arbitrary oligonucleotide primers**. *BioTechnology* **9**: 553–557

Caetano-Anolles G, Bassam B J, and Gresshoff P M. 1991b. **High-resolution DNA amplification fingerprinting (DAF): detection of amplification fragment length polymorphisms in soybean using very short arbitrary oligonucleotide primers**. *Soybean Genetics Newsletter* **18**: 279–283

Caetano-Anolles G, Bassam B J, and Gresshoff P M. 1994. **Multiple arbitrary amplicon profiling using short oligonucleotide primers**. In *Plant Genome Analysis*, edited by P M Gresshoff, pp. 29–45. Boca Raton: CRC Press

Cai W W, Chen R, Gibbs R A, Bradley A. 2001. **A clone-array pooled shotgun strategy for sequencing large genomes**. *Genome Research* **11**: 1619–1623

Carlson J E, Tulsieram L K, Glubitz J C, Luk V W K, Kuffeldt C, Tutledge R. 1991. Segregation of random amplified DNA markers in F1 progeny of conifers. Theoretical *and* Applied Genetics **83**: 194–200

Cerny J and Sasek A. 1996a. ***Bilkovinne Signalni Geny Psenice Obecne***. Praha: UZPI

Cerny J and Sasek A. 1996b. **Analysis of genetic structure of regional common wheat varieties using signal gliadin and glutenin genes**. *Scientific Agriculture, Bohemoslov* **27**: 161–182

Delourme R, Foisset N, Horvais R, Barret P, Champagne G, Cheung W Y, Landry B S, Renard M. 1998. Characterisation of the radish introgression carrying the Rfo restorer gene for the Ogu-INRA cytoplasmic male sterility in rapeseed (***Brassica napus*** L.). Theoretical and Applied Genetics **97**: 129–134

Echt C S, Erdah L A, and Mccoy T J. 1992. Genetic segregation of random amplified polymorphic DNA in diploid cultivated alfalfa. Genome **35**: 84–87

Edwards K J. 1998. **Randomly amplified polymorphic DNAs (RAPDs)**. In *Molecular Tools for Screening Biodiversity*, edited by A Karp, P G Isaac, and D S Ingram, pp. 171–175. London: Chapman and Hall

Ellsworth D L, Rittenhouse K D, and Honeycutt R L. 1993. **Artifactual variation in randomly amplified polymorphic DNA banding patterns**. *Biotechniques* **14**: 214–217

Epplen J T. 1992. In *Advances in Electrophoresis*, edited by A Chrambach, M J Dunn, and B J Radola, vol. 5, pp. 59–112. Cambridge: VCH

Giese H, Holm-Jensen A G, Mathiassen H, Kjaer B, Rasmussen S K, Bay H, Jensen J. 1994. Distribution of RAPD markers on linkage map of barley. Hereditas **120**: 267–273

Giovannoni J J, Wing R A, Ganal M W, Tanksley S. 1991. Isolation of molecular markers from specific chromosomal intervals using DNA pools from existing mapping populations. *Nucleic Acids Research* **19**: 6553–6558

Graner A, Streng S, Drescher A, Jin Y, Borovkova I, Steffenson B J. 2000. **Molecular mapping of the leaf rust resistance gene *Rph7* in barley**. *Plant Breeding* **119**: 389–392

Guidet F, Rogowsky P, Taylor C, Song W, Langridge P. 1991. **Cloning and characterization of a new rye-specific repeated sequence**. *Genome* **34**: 81–87

Gupta M, Chyi Y-S, Romero-Severson J, Owen J L. 1994. **Amplification of DNA markers from evolutionary diverse genomes using single primers of simple sequence repeats**. *Theoretical and Applied Genetics* **89**: 998–1006

Gupta P K, Varshney R K, Sharma P C, Ramesh B. 1999. **Molecular markers and their applications in wheat breeding**. *Plant Breeding* **118**: 369–390

Hallden C, Hansen M, Nilsson N O, Hejrdin A, Sall T. 1996. Competition as a source of errors in RAPD analysis. Theoretical and Applied Genetics 93: 1185–1192

Hamrick J L and Godt M J W. 1990. **Allozyme diversity in plants**. In *Plant Population Genetics, Breeding, and Genetic Resources*, edited by A H D Brown, M T Clegg, A L Kahler, and B S Weir, pp. 43–63. Sunderland, MA: Sinauer

Hatada I, Hayashizaki Y, Hirotsune S, Komatsubara H, Mukai T. 1991. **A genomic scanning method for higher organisms using restriction sites as landmarks**. *Proceedings of the National Academy of Sciences* **88**: 9523–9527

Hayashi K. 1992. **PCR-SSCP: a method for detection of mutations**. *Genetic Analysis, Techniques and Applications* **3**: 73–79

Hemmat M, Weeden N F, Manganaris A G, Lawson D M. 1994. **Molecular marker linkage map for apple**. *Journal of Heredity* **85**: 4–11

Henry R J. 1997. **Molecular techniques for the identification of plants**. In *Applied Plant Biotechnology*, edited by V L Chopra, V L Malik, and S R Bhat, pp. 269–284. New Delhi and Calcutta: Oxford and IBH Publishing Co. Pvt. Ltd.

Hirotsune S, Shibata H, Okazaki Y, Sugino H, Imoto H, Sasaki N, Hirose K, Okuizumi H, Muramatsu M, Plass C, Chapman C, Tamatsukuri S, Miyamoto C, Furuichi Y, Hayashizaki Y. 1993. **Molecular cloning of polymorphic markers on RLGS gel using the spot target cloning method**. *Biochemical and Biophysical Research Communications* **194**: 1406–1412

Ho J, Li G, Struss D, and Quiros C F. 1999. **SCAR and RAPD markers associated with 18-carbon fatty acids in rapeseed, *Brassica napus***. *Plant Breeding* **118**: 145–150

Hongtrakul V, Huestis G M, and Knapp S J. 1997. **Amplified fragment length polymorphisms as a tool for DNA fingerprinting sunflower** germplasm: genetic diversity among oilseed inbred lines. *Theoretical and Applied Genetics* **95**: 400–407

Hunt G J, and Page R E. 1992. Patterns of inheritance with RAPD molecular markers reveal novel types of polymorphism in the honey bee. Theoretical and Applied Genetics **85**: 15–20

Hunter R L and Markert C L. 1957. **Histochemical demonstration of enzymes separated by zone electrophoresis in starch gels.** *Science* **125**: 1294–1295

Jarvis P, Lister C, Szabo V, Dean C. 1994. **Integration of CAPS markers into the RFLP map generated using recombinant inbred lines of *Arabidopsis thaliana*.** *Plant Molecular Biology* **24**: 685–687

Jasieniuk M and Maxwell B D. 2001. **Plant diversity: new insights from molecular biology and genomics technologies.** *Weed Science* **49**: 257–265

Jeffreys A J, Wilson V, and Thein S L. 1985a. **Hypervariable minisatellite regions in human DNA.** *Nature* **314**: 67–73

Jeffreys A J, Wilson V, and Thein S L. 1985b. **Individual specific fingerprints of human DNA.** *Nature* **316**: 76–79

Joshi S P, Ranjekar P K, and Gupta V S. 1999. **Molecular markers in plant genome analysis.** *Current Science* **77**: 230–240

Karp A, Isaac P G, and Ingram D S (eds). 1998. *Molecular Tools for Screening Biodiversity*. London: Chapman and Hall

Karp A, Seberg O, and Buiatti M. 1996. **Molecular techniques in the assessment of botanical diversity.** *Annals of Botany* **78**: 143–149

Kaukinen J and Varvio S L. 1992. **Artiodactyl retroposons: association with microsatellites and use in SINE morph detection by PCR.** *Nucleic Acids Research* **20**: 2955–2958

Kawase M. 1994. **Application of the restriction landmark genomic scanning (RLGS) methods to rice cultivars as a new fingerprinting technique.** *Theoretical and Applied Genetics* **89**: 861–864

Kiss G B, Csanadi G, Kalman K, Kalo P, Okresz L. 1993. **Construction of a basic linkage map for alfalfa using RFLP, RAPD, isozyme and morphological markers.** *Molecular and General Genetics* **238**: 129–137

Klein P E, Klein R R, Cartinhour S W, Ulanch P E, Dong J M, Obert J A, Morishige D T, Schlueter S D, Childs K L, Ale M, Mullet J E, Dong J M. 2000. **A high-throughput AFLP-based method for constructing integrated genetic and physical maps: progress toward a sorghum genome map.** *Genome Research* **10**: 789–807

Koniecyzn A and Ausubel F M. 1993. **A procedure for mapping nomically important pathogen. Arabidopsis mutations using co-dominant ecotype-specific PCR-based markers.** *Plant Journal* **4**: 403–410

Koornneef M and Stam P. 2001. **Changing paradigms in plant breeding. Special issue: 75th anniversary. Conceptual breakthroughs in biology.** *Plant Physiology* **125**: 156–159

Kraic J, Horvath L, Gregova E, Zak I. 1995. **Standard methods for electrophoretic separation of wheat glutenins and gliadins by SDS-PAGE and A-PAGE.** *Rostlinna Vyroba* **41**: 219–223

Ledbetter D H. 1992. **Cryptic translocations and telomere integrity.** *The American Journal of Human Genetics* **52**: 451–456

Li-Wei M, Tang D Z, Wu W R, Lu H R. 2000. **A molecular map based on an indica/indica recombinant inbred population and its comparison with an existing map derived from indica/japonica cross in rice.** *Chinese Journal of Rice Science* **14**: 71–78

Lin X Y, Aken S, Kaul S, Creasy T H, Goodman H M, Somerville C R, Copenhaver G P, Preuss D, Nierman W C, White O, Eisen J A, Salzberg Ma Z Q, Steffenson B J, Prom L K, Lapitan N L V. 2000. **Mapping of quantitative trait loci for *Fusarium* head blight resistance in barley.** *Phytopathology* **90**: 1079–1088

Litt M and Luty J A. 1989. **A hyperverable microsatellite revealed by in vitro amplification of a dinucleotide repeat within the cardiac muscle action gene.** *The American Journal of Human Genetics* **444**: 397–401

Loveless M D and Hamrick J L. 1984. **Ecological determination of genetic structure in plant populations.** *Annual Review of Ecology and Systematics* **15**: 65–95

Mailer R J, Scarth R, and Fristenski B. 1994. **Discrimination among cultivars of rapeseed (*Brassica napus* L.) using DNA polymorphisms amplified from arbitrary primers.** *Theoretical and Applied Genetics* **87**: 697–704

Martin G B, Williams J G K, and Tanksley S D. 1991. **Rapid identification of markers linked to a *Pseudomonas* resistance gene in tomato by using random primers and near-isogenic lines.** *Proceedings of the National Academy of Sciences* **88**: 2336–2340

Matthes M C, Daly A, and Edwards K J. 1998. **Amplified fragment length polymorphism (AFLP).** In *Molecular Tools for Screening Biodiversity*, edited by A Karp, P G Isaac, D S Ingram, vol. 1, pp. 183–190. Cambridge: Chapman and Hall

Metakovsky E V. 1991. **Gliadin allele identification in common wheat II. Catalogue of gliadin alleles in common wheat.** *Journal of Genetics and Breeding* **45**: 325–344

Michelmore R W, Paran I, and Kesseli R V. 1991. **Identification of markers linked to disease-resistance genes by bulked segregant analysis: a rapid method to detect markers in specific genomic regions by using segregating populations.** *Proceedings of National Academy of Sciences* **88**: 9828–9832

Mullis K B and Faloona F A. 1987. **Specific synthesis of DNA** *in vitro* **via a polymerase-catalysed chain reaction.** *Methods in Enzymology* **155**: 335–350

Newton A C, Allnutt T R, Gillies A C M, Lowe A J, Ennos R A. 1999. **Molecular phylogeography, intraspecific variation and the** conservation of tree species. *Trends in Ecology and Evolution* **14**: 140–145

Obara-Okeyo P and Kako S. 1998. **Genetic diversity and identification of** *Cymbidium* **cultivars as measured by random amplified polymorphic DNA (RAPD) markers.** *Euphytica* **99**: 5–101

Ordon F, Schiemann A, Pellio B, Dauck V, Bauer E, Streng S, Friedt W, Graner A. 1999. **Application of molecular markers in breeding for resistance to the barley yellow mosaic virus complex.** Zeitschrift für Pflanzenkrankheiten und Pflanzenschutz **106**: 256–264

Ovesna J, Polakova K, and Leisova L. 2002. **DNA analyses and their applications in plant breeding.** *Czech Journal of Genetics and Plant Breeding* **38**: 29–40

Paglia G P, Olivieri A M, and Morgante M. 1998. **Towards second-generation STS linkage maps in conifers: a genetic map of Norway spruce (***Picea abies* **K.).** *Molecular and General Genetics* **258**: 466–478

Paran I, Kesseli R, and Michelmore R. 1991. **Identification of restriction fragment-length-polymorphism and random amplified polymorphic DNA markers linked to downy mildew resistance genes in lettuce, using near isogenic lines.** *Genome* **34**: 1021–1027

Parker P G, Snow A A, Schug M D, Booton G C, Fuerst P A. 1998. **What molecules can tell us about populations: choosing and using a molecular marker.** *Ecology* **79**: 361–382

Penner G A, Bush A, Wise R, Kim W, Domier L, Kasha K, Laroche A, Scoles G, Molnar S J, Fedak G. 1993. **Reproducibility of random amplified polymorphic DNA (RAPD) analysis among laboratories.** *PCR Methods and Applications* **2**: 341–345

Ranade S A, Farooqui N, Bhattacharya E, Verma A, Farooqui N, Bhattacharya E. 2001. **Gene tagging with random amplified polymorphic DNA (RAPD) markers for molecular breeding in plants.** *Critical Reviews in Plant Sciences* **20**: 251–275

Reiter R S, Williams J G K, Feldmann K A, Rafalksi J A, Tingey S V, Scolnik P A. 1992. **Global and local genome mapping in** *Arabidopsis thaliana* **by using recombinant inbred lines and random amplified polymorphic DNAs.** *Proceedings of the National Academy of Sciences* **89**: 1477–1481

Richardson T, Cato S, Ramser J, Kahl G, Weising K. 1995. **Hybridization of microsatellites to RAPD: a new source of polymorphic markers.** *Nucleic Acids Research* **23**: 3798–3799

Riedy M F, Hamilton W J, and Aquadro F C. 1992. **Excess of non-parental bands in offspring from known primate pedigrees assayed using RAPD PCR.** *Nucleic Acids Research* **20**: 918

Riesner D, Steger G, Zimmat R, Owens R A, Wagenhofer M, Hillen W, Vollbach S, Henco K. 1989. **Temperature-gradient gel electrophoresis of nucleic acids: analysis of conformational transitions, sequence variations, and protein-nucleic acid interactions.** *Electrophoresis* **10**: 377–389

Saiki R K, Gelfand D H, and Stoffel S. 1988. **Primer directed enzymatic amplification of DNA with thermostable DNA polymerase.** *Science* **239**: 487–497

Saiki R K, Scharf S, Faloona F, Mullis K B, Horn G T, Erlich H A, Arnhein N. 1985. **Enzymatic amplification of globin genomic sequences and restriction site analysis for diagnosis of sickle cell anemia.** *Science* **230**: 1350–1354

Schaal B A, Hayworth D A, Olsen K M, Rauscher J T, Smith W A. 1998. **Phylogeographic studies in plants: problems and prospects.** *Molecular Ecology* **7**: 465–474

Shariflou M R, Hassani M E, and Sharp P J. 2001. **A PCR-based DNA marker for detection of mutant and normal alleles of the Wx-DI gene of wheat.** *Plant Breeding* **120**: 121–124

Smithies O. 1955. **Zone electrophoresis in starch gels: group variation in the serum proteins of normal human adults.** *Biochemical Journal* **61**: 629–641

Sterling T M and Hou Y. 1997. **Genetic diversity of broom snakeweed (*Gutierrezia sarothrae*) and thread leaf snakeweed (*G. microcephala*).** *Weed Science* **45**: 674–680

Tanksley S D, Young N D, Paterson A H, Bonierbale M W. 1989. **RFLP mapping in plant breeding: new tools for an old science.** *BioTechnology* **7**: 257–264

Tingey S V, Rafalski J A, and Williams J G K. 1993. *Application of RAPD Technology to Plant Breeding*, edited by M Neff, pp. 3–8. Minnesota: ASHS Publishers

Tinker N A, Fortin M G, and Mather D E. 1993. **Random amplified polymorphic DNA and pedigree relationships in spring barley.** *Theoretical and Applied Genetics* **85**: 976–984

Torress A M, Weeden N F, and Martin A. 1993. **Linkage among isozyme, RFLP, and RAPD markers.** *Plant Physiology* **101**: 394–452

Vapa L and Radovic D. 1998. **Genetics and molecular biology of barley hordeins.** *Cereal Research Communications* **26**: 31–38

Vos P, Hogers R, Bleeker M, Reijans M, Lee T V N, Hornes M, Frijters A, Pot J, Peleman J, Kuiper M, Zabeau M. 1995. **AFLP: a new technique for DNA fingerprinting.** *Nucleic Acids Research* **23**: 4407–4414

Warwick S I. 1987. **Isozyme variation in proso millet.** *Journal of Heredity* **78**: 210–212

Warwick S I. 1990. **Allozyme and life history variation in five northwardly colonizing North American weed species.** *Plant Systematics and Ecology* **169**: 41–54

Warwick S I and Black L D. 1986a. **Genecological variation in recently established populations of *Abutilon theophrasti* (velvetleaf).** *Canadian Journal of Botany* **64**: 1632–1643

Warwick S I and Black L D. 1986b. **Electrophoretic variation in triazine-resistant and susceptible populations of *Amaranthusr etroflexus* L.** *New Phytologist* **104**: 661–670

Warwick S I and Black L D. 1993. **Electrophoretic variation in triazine-resistant and susceptible populations of the allogamous weed *Brassica rapa*.** *Weed Research* **33**: 105–114

Warwick S I and Thompson B K. 1989. **The mating system in sympatric populations of *Carduus nutans*, *C. acanthoides* and their hybrid swarms.** *Heredity* **63**: 329–337

Warwick S I, Bain J F, Wheatcroft R, Thompson B K. 1989. **Hybridization and introgression in *Carduus nutans* and *C acanthoides* re-examined.** *Systematic Botany* **14**: 476–494

Warwick S I, Thompson B K, and Black L D. 1984. **Population variations in *Sorghum halepense*, Johnson grass, at the northern limits of its range.** *Canadian Journal of Botany* **62**: 1781–1790

Warwick S I, Thompson B K, and Black L D. 1987a. **Life history and allozyme variation in populations of weed species *Setaria faberi*.** *Canadian Journal of Botany* **65**: 1396–1402

Warwick S I, Thompson B K, and Black L D. 1987b. **Genetic variation in Canadian and European populations of the colonizing weeds species *Apera spica-venti*.** *New Phytology* **106**: 301–317

Weber J L and May P E. 1989. **Abundant class of human DNA polymorphisms which can be typed using the polymerase chain reaction.** *The American Journal of Human Genetics* **44**: 388–396

Weising K, Nybom H, Wolff K, Meyer W. 1995. ***DNA Fingerprinting in Plants and Fungi***, edited by A Arbor, pp. 1–3. Boca Raton: CRC Press

Welsh J and McClelland M. 1990. **Fingerprinting genomes using PCR with arbitrary primers.** *Nucleic Acids Research* **18**: 7213–7218

Welsh J and McClelland M. 1991. **Genomic fingerprints produced by PCR with consensus tRNA gene primers.** *Nucleic Acids Research* **19**: 861–866

Werner K, Pellio B, Ordon F, Friedt W. 2000. **Development of an STS marker and SSRs suitable for marker-assisted selection for the BaMMV resistance gene *rym9* in barley.** *Plant Breeding* **119**: 517–519

White M B, Carvalho M, Derse D, O'Brien S J, Dean M. 1992. **Detecting single base substitutions as heteroduplex polymorphisms.** *Genomics* **12**: 301–306

Williams J G K, Hanafey M K, Rafalski J A, Tingey S V. 1993. **Genetic analysis using random amplified polymorphic DNA markers.** *Methods in Enzymology* **218**: 705–740

Williams J G K, Kubelik A R, Livark K J, Rafalski J A, Tingey S V. 1990. **DNA polymorphisms amplified by arbitrary primers are useful as genetic markers.** *Nucleic Acids Research* **18**: 6531–6535

Yong G, Glenn R, Buss G R, Saghai-Maroof M A. 1996. **Isolation of a superfamily of candidate disease-resistance genes in soybean based on a conserved nucleotide-binding site.** *Proceedings of the National of Academy of Sciences* **93**: 11751–11756

15

E-flora: the Future of Floristic Documentation

Gurcharan Singh

The Internet revolution in recent years has seen an all-round development in scientific interaction, with instant access to huge databases around the world. There have been cooperative efforts to bring together botanical literature in general and taxonomic literature in particular, with interconnected links to various electronic sites on the web. The past decade has seen significant improvement in documenting information about the plant wealth of different regions of the world, thereby eliminating the need for a researcher to sit with a pack of books to compare description, identify plants, decide correct nomenclature, and find distribution. A majority of the herbaria of the world have digitized their specimens, especially the types, and uploaded them in the form of virtual herbaria. These can be accessed instantly from any part of the globe. Collaborative efforts of various institutions have made lists of accepted names and their synonyms available online. Many important floras of the world, including the seven-volume *Flora of British India*, have been digitized and are available for download as PDF files. Online interactive keys go a long way in convenient identification of plants, and photographs of live plants are available through various official and social websites for easy comparison. The last few years have seen development of online floras with proper documentation of all relevant information regarding floristics of different regions. These online floras, appropriately called electronic floras or e-floras, are becoming increasingly user friendly with convenient links within the document and with other related e-Floras.

COMPONENTS OF A RELIABLE E-FLORA

A convenient and reliable e-flora needs to have elaborate citation of botanical names with author and full references, list of synonyms, including basionym (if exists), a key for the identification of genera, species and infraspecific taxa, detailed description, native habitat and distribution, illustrations in the form of line diagrams, and photographs. Working with e-floras can become convenient through the utilization of additional resources, as indicated below.

Nomenclature Sources

The process of setting nomenclature in order was initiated towards the end of the 19th century by the publication of two-volume *Index Kewensis* under the supervision of Sir J D Hooker, which was followed up by supplements almost every five years. At the beginning of the 1980s, the information was transferred to a computer database that continued to expand at a rate of 6000 records per year. The Global Strategy for Plant Conservation, adopted in 2002 by 193 governments who are parties to the Convention on Biological Diversity, set a target for the completion of a widely accessible working list of all known plant species by 2010, as a step towards a complete world flora, to coincide with the international year of biodiversity, a celebration of life on earth. A collaboration between the Royal Botanic Gardens, Kew, and Missouri Botanical Garden (MBG) helped create *The Plant List* by combining multiple checklist datasets held by these institutions and other collaborators. Version 1 of the list released in December 2010 contained 1244871 names, of which 298900 are accepted names, the rest being synonyms or unresolved names. The sooner the last category of names is resolved, the better it would be for the reliability of the database. The list at present contains vascular plants and bryophytes, and not algae and fungi. A simple search for each genus leads to a complete list of all published names within the genus differentiated as accepted names (green, bold italics), synonyms (green, italics), and unresolved (black, italics). There is also an indication of confidence level (1–3) and source database. The list includes links to the home pages of Royal Botanic Gardens, Kew; Global Compositae Checklist; International Legume Database and Information Service (ILDIS); International Organization for Plant Information (IOPI); New York Botanical Garden (NYBG), International Plant Names Index (IPNI); and Convention on Biodiversity. A click to the accepted name leads to full citation and list of all synonyms. This page contains links to websites that may have further information on the species: Kew World Checklist of Selected Plant Families (WCSP), Tropicos, Species 2000 Integrated Taxonomic Information System (ITIS) Catalogue of Life, Global Biodiversity Information Facility, New York Botanical Garden Virtual Herbarium, Jastor Plant Science, Herbarium Catalogue of Royal Botanic Gardens, Kew, National Centre for Biotechnology Information, GenBank,

Encyclopedia of Life, Plant Information Portal of Royal Botanic Gardens, Biodiversity Heritage Library, Wikispecies, and Google images.

Royal Botanic Gardens, Kew, has released an update to information resource discovery service, called ePIC—the electronic Plant Information Centre. From the ePIC interface, one can now search for plant information across six databases held at Kew, and also website, in one action. One can search for (1) plant names through the IPNI and WCSP; (2) bibliographic data in the Kew record of taxonomic literature, economic botany bibliography, library catalogue, and micromorphology bibliography; (3) collections from herbarium catalogues, economic botany collections, and living collection of plants; (4) species-level information about seed storage characteristics and Survey of Economic Plants for Arid and Semi-Arid Lands (SEPASAL); and (5) Flora Zambesiaca, a comprehensive descriptive account of the flowering plants and ferns native to and naturalized in Zambia, Malawi, Mozambique, Zimbabwe, Botswana, and the Caprivi Strip.

Global Compositae Checklist is a searchable integrated database of nomenclatural and taxonomic information for one of the largest plant families in the world, also known as the Asteraceae. Data sets concerning plant families across the world based on floras and distribution records have been integrated using the purpose-designed Checklist Integration software (C-INT, Landcare Research, New Zealand). Names are matched using a set of rules and a consensus name is generated with all original data sources linked to that name. Entries are complete to differing levels depending on the data contributed, as only few data sources have information regarding types. Broad distribution data, as derived from the data contributed, has been included using the TDWG Geographic Standard, although the distribution is not necessarily comprehensive.

Species 2000 Program was established by the International Union of Biological Sciences (IUBS), in cooperation with the Committee on Data for Science and Technology (CODATA) and the International Union of Microbiological Societies (IUMS) in September 1994, subsequently endorsed by UNEP Biodiversity Work Program 1996–97. Species 2000 aims at enumerating all known species of plants, animals, fungi, and microbes on earth as the baseline dataset for studying global biodiversity. Species 2000, in collaboration with ITIS, has developed *Catalogue of Life,* the latest annual checklist released in April 2011, incorporating 99 taxonomic databases covering 1 347 224 species. *Dynamic checklist* enables access to 26 taxonomic databases cached in 2005, covering 450 000 species. The project is developing an indexing system for all groups of organisms, with an ultimate goal of listing all known species on earth, operated by a federation of database organizations working closely with users, taxonomists, and sponsoring agencies. The *ITIS database* offers quality taxonomic information of flora and fauna of both aquatic and terrestrial habitats. ITIS is a product of the collaboration between federal agencies and systematists in the federal, state, and private sectors to provide taxonomic information. Geographic coverage will initially emphasize North American taxa.

ILDIS is an international project that maintains database concerning the family Fabaceae (Leguminosae). The database provides a taxonomic checklist plus basic factual data on distribution, common names, life forms, uses, literature references to descriptions, illustrations, and maps. This database can be searched online through the website or through ILDIS explorer. The ILDIS explorer, an innovative new program for browsing and searching, can be downloaded with database and software. One can also browse through the old draft checklist of world legumes created in 2001.

Global Plant Checklist Project, a part of Species 2000 database collection, is organised by the IOPI Checklist Committee. The checklist that will encompass about 300 000 vascular plant species and over 1 000 000 names is IOPI's first priority. It includes information on both vascular and non-vascular plants and is now operative in the form of a searchable database through Information Gateway of Ohio University Libraries. Entries include bibliographic source, family, protologue, and status (whether name is accepted or not).

IPNI, a database of the names and associated basic bibliographical details of all seed plants, is a result of the collaboration between The Royal Botanic Gardens, Kew; The Harvard University Herbaria; and the Australian National Herbarium. Over one million records have come from *Index Kewensis*, over 350 000 records from the *Gray Index* (originally the Gray Herbarium Card Index), and over 63 000 records from the *Australian Plant Names Index*. Until 1971, *Index Kewensis* did not include infraspecific names, although the other two indices included these names for their area, leading to many names below species level being missing from IPNI. IPNI intends to eliminate the need for repeated reference to primary sources for basic bibliographic information about plant names. Data are freely available and are gradually being checked and standardized. IPNI will be a dynamic resource, depending on direct contributions by all members of the botanical community.

TROPICOS is one of the several resources hosted by MBG. In addition to the angiosperm phylogeny website authored by Peter Stevens, the research page of the garden carries links to *TROPICOS, MOST, Image*

Index, and *Rare Books*. Although it was originally created for internal research, since its inception it has been made available to the world's scientific community. All of the nomenclatural, bibliographic, and specimen data accumulated in MBG's electronic databases during the past 25 years are publicly available here. W3TROPICOS, the latest version, provides a real-time look at the data of 1 253 419 scientific names, 3 957 305 specimens, 164 822 images, 47 518 publications, 121 419 references, and 52 576 common names in June 2011. It ensures improved access to the MBG's vast nomenclatural database and associated authority files. The searchable database provides information on names data; plant name and authors; group and family placement; place and date of publication; type information; basionym, with place and date of publication; next higher taxon, with place and date of publication; other uses of this name; synonyms of this name; references for alternative usage; homonyms; and infraspecific names for species. Reference can also be found with regard to author(s) of the publication, date of publication, title of the article, journal or book title, volume and page numbers, and keywords. Latest version of MOST (MOSs TROPICOS), W3MOST, is the moss database that is now integrated with TROPICOS. *Image Index* site carries links to images found at this site; the site is supported by the Institute of Museum and Library Services as part of a National Leadership Award for preservation and digitization of information. Images are now available under Open Archives Initiative Protocol for Metadata Harvesting (OAI-PMH). Image gallery includes line illustrations, photographs, and more usefully the images of type specimens. *Rare books* can similarly be searched through the database on the MBG website.

GRIN Taxonomy Home Page, maintained by Germplasm Resources Information Network of USDA's National Plant Germplasm System (NPGS), contains a record of economic plants. The database enables search of 94 887 names of species and infraspecies (56 046 accepted names), and 26 784 genera (14 180 accepted), with common names, geographical distributions, literature references, and economic impacts. Generally recognized standards for abbreviating author's names and botanical literature have been adopted in GRIN. The scientific names are verified, in accordance with the international rules of botanical nomenclature, by taxonomists of the National Germplasm Resources Laboratory, using available taxonomic literature and consulting with taxonomic specialists. Federal- and state-regulated noxious weeds and federally and internationally listed threatened and endangered plants are included in the GRIN taxonomy. The database can be searched for answers to simple and complex queries for species data. The database can also direct the searches to links of world economic plants, family and generic names, nomenclature of seed associations, federal and state noxious weeds, and rare and endangered plants. One can also download files on families, genera, and the species. GRIN is managed by the Database Management Unit (DBMU) of the National Germplasm Resources Laboratory, Agricultural Research Service, USDA.

Sorting Botanical and Common Names is a facility developed under the Multilingual Multiscript Plant Name Database (MMPND) by the University of Melbourne, Australia. The list of genera is gradually increasing and, at present, information on accepted names, synonyms, and common names is available in 70 languages (including 12 Indian languages) and in 25 scripts. Information regarding bamboos, fungi, palms, conifers, and medicinal plants can be accessed separately. It also provides link to the related ISTA Multilingual Glossary at ARS GRIN Database (USA), online references for all records and online bibliographies arranged by plant groups, languages, and photographs and botanical drawings.

Identification Keys

Taxonomic key is an integral part any flora, set out in the form of dichotomous keys: indented key (for shorter keys), parallel key (for longer keys), or serial key (suitable for both). These are conveniently used in e-floras with links to included taxa and their details. In recent years, taxonomists have seen the development of standalone or web-based interactive keys for convenient identification. Some of the common interactive keys are described here.

The DELTA System, an integrated set of programs based on the DELTA format (DEscription Language for TAxonomy), is a flexible and powerful method of recording taxonomic descriptions for processing by a computer. DELTA, a shareware program, has been adopted as a standard for data exchange by the International Taxonomic Databases Working Group. It enables generation and typesetting of descriptions and *conventional keys*, conversion of DELTA data for use by *classification programs*, and construction of *Intkey packages* for interactive identification and *information retrieval*. The system is developed during the Natural Resources and Biodiversity Program conducted by the Division of Entomology of the Commonwealth Scientific and Industrial Research Organisation (CSIRO), Australia, over a period of 20 years (by M J Dallwitz, T A Paine, and E J Zurcher). It is used worldwide for diverse kinds of organisms, including fungi, plants, and wood. The programs are continually refined and enhanced on the basis of the feedbacks from users.

The *DELTA program key* generates conventional identification keys. Depending on how well characters divide the remaining taxa, the characters are selected by the program for inclusion in the key. This information is then balanced against subjectively determined weights, which specify the ease of use and reliability of the characters.

DELTA data can readily be converted into the forms that are required by programs for phylogenetic analysis, for example, Paup, Hennig86, and MacClade. The characters and taxa for these analyses can be selected from the full data set. These programs cannot work with numeric characters; therefore, these characters are converted into multistate characters. Printed descriptions can be generated to facilitate checking of the data.

Intkey can also be used to access DELTA data and images over the Internet. For this data files (such as i-items, i-chars), intkey.ini, contents.ind (together with rtf files), and image files (optional) are converted into a zip file (or self-extracting zip file) and uploaded to the website along with a startup file (*.ink, which contains the information and the path of uploaded files of the project), intkey.ini, imagePath (optional), and InfoPath (optional). A dataset index file or link in WWW page must point to the special startup file (*.ink, not intkey.ini or intkey.ink). The startup file helps Intkey search for the dataset and its associated images on the website. When a person using an Internet browser clicks a link to an Intkey startup file, the browser activates Intkey and passes it a copy of the startup file. Then Intkey itself retrieves the actual dataset from the web, extracts its contents, and begins identification. However, for this web applicability, Intkey has to be installed on both the web server and each client PC, and an association of files has to be developed by the manager of web server, where the project files are located.

Intkey-based web applications are available for several families and genera from Flora of China, Families of the World (Watson and Dallwitz), Grass Genera of the World (Watson and Dallwitz), Grass Species of the World (RBG, Kew), Tree and Shrub Genera of Borneo (J K Jarvie and Ermayanti), identification facility for the vascular flora of western Australia, and is available in FloraBase. Additionally, interactive keys (using Intkey) to the families and genera of flowering plants in western Australia are soon to be added to the *FloraBase*. Specialist keys for certain significant genera are also in an advanced state of preparation.

NaviKey is a simple Java-based interactive identification key, a free program, which works on DELTA flat files (chars, items, and specs—present in your folder if you have developed a database ready for identification through Intkey, as detailed in preceding paragraphs). NaviKey v. 4 is developed in the frame of BIOTa Africa project (an international research network on biodiversity, sustainable use, and conservation) by Dieter Neubacher and Gerhard Rambold (University of Bayreuth, Germany), based on an earlier version (NaviKey v. 2.3 by Michael Bartley and Noel Cross, Harvard University Herbarium, Boston, USA). The program can be downloaded from www.NaviKey.net and can be used both as a standalone application and as a web application. After downloading and unzipping the files on your computer, add the three flat files of your project to this folder. For using it as a standalone application, click NaviKey.jar. Similar to Intkey, an identification window with four panels will open up: upper left window contains the character panel, right upper panel shows character states, lower right panel is the matching items panel that shows matching or remaining taxa (click any taxon to get its full description), and the lower left panel is the query criteria panel that displays previous (used) character state selections. NaviKey also allows checkbox matching options to (1) restrict view on used characters and character states of remaining items, (2) retain items unrecorded for the selected characters, (3) retain items matching at least one selected state of resp. characters, (4) use extreme interval validation, and (5) use overlapping interval validation. NaviKey does not display the list of excluded taxa, but the total number of taxa and the number remaining are displayed.

The use of software as a web application is very convenient. It involves only a few simple steps like filling in the title and subtitle of the project being developed in NaviKeyAppletWebpageTemplate.html using html editor (say Frontpage), uploading the whole folder to the website, and providing a link to NaviKey.html page. As this page opens, Java application gets loaded and the program is ready for interactive identification.

NaviKey identifications are available for several families and genera of Flora of China and genus *Arisaema* (Guy Gusman and Eric Gouda) and flowering plant families of Jamaica (Gerald Guala and Jimi Sadle). One key for the identification of families and genera covering the undergraduate courses of University of Delhi and several other universities is available at http://people.du.ac.in/~singhg45/.

Lucid Systems software (Lucid3) is a powerful, commercially used, and widely acclaimed Lucid Professional identification and diagnostic software developed by the Centre for Biological Information Technology, The University of Queensland, Brisbane, Australia. The Lucid3 system uses a builder and a player for creating and deploying effective and powerful identification and diagnostic keys. It helps create interactive, random-access keys that can be deployed over the world wide web or CD.

The key, when used for the identification of an unknown specimen, progressively eliminates entities that do not match the chosen features until only one or a few possible entities remain. Further information and images can be accessed to confirm the identification. The basic elements of a Lucid3 key include a list of entities, a list of features and states that may be used to describe those entities, a matrix of score data for the features associated with each of the entities for the features, and various attachments (images and web pages) for the entities and features, to provide extra information to users.

The Lucid3 builder provides all the tools necessary to create the entity and feature lists, encode the score data, and attach information files to items. The package, in addition, includes Lucid Phoenix, a computer-based dichotomous or pathway key builder and player that enables traditional paper-based identification keys to be published on the Internet or in a CD. Phoenix keys are interactive; they can be enhanced with multimedia and delivered across the Internet seamlessly. Additional Fact Sheet Fusion software facilitates rapid generation of standardized fact sheets in HTML (hypertext markup language) or XML (extensible markup language).

XID (Expert Identification Systems) Services Inc. produces commercial software with emphasis on biological sciences and is one of the leading providers of expert identification systems for major universities and botanical gardens in the USA. XID offers two identification packages: *Pankey*, a DOS-based identification program, and *XID Authoring Systems*, Windows-based databases and program for identification. The XID Authoring System allows authors to create their own "smart key" or random access expert system for the identification of plants, animals, or any other object. The simplicity of the XID system makes it extremely user friendly; it is as useful to a school teacher as to a professional scientist.

The XID system allows the user to select randomly the characteristics that are consistent with their specimen and skill level. If the users cannot decide upon a characteristic, they may query the program and obtain a list of suggestions in order of ease of use, effectiveness, and items remaining. In general, the program includes much more data about each item/species than is necessary to identify it. Thus, this abundance of data helps the user identify any of the items/species using the characteristics most obvious and easy to describe. With each characteristic entered by the user, the program eliminates all species that do not have that particular feature.

XID also offers *1000 Weeds of North America* CD ROM. This is the most comprehensive weed identification reference ever published in North America, containing 140 grass-like and 860 broadleaf weeds. The features include interactive key, coloured photos of all species, illustrated glossary of terms, page number references to over 40 weed reference books, searchable geographic data, and state-level distribution maps.

ActKey is a web-based interactive identification program developed by Hong Song of the MBG. This Java-based program uses MySQL as the database server and can handle data sets in DELTA, MS Excel, MS Access, and Lucid formats.

ActKey identification is available for the floras of China, North America, Madagascar, and Borneo at the Harvard University Herbaria Editorial Center, hosted at *e-Flora website*. Examples include several keys to the large- and medium-sized genera of China (also in Chinese), the genera of Brassicaceae of the world by Ihsan Al-Shehbaz, *Salix* (Salicaceae) of North America by George W Argus (also in Chinese), angiosperm families by B Hansen and K Rahn (also in Chinese and Spanish), Trilliaceae (*Trillium* and *Paris*) of the world by Susan B Farmer, generic tree flora of Madagascar by George Schatz, and trees and shrubs of Borneo by James K Jarvie and Ermayanti.

MEKA (pronounced "mecca") is an interactive multiple-entry key algorithm that enables rapid identification of biological specimens, now designed to run under Windows. The program, distributed free, is developed by Christopher Meacham at Jepson Herbarium, Berkeley, CA. The user chooses some character states present in the specimen from a list of possibilities. As the character states are scored by picking them, MEKA eliminates taxa that no longer match the list of scored character states. Different windows display different aspects of the underlying database. As the identification progresses, the windows are updated automatically. An index screen makes it easy to find and score particular classes of character states. MEKA does not restrict the user to a fixed stepwise progression through a series of questions; instead, the user can perform identifications by scoring character states in any order. Thus, unlike dichotomous keys, this program makes it possible to identify specimens that are much more fragmentary. New Windows version includes a conversion function that can convert any MEKA key to the SLIKS (Stinger's Light Weight Interactive Key Software) format developed by Gerald Guala for web-based identification. Thomas J Rosatti has developed many MEKA keys for the identification of California plants, and Professor Knud Ib Christensen (of the Botanic Garden of the University of Copenhagen) developed a key for Old World *Crataegus*.

SLIKS software, a small, free Javascript program that runs over the web or locally on your machine, facilitates the use of interactive keys. Users can download their own copy or use it from the website. It runs through the web browser; so it is essentially platform independent.

IdentifyIt is identification software of comprehensive commercial *Linnaeus II* multifunctional research tool developed by ETI BioInformatics for systematists and biodiversity researchers. It facilitates biodiversity documentation and species identification. Linnaeus II supports the creation of taxonomic databases, optimizes the construction of easy-to-use identification keys, expedites the display and comparison of distribution patterns, and promotes the use of taxonomic data for biodiversity studies. There are three "modules" of Linnaeus II: the "builder" to manage data and create an information system, the "runtime" engine to publish completed information systems on CD-ROM/DVD-ROM, and the "web publisher" to publish your completed project as a website.

The package offers three identification modules: *Text Key*™—an electronic version of written dichotomous keys, *Picture Key*™—similar to the text key but picture based, and *IdentifyIt*™—the most powerful identification tool. It is a multiple-entry key based on a matrix of taxa, characters, and character states. Unlike the Species and Higher Taxa that contain text descriptions of the taxa, in IdentifyIt taxa are described in a more structured format: as a series of character states. This allows one to easily obtain answers to specific questions: for example "Which species are red and/or white?"

Virtual Herbarium and Live Images

Virtual herbarium is a database consisting of images of herbarium specimens and the supporting text, available over the Internet. It is a huge advancement in herbarium use and design, coupling physical specimens directly with the Internet and integrating complete specimen data, with resources or information generation and retrieval. Although a virtual herbarium cannot exist without a physical herbarium, it enjoys several advantages over a physical herbarium: mainly (1) instant access, (2) no damage to specimens, (3) increased user interaction as numerous persons can work on same specimen simultaneously, and (4) convenient access to information on descriptive details, geographical distribution, photographs, illustrations, manuscripts, published work, microscopic preparations, gene sequences, and nomenclature through hyperlinks. Additionally, several virtual herbaria can be searched simultaneously.

Virtual herbaria with searchable database have been developed by many major organisations such as New York Botanical Garden (KE EMu),

Royal Botanic Gardens Melbourne (AVH), Fairchild Tropical Garden (e-FTG), Australian Virtual Herbarium (AVH), and Royal Botanic Gardens, Kew (ePIC).

The Digital Plant Research Center (DPRC) of the *New York Botanical Garden* has developed a system of interconnected digital resources to include the wide range of research endeavours, searchable through a common interface. The components of the DPRC include C V Starr Virtual Herbarium, electronic floras and monographs, economic botany collections, Pfizer laboratory, and living collections. In 1994, the herbarium began the development of a database containing information and images of 7.2 million specimens. The data of the virtual herbarium of the garden, initially developed as NYpc in 1994, was transferred to the new platform KE EMu in 2004 with additional search and display capabilities. It was renamed C V Starr Virtual Herbarium in 2007 in honour of Cornelius Vander Starr. The Starr Foundation (founded by C V Starr) has provided several major grants towards endowment and operating support for this important initiative. The digital collections of the virtual herbarium, comprising approximately 1.3 million herbarium specimens and 225 000 high-resolution specimen images are updated daily. Important links from the herbarium include vascular plants of the Osa Peninsula, Costa Rica; French Guianan e-Flora Project; Cyperaceae Pages; plants and lichens of Saba; and Barneby legume catalogue.

Fairchild Tropical Garden Virtual Herbarium (e-FTG) has a record of more than 100 000 specimens and more than 200 000 photographs (including data labels), including more than 20 000 high-resolution specimen photographs that can be zoomed in or out of the browser. Nearly 60 000 records are searchable online to obtain information regarding family, genus, collector, and other fields. e-FTG is the first truly virtual herbarium, as web portal of the herbarium allows simultaneous search through virtual herbaria of FTG, FLAS (Florida Museum of Natural History), MO (TROPICOS—MBG), NY Cassia (New York Botanical Garden), S (Linnean Herbarium, Swedish Museum of Natural History, Leiden), BM (British Museum of Natural History—including Clifford Herbarium), CAYM (National Trust for Cayman Islands), INB (Instituto Nacional de Biodiversidad, Costa Rica), and TAMU (Texas A&M University). Thus, the virtual herbarium of FTG includes not only specimens from Fairchild Herbarium but also from other herbaria. It also provides species lists, interactive keys, and photographs of living specimens in various databases and indices.

Australia's Virtual Herbarium (AVH) is a collaborative project of the state, Commonwealth, and territory herbaria, being developed under the auspices of the Council of Heads of Australian Herbaria (CHAH), which represents major Australian collections. It is an online botanical information resource accessible via the web, providing immediate access to the information on scientific plant specimens in each Australian herbarium.

Australian herbaria house over six million specimens that provide information regarding the classification and distribution of plants, algae, and fungi. These specimens are the working tools of scientists who contribute to our knowledge and understanding of biodiversity and conservation through the discovery, classification, and description of new species. The scope of knowledge will be enhanced by images, descriptive text, and identification tools.

The AVH is accessed via the website of any participating herbarium. A gateway at each of these herbaria links to the databases of all the other herbaria, consolidating the combined data into a nation-wide view of the botanical information. Most data related to specimens will be stored by the custodial institution, and some resources, such as the scientific names database (Australian Plant Names Index, APNI), will be common to all. More than 70% of the specimens housed in Australian herbaria have been included in the database, providing a comprehensive resource for accurate depiction of geographic distribution and occurrence, historical mapping, and information valuable for understanding the threatening processes of vegetation clearance and weed invasion. Flexible online search options allow customization of the data according to your requirements.

AVH provides descriptions of the flora dynamically linked to data and information from across the continent and distributed online as an electronic Australian Flora—a one-stop source of current information on the plants, algae, and fungi of the entire Australian continent. New observations can be released with minimal delay as they are confirmed and recorded in the database.

The Strong *ePIC database software of Royal Botanic Gardens*, Kew, also provides a window for digitized herbarium specimens. Since its inception in 2002, the herbarium's core digital collection program has grown at an increasing rate. The central herbarium catalogue has an image server and many project databases, containing information about specimens, which were built before the creation of the catalogue. These are being moved into the catalogue as resources permit. Label data from dry and spirit specimens of flowering plants, ferns, and gymnosperms held in Kew's herbarium are being uploaded. Information recorded includes the plant name, collection and determination data, locality, and type status.

JSTOR is another important online archive and research platform that allows faculty, researchers, students, and others to discover, use, and build upon a wide range of content including over 1000 academic journals, as well as conference proceedings, monographs, and other scholarly contents. It provides access to information on plant-type specimens, taxonomic structures, scientific literature, and related materials. A significant portion of the content available on JSTOR Plant Science has been contributed through an effort known as the Global Plants Initiative (GPI). It provides access to herbarium images from Africa (69 400), Europe (788 700), North America (296 500), and South America (44 000). In addition, miscellaneous collections from various sources, including Curtis Botanical Magazine (1200), Flora Capensis (17 500), Flora of Southern Africa (4000), Flora of Tropical Africa (13 200), Flora of Tropical East Africa (10 100), Flora of West Tropical Africa (9000), Flora Zambesiaca (10 400), Plants of Mascareignes (5000), Useful Plants of West Tropical Africa (4900), and huge collection of type specimens (159 000), can be accessed.

Additionally, official and private gardens and social organizations provide several online collections of images. *CalPhotos* is a huge database developed under a project of Biodiversity Sciences Technology (BSCIT) group, University of California, Berkeley. It contains 307 800 photographs of plants, animals, fossils, people, and landscapes from around the world, contributed by a variety of individuals and organizations. Plants, mostly Californian, are represented by 159 725 images. *Daves Garden* is perhaps the largest online plant identification guide, with *PlantFiles* having information on 186 308 different plants and their 296 966 images, which can be searched by common or botanical name. Currently, entries are from 431 families, 5099 genera, 42 734 species, and 115 065 cultivars. Each species page contains common names, botanical name with important synonyms, family, category, bloom time, bloom colour, foliage, other details, soil pH requirements, propagation methods, tips on seed collecting, and useful comments by various members, mostly gardeners. *Grass Images* (TAMU Bioinformatics Working Group) are represented in the TAMU-BWG Image Database. Image files available here are mostly derived from collaboration between the Hunt Institute for Botanical Documentation and the TAMU Press. *Photographic Collection of the Australian National Botanic Gardens* covers photos of species from over one million acres of the outer coastal plain in southern and central New Jersey. This website contains a collection of colour photographs of 45 pineland plants. The scientific and common names of the plant, the family to which it belongs, its approximate height at maturity, a brief description of the plant and its habitat, and the place where the photograph was taken are listed below each photograph. Plants Photo Gallery, USDA, USA, contains sample of plant images being integrated into the PLANTS database for North American plants. These images include photographs of plants and plant habitats, with vernacular and scientific names, family name, photographer's name, and location. *Smithsonian Catalogue of Botanical Illustrations* is a database of the Department of Botany of the Smithsonian Institution's National Museum of Natural History, USA, with more than 3000 botanical illustrations curated by Alice Tangerini, the scientific illustrator of the department. As part of a long-term project to make an online catalogue of these illustrations available to their staff and others needing access to this information, the department offers 500 images in the three families that have been completed: Bromeliaceae, Cactaceae, and Melastomataceae. *Vascular Plant Image Library* was originally developed with the support from Texas Higher Education Coordinating Board as a part of Digital Flora of Texas. Links are provided family-wise to the images of plant species in databases including Flowers of India, CalPhotos, Flora of Chile, Missouri Plants, Floral Images, Plants of Hawaii, Oregon Flora image project and several individual image collections.

The last few years have seen a spurt in Internet-based exchange of information in India. Google *eGroup efloraofindia* (earlier indiantreepix) with more than 1600 members is the largest Google e-group in the world in this field, and the largest nature-related e-group in India. It has a database of more than 4500 species based on nearly 70000 messages containing photographs and discussions on Indian flora. It is devoted to creating awareness, helping in identification, and discussion and documentation of Indian flora. The group aims at compiling a database of photographs, nomenclature, relevant information, and local names in different languages. The database is building up at a rapid pace and new taxa are being added after confirmation from various experts.

Flowers of India is a very useful website for the identification of Indian plants. It contains a species page with one or more photographs, common names of the species in English and Indian languages, common synonyms, short description, relevant information, and the place of photography (for example, *Abelia X grandiflora*, Figure 1). These species pages can be accessed through various links such as botanical names (sorted alphabetically or family-wise) or common names. There are also special links for flowering trees, orchids, grasses, medicinal plants, flowers in ancient literature, cacti and succulents, vines and creepers, bulbous plants, garden flowers, and Himalayan flowers. The plants are also arranged by colour of flowers and their fragrance. In addition, there is provision for the identification of trees by the type and shape

Figure 1 Flowers of India website: a platform for identification of Indian plants
Source <www.flowersofindia.in/catalog/slides/Glossy%20Abelia.html>

of leaves. Photographs of additional species are being added regularly by contributors. At present, the website has information on more than 5000 Indian flowering plant species.

Centre for Conservation of Natural Resources (CCNR), under the Institute of Ayurveda and Integrative Medicine (I-AIM), an expression of FRLHT in Bangalore, has developed an internationally accredited herbarium of medicinal plants, with a collection of more than 70% of medicinal plants used by the codified Indian system of medicine. The digitized herbarium, known as *FRLH-Herbarium and Raw Drug Repository*, houses more than 35 000 accessions of nearly 2800 medicinal plants. The exclusive and innovative search-based database stores 7637 botanical names (6198 medicinal plants species) with 101 745 vernacular names from 12 languages across India. Thus, these can be searched by botanical name or vernacular name. Nearly 998 high-resolution plant images are also available in the database. Information and images of nearly 100 red-listed medicinal plants of southern India are available here.

Ayurvedic Medicinal Plants is another useful website with photographs of nearly 500 Indian plants that are arranged by botanical names, Sanskrit names, Malayalam names, and whether they are endangered medicinal plants or not. The website also provides information on nearly 150 medicinal plants that are difficult to find and their herbal remedies. It also has a collection of several useful photographs by individuals. J M Garg, the owner of efloraofindia, has a collection of photographs of more than 1000 species of plants on Wikimedia commons grouped under different categories: climbers, ferns, grasses, herbs, shrubs, succulents and trees, further arranged alphabetically. Each photograph contains information on species botanical name, common name, and place of taking photograph. Dinesh Valke's photostream on Flickr from Yahoo hosts photographs of more than 1000 species under the category Flora of India, including photographs of more than 600 species from Western Ghats and nearly 300 cultivated plants, in addition to about 120 grasses and sedges and few lower Himalayan plants. Each photograph contains information on botanical name, synonyms, common names in different languages in regional scripts, native region, and important references about the species.

The personal page of Dr Gurcharan Singh in the website of the University of Delhi (Figure 2) and the website of SGTB Khalsa College have link to the digital herbarium containing nearly 10 000 digital images. These include digital images of herbarium specimens collected

Figure 2 Dr Gurcharan Singh's personal page in the University of Delhi website
Source <http://singhg45.jalbum.net/Abrus%20precatorius/>

by the author from Kashmir, Ladakh, and other parts of India and deposited in the Herbarium of Botany Department, Sri Guru Tegh Bahadur Khalsa College, University of Delhi. The collection also includes images of herbarium specimens deposited by staff and students of the botany department. The project aims at displaying digital images of live plants (photographed by the author) mainly from Delhi, Himalayas, and other areas. The digital images are being uploaded progressively.

ELECTRONIC FLORAS

The last few years have seen the online availability of digitized form of many popular floras. These online floras (e-floras) help users browse and search for floristic treatments and work dynamically on these treatments. One such effort by MBG has resulted in the publication of <www.eFloras.org>, combining together the information from several floras, including Flora of Chile, Flora of China (Figure 3), Flora of Missouri, Flora of North America, Flora of Pakistan, Moss Flora of China, Trees and Shrubs of Andes and Ecuador, as also the Annotated Checklist of Flowering Plants of Nepal. These floras can be searched through a common search engine to obtain relevant information. The hyperlinks to families, genera, and species are very handy in identifying and retrieving information. The website also hosts the interactive *ActKey* provided by the Harvard University Herbarium, allowing visitors to locate and use a key for identifying an unknown specimen. The keys for families of angiosperms by Bertel Hansen and Knud Rahn, families of dicotyledons

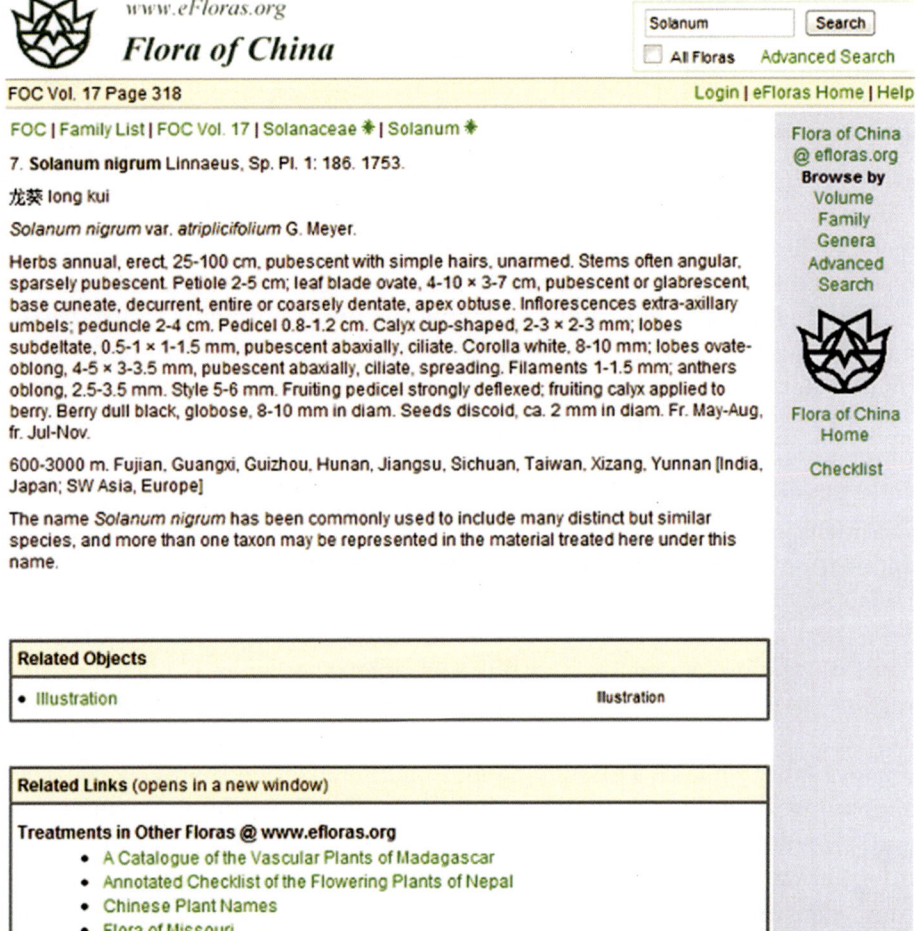

Figure 3 Information on Flora of China is found in www.efloras.org

Source <eFloras.org>

of the Western Hemisphere, South of the United States, generic tree flora of Madagascar, key to taxa of China in ActKey, trees and shrubs of Borneo, and weeds of rain-fed lowland rice fields of Laos and Cambodia are already incorporated in a user-friendly interface.

Royal Botanic Gardens, Kew, has hosted e-Flora *Flora Zambesiaca*, which provides not only an easy way of searching the information but also an identification tool. This website allows a search for a plant name across the whole flora, which would otherwise entail looking up separate indexes. It also allows listing of endemics, species from a particular division or country, species that match a particular habit, and species that occur at a specific altitude. As far as possible, no changes have been made to the information existing in the original text and the information is presented in the same way as in the original.

Collaborative efforts of several organizations have resulted in the development of *e-Flora BC*, a volunteer-driven GIS-based biogeographic atlas of vascular plants, fungi, algae, bryophytes, and lichens of British Columbia, having more than 8000 individual atlas pages. e-Flora BC includes a significant citizen science component. In addition to collection records, this website also presents observation records and photographs contributed by citizens. For each species, information is available on habitat, taxonomy, invasive, and poison status. Links are also provided to related databases on taxonomy, nomenclature, and distributional range. The valuable work *The Illustrated Flora of British Columbia* in eight volumes, containing detailed species descriptions, illustrations, maps, and range information, is provided online by the e-Flora BC project.

Smithsonian research botanist John Cress has created a smartphone application *Leafsnap*, in collaboration with engineers from Columbia University and the University of Maryland. Originally conceived in 2003 as a high-tech aid for scientists to discover new species in unknown habitats, the project evolved with the emergence of smartphones as a new way for citizens to contribute to research even without any formal training. The mobile application helps in the identification of plants by simply photographing a leaf. This iphone and ipad application instantly searches a growing library of leaf images amassed by Smithsonian institution, and provides names of likely species, high-resolution photographs, and information on the tree's flowers, fruits, and bark. The user makes the final identification based on inputs from the application and shares the findings with application's growing database. Developed only in May 2011, it covers trees in New York's Central Park and Washington Rock Creek Park. During this summer, it will include all the trees of the northeast and eventually will cover all the trees of North America.

BIBLIOGRAPHY

Aguilar R, Cornejo X, Bainbridge C, Tulig M, Mori S A. 2008. **Vascular plants of the Osa Peninsula, Costa Rica**. Bronx, New York: The New York Botanical Garden. Details available at <http:sweetgum.nybg.org/osa>

Biodiversity Heritage Library. Details available at <www.biodiversitylibrary.org>, last accessed on 14 June 2011

Bisby F A, Roskov Y R, Orrell T M, Nicolson D, Paglinawan L E, Bailly N, Kirk P M, Bourgoin T, Baillargeon G (eds). 2010. **Species 2000 and ITIS** *catalogue of life: 2010 annual checklist*. Reading, UK: Species 2000. Details available at <www.catalogueoflife.org/annual-checklist/2010>

Brach A R and Song H. 2006. **eFloras: new directions for online floras exemplified by the Flora of China Project**. *Taxon* **55**(1): 188–192

Klinkenberg B (ed). 2010. **E-Flora BC: electronic atlas of the plants of British Columbia**. Vancouver: Lab for Advanced Spatial Analysis, Department of Geography, University of British Columbia.

Missouri Botanical Garden and Harvard University Herbaria. 2008. **e-Floras** St Louis, MO: Missouri Botanical Garden and Cambridge, MA: Harvard University Herbaria. Details available at: <www.efloras.org>, last accessed on 15 May 2011

Mori S A, Tulig M, de Granville J J, Gonzalez S, Guerin V. 2007. **French Guiana e-Flora project**. The New York Botanical Garden and the Institut de Recherche pour le Développement. Details available at <http://sweetgum.nybg.org/fg>

The herbarium catalogue. 2006. Details available at <http://www.kew.org/herbcat>, last accessed on 14 June 2011

The plant list. 2010. Version 1. Details available at <www.theplantlist.org>, last accessed on 5 June 2011

Royal Botanic Gardens, Kew. 2002. **Electronic Plant Information Centre**. Details available at <http://epic.kew.org/epic>, last accessed on 14 June 2010

World checklist of selected plant families. Kew: The Board of Trustees of the Royal Botanic Gardens. Details available at <www.kew.org/wcsp/>, last accessed on 8 June 2011

Zonker B. 2011. **What is that tree? Try Smithsonian's new app to see**. Details available at <www.mercurynews.com/business/ci_18258883?IADID>

WEBSITES

Encyclopedia of life. <www.eol.org>

GenBank—nucleotide alphabet of life. <www.ncbi.nlm.nih.gov>

Jstor plant science. <http://plants.jstor.org>

Royal Botanic Gardens, Kew. <www.kew.org>

Tropicos.org. Missouri Botanical Garden. <www.tropicos.org>

Wikispecies Contributors. <http://species.wikimedia.org/w/index.php?title=Main_Page&oldid=1110094>

About the Editor

Dr Rajni Gupta is Assistant Professor in the Department of Botany, Kirori Mal College, University of Delhi, New Delhi. She obtained her MSc and PhD in Botany from Agra University, Agra, and completed her postdoctoral work in the Department of Botany, under the guidance of Professor K G Mukerji.

Dr Gupta has published more than 45 research papers in various national and international journals. She has edited the book *Advances in Microbial Biotechnology* and authored *Microbial Technology* and *The Fungi*. Currently, she is working on a University Grants Commission sponsored project related to vesicular–arbuscular mycorrhiza association in lower plants.